DIFFUSION OF REACTIVE MOLECULES IN SOLIDS AND MELTS

DIFFUSION OF REACTIVE MOLECULES IN SOLIDS AND MELTS

ROBERT H. DOREMUS

A Wiley-Interscience Publication
JOHN WILEY & SONS, INC.

This book is printed on acid-free paper. ♾

Published simultaneously in Canada.

For ordering and customer service, call 1-800-CALL-WILEY.

Library of Congress Cataloging-in-Publication Data Is Available

ISBN 0-471-38545-X

Printed in the United States of America.

10 9 8 7 6 5 4 3 2 1

CONTENTS

Preface xi

1 Introduction 1

1.1 Importance 1
1.2 Defects 2
1.3 Diffusion–Reaction Mechanisms 3
1.4 Outline of the Book 4
1.5 Omissions 7
References 8

2 Diffusion Without Chemical Reaction 9

2.1 Introduction 9
2.2 Fundamental Diffusion Equations 9
2.3 Infinite and Semi-Infinite Geometries 12
2.4 Plane Sheet 14
2.5 Permeation Through a Membrane 16
2.6 Diffusion Coefficient a Function of Concentration 19
References 22

3 Diffusion and Reaction with Local Equilibrium 23

3.1 Introduction 23
3.2 Reaction Experiments 24
3.3 Exchange Experiments 27
3.4 Pressure Effects 29

3.5 Moving Boundary-Tarnishing Reactions 30
References 35

4 Diffusion and Nonequilibrium Reactions 36

4.1 Introduction 36
4.2 No Local Equilibrium, Forward Reaction Only 37
4.3 No Local Equilibrium, First-Order Forward and Reverse Reactions 42
4.4 No Equilibrium, First-Order Forward Reaction, Second-Order Reverse Reaction 49
4.5 No Local Equilibrium, Tracer Diffusion 52
4.6 Time Ranges of Validity of Equations 54
4.7 Surface Exchange 55
References 57
Appendix: Error Functions of Complex Argument 57

5 Structure and Sizes of Atoms and Molecules 59

5.1 Introduction 59
5.2 Electron Distributions Around Atoms and Molecules 60
5.3 Sizes of Gaseous Atoms and Molecules Dissolved in Solids and Melts 62
5.4 Sizes of Ions in Solids and Melts, and Openness of Solid Structures 68
5.5 Monoxides 70
References 72

6 Diffusion and Reaction of Water in Silica Glass 74

6.1 Introduction 74
6.2 Structure of Silica Glass 76
6.3 Gas Solubility in Silica Glass 78
6.4 Experiments of Roberts and Co-Workers at Temperatures of 700 to 1200°C 80
6.5 Experiments of Spitzer and Ligenza at 650°C, and Rigo *et al.* from 400 to 900°C 86
6.6 Experiments of Wakabayashi and Tomozawa at 350 and 450°C 90
6.7 Experiments of Rauch *et al.* and Lanford *et al.* at Temperatures of 200 to 90°C 91
6.8 Discussion 93
References 94

7 Diffusion, Reaction, and Exchange of Water in Quartz 96

7.1 Introduction 96
7.2 Structure of Quartz 97

7.3 Experiments of Kats on Hydrogen–Deuterium Exchange 98
7.4 Experiments on ^{18}O–^{16}O Exchange 101
7.5 Experiments of Dersch and Rauch at Temperatures of 200°C and Lower 104
7.6 Discussion 106
References 107

8 Diffusion and Reaction of Water in Rhyolite Glass (Obsidian) 108

8.1 Introduction 108
8.2 Diffusion–Reaction Model 109
8.3 Analysis of High-Temperature Diffusion Profiles 111
8.4 Analysis of Low-Temperature Profiles 114
8.5 Discussion 117
References 118
Appendix 118

9 Diffusion and Reaction of Water in Oxides 120

9.1 Introduction 120
9.2 Equations 121
9.3 Diffusion of Water in Pyrope and Calcite 122
9.4 Comparison of Experimental Activation Energies for Diffusion of Water in Oxides with Ionic Porosity 124
References 127

10 Speciation of Water in Silicate Glasses and Melts 129

10.1 Introduction 129
10.2 Langmuir Model for Speciation 130
10.3 Comparison with Experiment 132
10.4 Solubility of Water 136
10.5 Partial Molar Volume 140
10.6 Discussion 145
References 145

11 Hydrogen in Oxides 148

11.1 Introduction 148
11.2 Hydrogen in Silica Glass 149
11.3 Hydrogen in Fiber Optic Cables of Silica 154
11.4 Hydrogen in Multicomponent Silicate Glasses 158
11.5 Hydrogen in Crystalline Oxides 159
References 160

12 Diffusion of Oxygen in Nominally Dry Oxides 163

12.1 Introduction 163
12.2 Diffusion of Oxygen in Silica Glass 164

12.3 Molecular Transport of Oxygen 170
12.4 Oxygen Transport by Ions 172
12.5 Experimental Activation Energies 175
12.6 Further Considerations of Mechanisms 177
12.7 Diffusion of Oxygen in Nominally Dry Silicate Melts 180
References 182

13 Oxidation State 185

13.1 Introduction 185
13.2 Dissolved Gases 186
13.3 Metals and Oxygen 193
13.4 Ions with Multiple Valence States and the Phase Rule 193
13.5 Oxidation with Transport of Cations 197
References 198

14 Diffusion in Silicon: Fast and Intermediate Diffusers 199

14.1 Introduction 199
14.2 Classification of Diffusing Atoms 200
14.3 Interstitials, Vacancies, and Fast Diffusers in Silicon 202
14.4 Exchange or Kick-Out Mechanism and Diffusion of Intermediate Diffusers in Single-Crystal, Dislocation-Free Silicon 205
14.5 Influence of Dislocations in Silicon on Intermediate Diffusers 214
14.6 Diffusion in Amorphous Silicon 215
14.7 Summary and Subsequent Chapters 216
References 217

15 Diffusion of Hydrogen in Silicon 219

15.1 Introduction 219
15.2 Mechanisms of Hydrogen Diffusion in Silicon 220
15.3 Equations 223
15.4 Comparison with Experiments 229
15.5 Discussion 237
References 239

16 Diffusion of Oxygen and Silicon in Silicon: Silicon Monoxide Model 241

16.1 Introduction 241
16.2 Structure of Dissolved SiO 244
16.3 Infrared Spectra and Dichroism 245
16.4 Diffusion of Oxygen, Infrared Dichroism, and Internal Friction 246
16.5 Fast Diffusion of Oxygen 247
16.6 Diffusion of Silicon 248

16.7 Conclusions 250
References 250

17 Diffusion of Group III, IV, and V Elements in Silicon 253

17.1 Introduction 253
17.2 Diffusion Coefficients and Activation Energies for Group III, IV, and V Elements 254
17.3 Mechanisms of Diffusion and Reaction 258
17.4 Concentration Dependence of Effective Diffusion Coefficients of Group III and V Elements in Silicon 262
17.5 Internal Friction of Silicon 267
17.6 Influence of Other Factors on Dopant Diffusion in Silicon 268
References 270

18 Diffusion in Germanium 273

18.1 Introduction 273
18.2 Measurements of Diffusion Coefficients in Germanium and Silicon 274
18.3 Diffusion Profiles in Germanium 275
18.4 Mechanisms of Diffusion in Germanium 279
References 281

Appendix 283

Author Index 285

Subject Index 291

PREFACE

A few years ago I spent a sabbatical leave at the University of New Mexico in the Department of Earth and Planetary Sciences. Professor Rodney Ewing and his students and postdoctoral associates provided a stimulating atmosphere for my visit. This experience convinced me of the close relationship of many problems in geology and materials science. Diffusion, dissolution, and chemical reactions of gases such as water in minerals, crystals, and melts are of great importance in understanding magmas, volcanos, and growth of minerals. Materials scientists are also interested in these processes in many different applications, such as deterioration of silica optical fibers and mechanical properties of crystalline ceramics and glasses. Thus a detailed description of a model of diffusion and chemical reaction in solids and its application to specific systems should facilitate interaction between these two communities.

Initially, my emphasis was on diffusion and reaction of water in silicates, but I soon found that similar processes occur in other crystalline oxides and also for hydrogen. To my surprise I discovered that the diffusion of hydrogen in silicon showed phenomenology very similar to that of the diffusion of water in oxides; this finding led me to examine diffusion of many other elements in silicon, most of which react or exchange with the silicon. Therefore, the diffusion–reaction model is valid for a disparate collection of systems, from water and hydrogen in oxides to hydrogen, gold, and many dopants in silicon.

Mechanisms of diffusion and reaction that are much different from widely accepted ones are proposed in this book. The purpose of these proposals is to stimulate discussion and experiments to test both old and new ideas. Some examples of fresh proposals are: the importance of residual water in the diffusion of oxygen, the diffusion of silicon monoxide to transport oxygen and silicon in silicon, and the diffusion of molecular hydrogen in silicon. Several subjects related to diffusion and reaction are also discussed in the book, including hard-sphere models of atoms, oxidation states of solids and melts, openness of crystals, glasses and melts, and the structure of oxygen in silicon. Some of these discussions are original and different from many previous ideas about these subjects.

I have benefited from many discussions with colleagues about subjects in this book, especially Professor Minoru Tomozawa of Materials Science and Engineering and Professor Bruce Watson of Earth and Environmental Sciences at Rensselaer. This is the third book that Mrs. Kathy Curtice has typed for me. She has devoted much care and patience to the preparation of this volume.

Robert H. Doremus

Troy, New York

DIFFUSION OF REACTIVE MOLECULES IN SOLIDS AND MELTS

1

INTRODUCTION

1.1 IMPORTANCE

Chemical reactions in solids and melts can influence many of their properties. Water in oxides can break up the oxide structure, leading to lower strength, shorter fatigue failure, more rapid creep of solids, and lower viscosity of melts. Outgassing of water in magmas can change their viscosity and cause explosive eruptions, so geologists have been especially interested in the reactions of water in minerals and their melts. Water or hydrogen in fiber optic cables of silica glass can reduce their optical transmission, leading to unacceptable performance. Mechanical fatigue in oxides results from the stress-enhanced reaction of water at crack tips at the oxide surface.

The state of oxidation of a solid or melt is determined by the solubility and reactivity of gases such as water, hydrogen, carbon monoxide, and oxygen in the solid or melt. The state of oxidation determines the valency and consequently the optical absorption of transition metal ions such as iron, copper, and manganese. In silicon the exchange of dopant atoms such as boron, phosphorus, and arsenic with network silicon atoms leads to generation of electrical carriers required for electronic devices. Hydrogen can react with dangling bands in silicon to improve its properties. Chemical reactions in these and many other solids influence properties, especially mechanical, optical, and chemical properties. Many of these effects are just being recognized as new materials and applications are developed. These reactions are also of fundamental importance in a more complete characterization of structure and chemistry of solids and melts.

1.2 DEFECTS

In the years following World War II, an understanding of the role of structural defects on the properties of solids burgeoned. Concentrations of lattice vacancies in metals were measured and found to be directly related to rates of diffusion; the mechanism of vacancy diffusion of substitutional atoms in metals and alloys is firmly established (see, e.g., Refs. 1 and 2). Small atoms such as carbon, nitrogen, oxygen, and hydrogen were found to dissolve and diffuse interstitially in metals, especially those with body-centered cubic structure, which is more open than face-centered cubic structure [3, 4]. Diffusion by jumping of ions into lattice vacancies also explained experimental diffusion coefficients in alkali metal halides and other ionic solids [5].

The great success of these mechanisms involving lattice vacancies in metals and ionic solids led naturally to the application of vacancy mechanisms to explain diffusion of substitutional atoms and ions in covalently bonded solids. Thus the diffusion of water in oxides was explained by the jumping of hydroxyl ions (OH^-) from one site to another in the solid, or by the motion of other lattice constituents, such as oxygen ions. However, interstitial diffusion of water molecules in silica glass was also proposed [6]. When I suggested the dissolution and diffusion of molecular water and its subsequent reaction with the silicon–oxygen network to explain the diffusion of water in silica glass [7], this mechanism was strongly resisted by adherents of ionic and vacancy diffusion in the silica. However, this diffusion–reaction mechanism has gradually been accepted for water in silica glass, because it explains the concentration dependence of the effective diffusion coefficient of the water and its pressure dependence (see Section 1.3).

In covalently bonded solids such as oxides and semiconductors, the energy to break the primary bonds is large, so that vacancy formation requires a large energy and is unfavorable. To move vacancies and lattice elements in these solids also requires high energies because several strong covalent bonds must be broken and re-formed in this motion, so their rate of diffusion should be low. In metals and ionic solids the bonds are not directional and are usually weaker than in covalent solids, so the formation and motion of vacancies are more favorable. Furthermore, most covalently bonded solids have relatively open structures, so the dissolution and motion of interstitial atoms and molecules are favored. An important purpose of this book is to explore the extension of the diffusion–reaction mechanism to other systems and diffusants beyond silica glass and water.

In silicon and other semiconductors, such as germanium, diffusion was first explained in terms of lattice vacancies. However, the realization that many elements, including copper and iron, diffuse very rapidly in silicon led to the adoption of interstitial mechanisms of diffusion in silicon. The elements of groups III and V of the periodic table appear to diffuse much more slowly in silicon, so their diffusion mechanisms were not as clear. Various combinations of interstitial and vacancy diffusion for these dopant elements in silicon are often suggested [8]. In this book I have taken the view that lattice vacancies in silicon are effectively immobile, and that diffusion in silicon is always interstitial, because of its very open structure. The kick-out or exchange mechanism is therefore predominant for dopants in silicon, in which interstitially dissolved dopant atoms exchange with silicon atoms in the silicon structure to give interstitial silicon atoms (self-interstitials). For diffusion of hydrogen in silicon, the traditional mechanism involves diffusion of hydrogen atoms or ions that are bonded to silicon; molecular hydrogen is immobile. I propose that the mobile species is interstitial molecular hydrogen, and that bonded hydrogen atoms and ions are immobile.

One important conclusion of the discussions in this book is that interstitial diffusion plus chemical reaction is a more usual transport mechanism in covalent solids than is diffusion of lattice vacancies or lattice elements. Exploration of the applicability of the diffusion–reaction mechanism to covalently bonded systems not discussed in this book should therefore be profitable.

1.3 DIFFUSION–REACTION MECHANISMS

In a diffusion–reaction mechanism, a molecule dissolves and diffuses interstitially in a solid or melt. An example is the diffusion of water in an oxide, for example a silicate; the reaction is

$$H_2O + Si\text{—}O\text{—}Si = 2SiOH \tag{1.1}$$

In a similar way, hydrogen can dissolve in silicon and react with dangling silicon–silicon bonds, especially in amorphous silicon:

$$H_2 + 2Si = 2SiH \tag{1.2}$$

The binary reverse reactions in these equations result in an effective diffusion coefficient that is directly proportional to the concentration of reaction groups (see Chapter 3). It is the concentration of these reacted groups that is actually measured in an experiment. Since the concentration of reacted groups is usually much greater than the concentration of interstitially dissolved molecules (H_2O or H_2), the effective diffusion coefficient calculated from the profiles of reactants is much smaller than the diffusion coefficient of the diffusing interstitial atoms. The reacted groups absorb dissolved interstitial atoms, reducing their concentration and their diffusion distance.

Because the effective diffusion coefficient is proportional to reactant concentration, the measured profile of reactant is different from that expected for a diffusion coefficient independent of concentration. Another result is that the effective diffusion coefficient is proportional to the pressure of the water vapor or hydrogen that supplies the diffusing material. Furthermore, the diffusion is more rapid for sorption than for desorption. *Sorption* means diffusion into a material that contains no reactant; *desorption* means diffusion out of a material containing reactant (e.g., OH groups). These properties of the measured diffusion coefficients are signs that the diffusion–reaction mechanism is operative.

At lower temperatures it is possible that Eqs. 1.1 and 1.2 do not reach local equilibrium (see Chapter 4 for details). *Local equilibrium* means that at each point in the solid or melt the participants in Eq. 1.1 or 1.2 follow the equilibrium constant K:

$$K = \frac{S^2}{C}. \tag{1.3}$$

In Eq. 1.3, S is the concentration of reactant (SiOH or SiH) and C is the concentration of interstitial molecules (H_2O or H_2). It is assumed that the concentrations of structural groups Si–O–Si or Si are constant. If local equilibrium is not reached, the concentration of reactant in sorption is lower than expected; for example, it is lower than the equilibrium value of Eq. 1.3 at the solid surface in contact with a gaseous source (H_2O or H_2). In a diffusion couple with two parts containing different reactant concentrations, lack of local equilibrium can lead to a concentration step at the interface between the two parts. Furthermore, effective diffusion coefficients can become functions of time if local equilibrium does not hold. Again, these features are evidence of the diffusion–reaction mechanism.

1.4 OUTLINE OF THE BOOK

This book has three distinct parts. In the first part (Chapters 2 to 5) pertinent diffusion equations and structural factors are described, the second part (Chapters 6 to 13) covers experimental results on diffusion and structural and chemical features in oxides, and diffusion in silicon and germanium is described in the third part. More details about each chapter are given next.

In Chapter 2 the differential equations for one-dimensional diffusion are derived, and solutions for semi-infinite and slab geometries with a constant diffusion coefficient are given. Then some examples of solutions with a diffusion coefficient directly proportional to concentration, which arises in the diffusion–reaction model, are given. This model is introduced in Chapter 3 for conditions of local equilibrium, as defined previously. Diffusion–exchange conditions, where there is no net chemical reaction, are also described in Chapter 3; in silicon interstitial diffusion plus exchange is sometimes called the kick-out mechanism. In Chapter 4 the diffusion–reaction model without local equilibrium is treated. This condition can lead to steps in the interfacial concentrations of diffusant, and time-dependent apparent diffusion coefficients; these effects are often explained with complicated models of defects and traps, but they arise naturally in the diffusion–reaction model. In Chapter 5, I describe atom sizes and crystal structures. The hard-sphere model is valuable as an approximation for atoms, but can be misleading, especially for covalently bonded solids. The openness of solid structures is described and is closely related to values and activation energies of interstitial diffusion coefficients discussed in subsequent chapters.

The diffusion and reaction of water in a series of specific oxide systems is discussed in Chapters 6 to 9. In Chapter 6 the extensive data on diffusion of water in silica glass are compared to the diffusion–reaction model. The effective diffusion coefficients show puzzling effects of water concentration, sorption and desorption, temperature, and diffusion time (at temperatures below about 600°C). When these results are analyzed with the diffusion–reaction model, the resulting diffusion coefficients of molecular water are independent of concentration, time, and type of diffusion experiment, and fit on a single Arrhenius curve from 200 to 1200°C. These results are strong evidence for the diffusion–reaction model.

In Chapter 7 it is shown that diffusion of water in crystalline quartz takes place by molecular dissolution and diffusion of water in the quartz, and reaction with the silicon bonds in the quartz structure by Eq. 1.1, even though the quartz structure is more dense than that of vitreous silica. Especially important evidence for the diffusion–reaction model in quartz is the proportionality of the effective diffusion coefficients to the water vapor pressure, as found by Farver and Yund [9].

The diffusion–reaction model fits diffusion experiments of water in rhyolite glass (obsidian), as described in Chapter 8. A survey of activation energies of diffusion of water in a wide variety of crystalline and amorphous oxide is provided in Chapter 9. The activation energy is directly related to the ionic porosity or openness of the oxide structure, which is good evidence for the interstitial molecular diffusion and dissolution of water molecules in the oxide structures. Even the dense mineral pyrope (a garnet) shows a direct proportionality of the effective diffusion coefficient on water concentration, as expected from the diffusion–reaction model [10].

In Chapter 10 the speciation of water in some mineral glasses in shown to fit a Langmuir model of limited reaction sites. *Speciation* is the distribution or ratio of water between interstitially dissolved molecules and the reacted SiOH groups. The concentration of these groups saturates at high total water concentrations, as expected from the limited site model. In Chapter 11 the diffusion of hydrogen molecules in oxides is discussed. There are many data on hydrogen diffusion in silica glass, recently because of the potential for degradation of the optical transmission of fiber optic cables of silica glass from formation of SiOH groups by reaction with hydrogen. The results on diffusion of hydrogen in silica glass are consistent with the diffusion–reaction model. There are also data on hydrogen diffusion and reaction in other silicate glasses and in crystalline oxides such as titania and diopside.

In the geological literature a distinction is made between the diffusion of hydrogen and oxygen from wet atmospheres into oxides. In my view both of these processes are controlled by the diffusion and reaction of interstitially dissolved water. The apparent diffusion coefficient of hydrogen from water vapor is much larger than the apparent diffusion coefficient of oxygen, because the oxygen isotope in water can exchange with all the structural oxygen in the oxide, whereas the concentration of reacted or network hydrogen is much lower than that of structural oxygen. See Chapter 6 for data on silica glass that establish this explanation.

The mechanism of diffusion of oxygen in oxides from nominally dry atmospheres is uncertain, as described in Chapter 12. There is good evidence [11] that residual water controls the diffusion of oxygen in silica glass from nominally dry oxygen. The extent to which this residual water is important in other oxides is explored in Chapter 12, where it is shown that the activation energies for diffusion of oxygen in nominally dry oxides is correlated with the openness or ionic porosity of the oxide. This result suggests that the diffusing entity is an interstitial molecule, possibly residual water. There is also a section in Chapter 12 on diffusion of oxygen in oxide melts. The most likely carriers of oxygen are found to be residual water or silicon monoxide (SiO).

Chapter 13 is a discussion of the oxidation state of solids and melts. The oxidation state is controlled by gases dissolved in the solid, mainly water, hydrogen, and carbon monoxide, and possibly molecular oxygen and carbon dioxide. The times required to reach equilibrium with these gases is estimated in this chapter; it is found that often, even at high temperatures, the oxidation state of a solid or melt is not uniform. The influence of oxidation state on the valency of transition metal ions dissolved substitutionally in a solid or melt is treated in Chapter 13 with the help of the thermodynamic phase rule.

In the third part of this book (Chapters 14 to 18) I discuss diffusion in silicon and germanium. There is an enormous literature on this subject because of its commercial importance and because of many elusive effects and poor reproducibility. There is general agreement that diffusing elements such as copper, lithium, and iron dissolve and diffuse interstitially in silicon, because they diffuse very rapidly with low activation energy, as described in Chapter 14, which includes a survey of diffusion in silicon. Diffusion experiments on gold in silicon show unusual U-shaped profiles, which can be explained with a diffusion–exchange or kick-out mechanism (see Chapter 14). This mechanism results when an interstitially diffusing atom exchanges with, or "kicks out", a silicon atom from the silicon structure to become an interstitial silicon atom (self-interstitial).

The mechanisms of diffusion of hydrogen and oxygen in silicon have been considered to be transport of hydrogen or oxygen covalently bonded to silicon in the silicon structure. I suggest that such mechanisms are unlikely because of the high energy levels required to break the covalent bonds. I suggest that diffusion and reaction of molecular hydrogen (Chapter 15) and silicon monoxide, SiO (Chapter 16), are more likely mechanisms for transport of hydrogen and oxygen in the open silicon structure.

Some years ago the diffusion of substitutional dopant atoms from groups III and V of the periodic table was thought to involve lattice vacancies and substitutional diffusion in silicon. However, more recently the greater importance of interstitial diffusion of groups III and V atoms and their subsequent exchange with network silicon atoms (the kick-out mechanism) has been recognized. Extensive data on diffusion of group III, IV, and V elements in silicon are summarized in Chapter 17, and it is shown that diffusion–reaction and diffusion–exchange mechanisms are appropriate. In Chapter 18 diffusion in germanium is compared to diffusion in silicon.

1.5 OMISSIONS

Fast chemical reactions in liquid solution can be limited by diffusion of reactants to a reaction site or molecule. Examples are oxygenation of red blood cells, quenching of fluorescence, photochemical reactions, and polymerization. It might seem that these diffusion-limited reactions are closely related to the diffusion–reaction mechanisms described in this book. However, there are at least three differences between diffusion-limited reactions and the diffusion–reaction mechanisms.

1. In the diffusion–reaction mechanisms described in this book, the emphasis is on the modification of a transport or diffusion process by a chemical reaction. Thus the diffusion of interstitial water into a solid or molten oxide is slowed by the reaction of the water with the oxide, or the diffusion of phosphorus in silicon is slowed by the exchange of the phosphorus with a structural silicon atom. In diffusion-limited reactions the emphasis is on the slowing of a chemical reaction because of slow diffusion together with the reactants; the results are reported in terms of a changed rate of a chemical reaction, whereas in this book the results are reduced apparent diffusion coefficients.
2. The present interest is in diffusion in solids and viscous melts, whereas diffusion-limited reactions are studied in fluid liquids, where diffusion coefficients are in a range of about 10^{-5} to 10^{-7} cm^2/s. Usually, the diffusion coefficients in solids and even viscous melts are much smaller than these, although there is a curious overlap with interstitial diffusion of a few elements (e.g., copper) in silicon at high temperatures, which can be as high as 10^{-5} cm^2/s at 1150°C.
3. In recent years the study of diffusion-limited reactions has led to extensive interest in complicated mathematical solutions of a wide variety of possible configurations. For example, various wave patterns have been found experimentally and lead to complicated solutions of the appropriate differential equations.

Thus the subject of diffusion-limited reactions is not treated in this book, although some of their mathematical treatments have special cases similar to solutions considered here. There is an excellent book on diffusion-limited reactions by Stephen Rice [12] which begins with a clear and careful explanation of diffusion-limited reactions with many specific examples. The author describes the Smoluchowski theory of diffusion-limited reactions and builds to more complicated mathematical cases. Especially insightful is Rice's Chapter 8, entitled "A Critique of the Diffusion Equation and Molecular Pair Treatments," where he analyzes the limits of validity of the diffusion equation, especially at short diffusion times. Two additional, more recent books on diffusion-limited reactions are Refs. 13 and 14; they both have a strong mathematical emphasis.

I have tried to discuss all usable and pertinent experimental data. For oxides, most data are accessible from tables or graphs of concentration profiles. The geological literature often includes tables of data, which are highly valuable for testing ideas. Many of the experimental data on diffusion in silicon are inaccessible. Most papers are short and fragmentary, with inadequate descriptions of experimental techniques and one or two concentration profiles. Thus for most diffusion studies in silicon it was not possible to examine profiles over a wide range of experimental conditions and compare measured profiles with calculated profiles. It is very sad that many valuable experimental data are lost because authors publish short papers quickly, rather than providing a complete description of all of their results.

REFERENCES

1. H. B. Huntington and F. Seitz, *Phys. Rev.* **61**, 315, 325 (1942).
2. A. H. Cottrell, *Theoretical Structural Metallurgy*, Edward Arnold, London, 1948, Chap. XII.
3. R. W. Powers and M. V. Doyle, *J. Appl. Phys.* **30**, 514 (1959).
4. D. N. Beshers, in *Diffusion*, American Society for Metals, Metals Park, OH, 1973, p. 209.
5. A. B. Lidiard, in *Encyclopedia of Physics*, S. Flügge, ed., Springer-Verlag, Berlin, 1957, p. 246.
6. W. G. Spitzer and J. R. Ligenza, *J. Phys. Chem. Solids* **17**, 196 (1961).
7. R. H. Doremus, in *Reactivity of Solids*, J. N. Mitchell, R. C. DeVries, R. W. Roberts, and P. Cannon, eds., Wiley, New York, 1969, p. 66.
8. P. M. Fahey, P. B. Griffin, and J. D. Plummer, *Rev. Mod. Phys.* **61**, 289 (1989).
9. J. R. Farver and R. A. Yund, *Chem. Geol.* **90**, 55 (1991).
10. L. Wang, Y. Zhang, and E. J. Essene, *Am. Mineral.* **81**, 706 (1996).
11. R. H. Doremus, *J. Electrochem. Soc.* **143**, 1992 (1996).
12. S. A. Rice, *Diffusion-Limited Reactions*, Chemical Kinetics, Vol. 25, Elsevier, Amsterdam, 1996.
13. P. Grindrod, *The Theory and Applications of Reaction–Diffusion Equations: Patterns and Waves*, Oxford University Press, Oxford, 1996.
14. E. Kotomin and V. Kuzovkov, *Modern Aspects of Diffusion-Controlled Reactions: Cooperative Phenomena in Bimolecular Processes*, Chemical Kinetics, Vol. 34, Elsevier, Amsterdam, 1996.

2

DIFFUSION WITHOUT CHEMICAL REACTION

2.1 INTRODUCTION

In this chapter the fundamental diffusion equations are described, and a few of their solutions are given for comparison with reactions that are presented in later chapters. Many more solutions are given by Crank [1] for different geometries, diffusion coefficient as a function of concentration, and diffusion plus chemical reaction. Carslaw and Jaeger [2] give solutions for heat conduction which are also applicable to diffusion problems. In an unusual treatment, Mikhailov and Özizik have combined solutions for different geometries in a unified treatment and include diffusion plus chemical reaction in their class VII solutions [3].

2.2 FUNDAMENTAL DIFFUSION EQUATIONS

The chemical potential μ_i of a substance i in a mixture is the partial molar Gibbs free energy:

$$\mu_i = \left(\frac{\partial G}{\partial n_i}\right)_{T,p,n} \tag{2.1}$$

in which G is the Gibbs free energy, n_i the number of moles or molecules of substance i, T and p the temperature and pressure, and n the number of moles or molecules of all other components in the mixture. The chemical potential gives a criterion for equilibrium in a thermodynamic system: The chemical potential μ_i of component i is uniform in a phase at uniform temperature, and the chemical potentials of i in different phases α, β, γ, ... are the same at equilibrium:

$$\mu_\alpha = \mu_\beta = \mu_\gamma \tag{2.2}$$

The chemical potential μ_i can also be related to the thermodynamic activity a_i of i:

$$\mu_i = \mu_0 + RT \ln a_i \tag{2.3}$$

in which μ_0 is a constant depending only on temperature, R the gas constant, and T the temperature. If there is a change of chemical potential along the one-dimensional coordinate x, a derivative of Eq. 2.3 gives

$$\left(\frac{\partial \mu}{\partial x}\right)_t = \frac{RT}{c}\frac{\partial c}{\partial x}\frac{\partial \ln a}{\partial \ln c} \tag{2.4}$$

where c is the concentration of substance i. Subscripts i are omitted because only one diffusion component is being considered. The diffusive flux J (amount of material flow per unit area) is assumed to be proportional to the gradient of the chemical potential:

$$J = -D\frac{\partial c}{\partial x}\frac{\partial \ln a}{\partial \ln c} \tag{2.5}$$

in which D is a diffusion coefficient, which can be a function of concentration. Equation 2.5 can be considered as a definition of the diffusion coefficient D. The SI units for D are m^2/s and are being used increasingly, but as much of the present literature on diffusion is in cm^2/s, the latter units are used in this book.

If the activity coefficient a/c is constant,

$$\frac{\partial \ln a}{\partial \ln c} = \frac{c}{a}\frac{\partial a}{\partial c} = 1 \tag{2.6}$$

and this condition is assumed throughout the book unless a concentration dependence of a/c is stated explicitly. Then

$$J = -D\frac{\partial c}{\partial x} \tag{2.7}$$

which is known as *Fick's first law of diffusion*.

A mass balance in a volume element $2dx$, $2dy$, $2dz$ (Fig. 2.1) can be derived from the fluxes that enter and leave this element. Let the concentration of diffusing substance be c at the center P of the element. If J is the flux at P, in the x direction, then at $x - dx$ or face $ABCD$, the rate at which diffusing substance enters the element is

$$4\,dy\,dz\left(J - \frac{\partial J}{\partial x}dx\right) \tag{2.8}$$

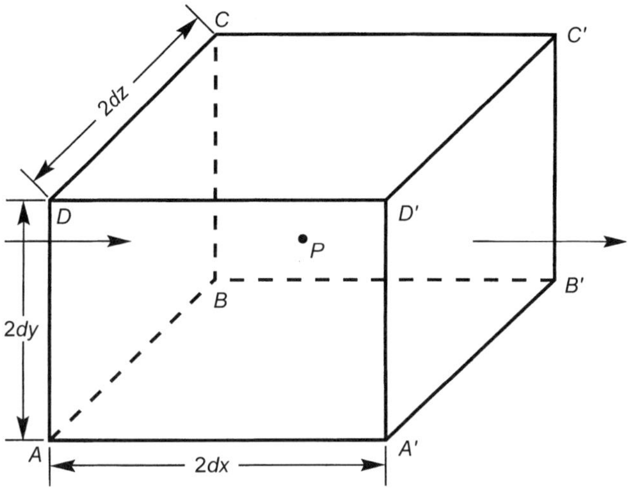

Figure 2.1 Element of volume 8 *dx dy dz* with fluxes of diffusion substance entering and leaving the element. (After Ref. 1, p. 3.)

and the rate of transfer out of the element at face $A'B'C'D'$ is

$$4\,dy\,dz\left(J + \frac{\partial J}{\partial x}dx\right) \tag{2.9}$$

The net rate of increase of concentration in the volume element is the difference between these two equations, or

$$-8\,dx\,dy\,dz\left(\frac{\partial J}{\partial x}\right)_t \tag{2.10}$$

This rate of increase is also

$$8\,dx\,dy\,dz\left(\frac{\partial c}{\partial t}\right)_x \tag{2.11}$$

where t is the time, so

$$\left(\frac{\partial c}{\partial t}\right)_x = \left(\frac{\partial J}{\partial x}\right)_t \tag{2.12}$$

for diffusion in the x direction only. From Eqs. 2.7 and 2.12,

$$\left(\frac{\partial c}{\partial t}\right)_x = \frac{\partial}{\partial x}\left(D\frac{\partial c}{\partial x}\right)_t \tag{2.13}$$

This is the *equation of continuity* or *Fick's second law*. If D is not a function of concentration,

$$\left(\frac{\partial c}{\partial t}\right)_x = D\left(\frac{\partial^2 c}{\partial x^2}\right)_t \tag{2.14}$$

2.3 INFINITE AND SEMI-INFINITE GEOMETRIES

A common type of diffusion experiment is to join two pieces of material with different uniform concentrations of diffusing substance, as shown in Fig. 2.2. The extent of diffusion in either direction from the interface $x = 0$ is much less than the thicknesses of the regions, making them essentially infinite. The solution to Eq. 2.14 for this initial condition and geometry is

$$c(x, t) = \frac{1}{2} c_i \,\mathrm{erfc}\left(\frac{x}{2\sqrt{Dt}}\right) \tag{2.15}$$

In this equation c_i is the uniform initial concentration for $x < 0$; initially, $c = 0$ for $x > 0$. Equation 2.15 is plotted in Fig. 2.2. The complementary or conjugate error function erfc is related to the error function erf by

$$\mathrm{erfc}(u) = 1 - \mathrm{erf}(u) = 1 - \frac{2}{\sqrt{\pi}} \int_0^u e^{-\lambda^2} d\lambda = \frac{2}{\sqrt{\pi}} \int_u^\infty e^{-\lambda^2} d\lambda \tag{2.16}$$

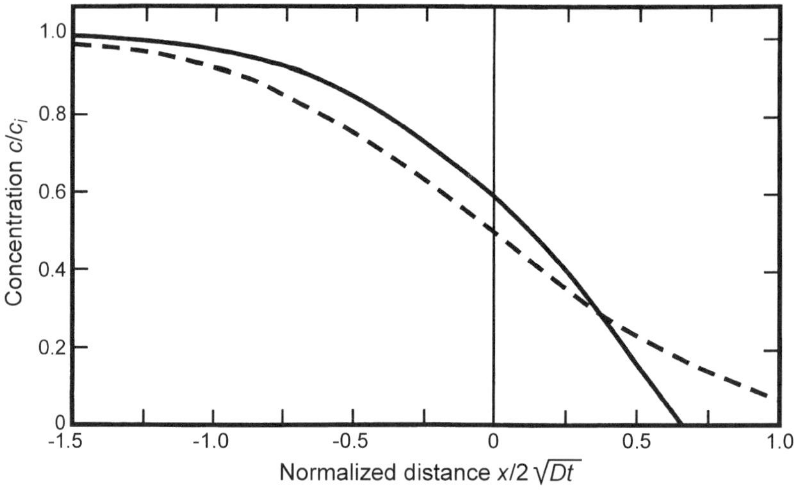

Figure 2.2 Profiles of diffusion in infinite geometry, or joining two regions of different uniform initial concentrations. Initial boundary conditions, $c = c_i$ at $x < 0$ and $c = 0$ at $x > 0$. Dashed line, diffusion coefficient not a function of concentration, from Eq. 2.15; solid line, diffusion coefficient directly proportional to concentration.

These functions have the properties erf(0) = 0, erf(∞) = 0, erf(− u) = u, erfc(0) = 1, erfc(∞) = 0, erfc(− u) = 2 − erfc(u). These functions are tabulated in Refs. 1 and 2. In Eq. 2.15, $c(\infty, t) = 0$ and $c(-\infty, t) = c_0$, as required by the condition of infinite geometry. At $x = 0$, $c = c_0/2$ for all $t > 0$. In this chapter solutions to the diffusion equation are stated without giving methods of derivation, which are described fully in Refs. 1 and 2.

The configuration of a substance whose surface is held at some boundary condition and whose thickness is much greater than the diffusion distance is called *semi-infinite geometry.* If the concentration of diffusing substance is held constant at c_0 at the sample surface, the boundary conditions are $x(0, t) = c_0$ for $x = 0$, $t > 0$, and $c(x, 0) = 0$ for $t = 0$, and the solution to Eq. 2.14 is

$$c(x, t) = c_0 \operatorname{erfc}\left(\frac{x}{2\sqrt{Dt}}\right) \tag{2.17}$$

This function is plotted in Fig. 2.3. If the surface is maintained at concentration zero and the initial concentration is c_i, then

$$c(x, t) = c_i \operatorname{erf}\left(\frac{x}{2\sqrt{Dt}}\right) \tag{2.18}$$

At $x = 0$, $c = 0$; as $x \to \infty$, $c \to c_i$; as $t \to \infty$, $c \to 0$ for all x.

The total amount M (units of mol/cm^2) of material that has diffused into or out of the semi-infinite space in time t is proportional to the square root of time:

$$M = 2c_i\sqrt{\frac{Dt}{\pi}} \tag{2.19}$$

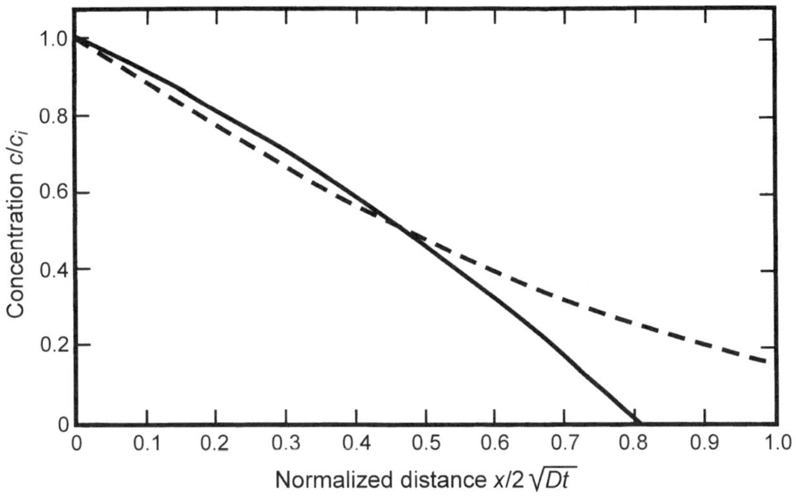

Figure 2.3 Profiles of diffusion in semi-infinite geometry, sorption. Boundary conditions $c = c_i$ at $x = 0$ for $t > 0$; $c = 0$ for $x > 0$, $t = 0$. Dashed line, constant diffusion coefficient; solid line, diffusion coefficient directly proportional to concentration.

If a total quantity of material M is deposited on the surface of a semi-infinite solid at $t = 0$, the resulting profiles are

$$c = \frac{M}{\sqrt{\pi D t}} \exp\left(-\frac{x^2}{4Dt}\right) \tag{2.20}$$

This type of experiment is often used for a tracer measurement, with a high amount of tracer being deposited on the sample surface. Equation 2.20 can be plotted as log c versus x^2 to give a straight line of slope $-1/4\,Dt$; thus one does not need to know M to calculate a diffusion coefficient.

2.4 PLANE SHEET

Diffusion in a plane sheet or membrane under steady-state conditions, $(\partial c/\partial t)_x = 0$, is important in diffusion-controlled oxidation of a metal or silicon, when the growth rate of oxide on the sample surface is slow enough to establish the steady state, and after long times of diffusion. This condition of steady-state diffusion through a membrane is also important in permeation of a gas through the membrane, as discussed below. If the concentration of diffusing substance is held at concentration c_i at the outer oxide surface $(x = 0)$ and at zero at the oxide–silicon or metal interface $(x = L)$, the concentration gradient is constant:

$$c(x, t) = c_i\left(1 - \frac{x}{L}\right) \tag{2.21}$$

The thickness L of oxide for diffusion control (parabolic growth) is

$$L^2 = \frac{c_i D t}{\rho} \tag{2.22}$$

in which ρ is the density of the oxide in concentration units and D is the diffusion coefficient of oxidant (water or oxygen) in the oxide. Often, especially for silicon, the growth of oxide is linear parabolic:

$$L^2 + AL = \frac{c_i D t}{\rho} \tag{2.23}$$

in which A is a coefficient for linear growth.

The complete solution for non-steady-state diffusion in the slab is an infinite series. If the concentration is initially c_i in the region $-L < x < L$, and the surfaces are held at concentrations c_0, then [1, p. 47]

$$\frac{c_i - c}{c_i - c_0} = 1 - \frac{4}{\pi}\sum_{n=0}^{\infty}\frac{(-1)^n}{2n+1} \exp\left[-\frac{D(2n+1)^2\pi^2 t}{4L^2}\right] \cdot \left[\cos\frac{(2n+1)\pi x}{2L}\right] \tag{2.24}$$

These concentration distributions are plotted in Ref. 1, p. 50, and Ref. 2, p. 101. If M_t is the total amount of diffusing substance that has entered the sheet at time t, and M_∞ at long time, then

$$\frac{M_t}{M_\alpha} = 1 - \sum_{n=0}^{\infty} \frac{8}{(2n+1)^2 \pi^2} \exp\left[-\frac{D(2n+1)^2 \pi^2 t}{4L^2}\right] \tag{2.25}$$

Solutions useful for short time periods are

$$\frac{c_i - c}{c_i - c_0} = \sum_{n=0}^{\infty} (-1)^n \text{erfc}\left[\frac{(2n+1)L - x}{2\sqrt{Dt}}\right] + \sum_{n=0}^{\infty} (-1)^n \text{erfc}\left[\frac{(2n+1)L + x}{2\sqrt{Dt}}\right] \tag{2.26}$$

and

$$\frac{M_t}{M_\alpha} = 2\left(\frac{Dt}{L^2}\right)^{1/2} \left[\frac{1}{\sqrt{\pi}} + 2\sum_{n=1}^{\infty} (-1)^n i\ \text{erfc}\left(\frac{nL}{\sqrt{Dt}}\right)\right] \tag{2.27}$$

in which i erfc is the integral of the conjugate error function:

$$i\,\text{erfc}(u) = \int_x^{\infty} \text{erfc}\,(\lambda)\, d\lambda \tag{2.28}$$

Jain [4] has suggested a way to determine an approximate solution for the plane sheet that is quite accurate to times of about $Dt/L^2 = 0.4$. The approximate equation for the concentration c/c_i, with initial concentration of c_i and concentrations at $x = \pm L$ held at zero is

$$\frac{c}{c_i} = \text{erf}\left(\frac{L+x}{2\sqrt{Dt}}\right) + \text{erf}\left(\frac{L-x}{2\sqrt{Dt}}\right) - 1 \tag{2.29}$$

Concentration profiles from this equation are given in Table 2.1 and Fig. 2.4, as compared with those from Eq. 2.24. The approximation of Eq. 2.29 is quite accurate for values of $Dt/L^2 < 0.4$. At $x = \pm L$ the Jain concentration is less than zero, which of course is not physically correct, but most of the curve fits the rigorous solution well for $Dt/L^2 = 0.4$. At $Dt/L^2 = 0.6$ the approximation is not as accurate, but it is still useful to obtain a reasonably close estimate of the profile. This method can also be used for a diffusion coefficient that is a function of concentration, as shown in Chapter 15.

Table 2.1 Comparison of an Approximate Solution and the Complete Solution for Concentration in a Plane Sheet[a]

	Concentration c/c_i	
$\pm x/L$	Complete (Eq. 2.24)	Approximate (Jain [4] and Eq. 2.29)
	$Dt/L^2 = 0.4$	
1.0	0	−0.028
0.8	0.144	0.136
0.6	0.384	0.274
0.2	0.478	0.450
0	0.478	0.478
	$Dt/L^2 = 0.6$	
1.0	0	−0.070
0.8	0.086	0.060
0.6	0.172	0.15
0.4	0.236	0.22
0.2	0.278	0.266
0	0.288	0.280

[a]Initial concentration c_i and surfaces $x = \pm L$ held at zero.

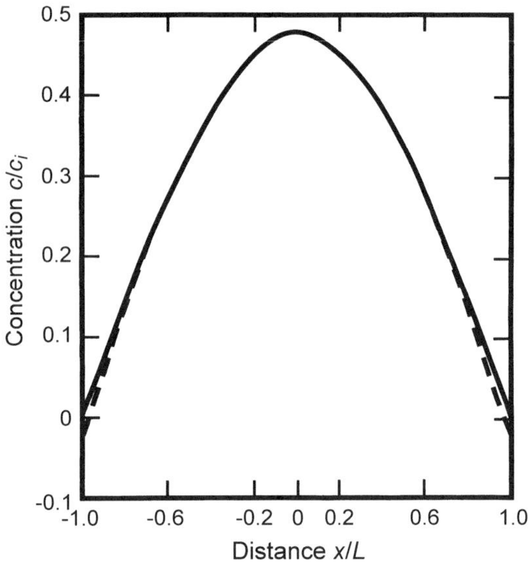

Figure 2.4 Concentration profiles in a plane sheet. Boundary conditions $c = c_i$ at $t = 0$ for $-L > x < L$; $c = 0$ at $x = \pm L$ for $t > 0$. $Dt/L^2 = 0.4$ (see Table 2.1). Solid line, complete solution from Eq. 2.21; dashed line, approximation from Eq. 2.29.

2.5 PERMEATION THROUGH A MEMBRANE

The permeation of a gas through a membrane is often used to measure the diffusion coefficient of the gas in the membrane material. The membrane is mounted between two chambers which are initially pumped out; then the permeating gas is introduced into one chamber, and the amount of gas that appears on the other side of the membrane is measured as a function of time. The amount of gas that permeates is kept so low that the drop in pressure at the inlet side is negligible and the pressure on the outlet side is small. Then the boundary conditions are: $c = c_i$ at $x = 0$, for all t; $c = 0$ at $x = L$ for all t; $c = 0$ for $t = 0$, all x; for a membrane thickness L. These are the same conditions as those for the steady-state result of Eq. 2.21. The concentration c_i in the solid membrane at $x = 0$ is maintained by dissolution from the gas phase; there is a distribution coefficient of gas between the gas phase and the solid. For example, in silica glass the ratio c_i/c_g of the concentration of gas c_i dissolved molecularly in the solid to the concentration c_g in the gas phase ranges from about 0.01 to 0.03 and is nearly independent of temperature [5]. The flux J of diffusing gas that passes through a unit area of membrane of thickness L in unit time for the boundary conditions above is [1, p. 49ff.]

$$J = \frac{Dc_i}{L}\left[1 + 2\sum_{n=1}^{\infty} \cos n\pi \, \exp\left(\frac{-Dn^2\pi^2 t}{L^2}\right)\right] \tag{2.30}$$

where t is the time after emission of the gas. Holstein (in Ref. 6) has shown that this equation can be transformed to

$$J = 2c_i\left(\frac{D}{\pi t}\right)^{1/2} \sum_{n=0}^{\infty} \exp\left[\frac{-L^2(2n+1)^2}{4Dt}\right] \tag{2.31}$$

where the series converges rapidly for small t. For long times $t \gg D/L^2$, these equations converge to $J = Dc_i/L$, which is the steady-state solution with the profile of Eq. 2.21. The measured pressure p of gas on the outlet side is related to J by the equation

$$J = \frac{V}{ART_p}\frac{dp}{dt} \tag{2.32}$$

where V is the volume on the outlet side where p is measured, A the surface area of the glass membrane, R the gas constant, and T_p the temperature at which the pressure is measured. The results of permeation measurements in the steady state are usually expressed in terms of a permeation velocity K, which is defined as

$$K = \frac{VL}{\Delta p A}\frac{dp}{dt} = \frac{LRT_pJ}{\Delta p} \tag{2.33}$$

where Δp is the pressure difference across the membrane. If the outlet pressure is negligible, $\Delta p = c_g RT$, where c_g is the concentration of gas on the inlet side in mol/cm^3 and T is the temperature of the diffusion experiment. Then, since $J = Dc_i/L$, $K = Dt_p c_i/c_g T$. In this book the solubility of gas in a solid or melt is defined as

$$\alpha = \frac{c_i}{c_g} \tag{2.34}$$

Thus K is proportional to this solubility, and the units of K are cm^2/s, since $T_p c_i/c_g T$ is dimensionless. A discussion of the units of K and solubility is given by Rogers *et al.* [6] in an appendix to their article.

In the early stages of permeation, D can be calculated from the measured change in pressure with time (dp/dt) in the outlet side. For this purpose, Eq. 2.31 is useful, since for short times only the first term of the series is needed, so that a plot of $\log(t^{1/2}dp/dt)$ against $1/t$ is linear, because J is proportional to dp/dt; the slope of the line is equal to $L^2/4D$, from which D can be calculated since L is known. This method, suggested by Rogers *et al.* [6], gives the most accurate values of D from permeation data. The intercept of this line with the $\log(t^{1/2}dp/dt)$ axis ($1/t \to 0$) can also be used to calculate the solubility if D is known. When the outlet pressure p_o is plotted against time, the curve becomes linear at long times (this is also the steady-state region), and the diffusion coefficient can also be found from the intercept t_c of this line with the pressure axis from the relation $D = L^2/6t_c$, as shown by Barrer [7].

If the gas on the inlet side is pumped out rapidly after the steady state is attained, the diffusion coefficient can be calculated from the rate at which the dissolved gas diffuses out of the membrane. The initial concentration for this desorption is given by $c = c_i(1 - x/L)$, and the boundary conditions are that $c = 0$ at both surfaces. The flux at the outlet is then [1]

$$J = \frac{2c_i D}{L} \sum_{n=0}^{\infty} \cos n\pi \exp\left(\frac{-Dn^2\pi^2 t}{L^2}\right) \tag{2.35}$$

Thus the diffusion coefficient and solubility can be calculated from permeability data in a variety of ways, and comparisons between the various calculations should give some idea of the reliability of the results.

In the powder method, the diffusion coefficient is calculated from the amount of gas absorbed or desorbed from a powdered sample, and the solubility is found from the total amount of gas absorbed or desorbed per unit weight of solid. For desorption from spheres of radius R, the amount of gas desorbed Q is [8, 9]

$$\frac{Q}{Q_\infty} = \frac{6(Dt)^{1/2}}{\pi R} - \frac{3Dt}{R^2} \tag{2.36}$$

where Q_∞ is the total amount of gas desorbed. A comparison of the experimental curve with Eq. 2.36 gives a measure of D when t and Q_∞ are known.

2.6 DIFFUSION COEFFICIENT A FUNCTION OF CONCENTRATION

Crank [1] has given many methods for solving the diffusion equations when diffusion coefficients are a function of concentration and presents equations and graphs for these solutions. In this section solutions for a diffusion coefficient directly proportional to concentration are given, because this functionality is found in several different diffusion–reaction equations. These solutions are given in Crank [1, p. 119ff.); the original solutions were given in Refs. 10 and 11.

Profiles for different boundary conditions and geometries are given in Table 2.2 and Figs. 2.2, 2.3, and 2.5 for

$$D_e = CD_i \tag{2.37}$$

in which D_e is the effective diffusion coefficient that is directly proportional to a dimensionless concentration $C = c/c_i$; D_i is the diffusion coefficient at concentration c_i. For semi-infinite geometry and sorption ($C = 0$ for all x at $t = 0$, and $c = c_i$ at the surface $x = 0$), profiles are given in Table 2.2 and Fig. 2.3. The profile intersects the abscissa ($c = 0$) at a finite value $x/2\sqrt{D_i t} = 0.81$ with a finite slope. Thus the advancing velocity of the diffusing material at $c = 0$ is $dx/dt = 1.31 D_i/x = 0.81\sqrt{D_i/t}$.

Table 2.2 Profiles for Diffusion Coefficient Directly Proportional to Concentration[a]

Sorption, Semi-infinite [1, Fig. 7.5]		Desorption, Semi-infinite [1, Fig. 7.7]		Infinite [1, Fig. 7.6]	
$c=0,\ t=0$	$c=c_i,\ x=0$	$c=c_i,\ t=0$	$c=0,\ x=0$	$c=0,\ x>0\ t=0$	$c=c_i,\ x<0,\ t=0$
1.0	0	0	0	1.0	−2.0
0.9	0.115	0.27	0.1	1.0	−1.6
0.8	0.218	0.40	0.2	0.99	−1.6
0.7	0.310	0.57	0.4	0.97	−1.0
0.6	0.398	0.65	0.5	0.94	−0.8
0.5	0.477	0.69	0.6	0.89	−0.6
0.4	0.549	0.78	0.8	0.81	−0.4
0.3	0.619	0.86	1.0	0.71	−0.2
0.2	0.685	0.94	1.5	0.59	0
0.1	0.749	—	—	0.45	0.2
0	0.810	—	—	0.27	0.4
				0.07	0.6
				0	0.66

Source: Ref. 1, p. 119ff.

[a]Columns are c/c_i; $x/2\sqrt{D_i t}$ for each condition.

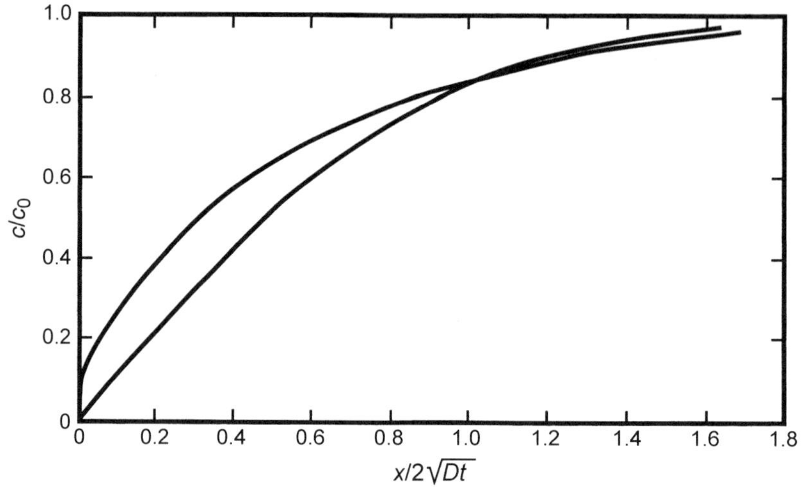

Figure 2.5 Concentration profiles in semi-infinite geometry, desorption. Boundary conditions $c = 0$ at $x = 0$ for $t > 0$; $c = c_i$ for $x > 0$, $t = 0$. Lower line, diffusion coefficient constant; upper line, diffusion coefficient directly proportional to concentration.

Barber [12] has described an analytical solution for the diffusion coefficient of Eq. 2.37 with an amount of material Q deposited at $x = 0$ and $t = 0$. In Barber's treatment C in Eq. 2.37 is a dimensionless concentration and Q has dimensions of length. His solution for semi-infinite geometry is:

$$C = \frac{1}{2}\left(\frac{3Q^2}{D_i t}\right)^{1/3} - \frac{x^2}{6D_i t} \tag{2.38}$$

Concentration profiles from this equation are plotted in Fig. 2.6 from Table 2.3. This profile goes to zero concentration at a finite value of $x = x_0$, which is

$$x_0 = (9QD_i t)^{1/3} \tag{2.39}$$

Thus the intercept increases proportional to $t^{1/3}$ for this boundary condition instead of $t^{1/2}$ for constant surface concentration. The velocity dx/dt of the intercept is

$$\frac{dx}{dt} = \frac{1}{3}\frac{(9QD_i)^{1/2}}{t^{2/3}} \tag{2.40}$$

instead of being inversely proportional to $t^{1/2}$.

In infinite geometry with the diffusion coefficient of Eq. 2.37 and with boundary conditions $c = c_i$ at $t = 0$ and $x < 0$ and $c = 0$ for $x > 0$, the profiles are as shown in Table 2.2 and Fig. 2.2. Again the profile ends at a finite value of $x/2\sqrt{D_i t} = 0.66$. The value of c/c_i at $x = 0$ is 0.59 instead of 0.5 as it is for this geometry and a constant diffusion coefficient. The profile for desorption in semi-infinite geometry

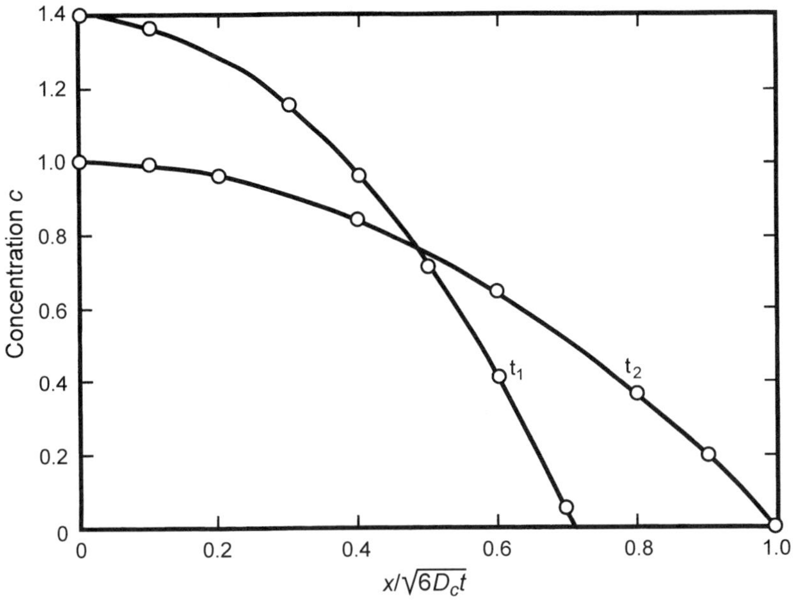

Figure 2.6 Profiles of diffusion in semi-infinite geometry for the diffusion coefficient proportional to concentration (Eq. 2.37) when an amount Q of diffusing material is deposited at $x=0$ and $t=0$. Lines from Eq. 2.38 for two different times $t_2=2.74t_1$. From Barber [12].

Table 2.3 Concentration Profile from a Diffusion Coefficient Directly Proportional to Concentration (Eq. 2.37) with an Amount Q of Diffusing Deposited at $x=0$ and $t=0$, from Eq. 2.38

	Concentration C	
$x/\sqrt{6Dt}$	t_1, $C_0=1.4$	t_2, $C_0=1.0$
0	1.400	1.0
0.1	1.373	0.990
0.2	1.290	0.960
0.3	1.153	0.910
0.4	0.961	0.840
0.5	0.714	0.750
0.6	0.412	0.640
0.7	0.055	0.510
0.8	—	0.360
0.9	—	0.190
1.0	—	0
0.714	0	—

($c = c_i$ for all x at $t = 0$; $c = 0$ at $x = 0$ for $t > 0$) is quite different from the profile for sorption and is somewhat similar to that for a constant diffusion coefficient except near $c = 0$. The finite intersections at $c = 0$ for sorption result from the zero diffusion coefficient at this concentration, meaning that no flux can pass the abscissa intersection points. In desorption the slope dc/dx goes to infinity at $x = 0$ ($c = 0$); thus there still can be a finite flux desorbing at the surface even though the effective diffusion coefficient is zero there. The total amount of material M that has diffused into or out of the solid is proportional to the square root of time:

$$M = 0.94 c_i \sqrt{D_i t} \tag{2.41}$$

in which D_i is the diffusion coefficient from Eq. 2.37 at concentration c_i ($x = 0$) for sorption and $t = 0$ for desorption (Table 7.6 in Ref. 1).

It is possible to calculate diffusion coefficients that are a function of concentration from a concentration profile by the *Matano method* [1, p. 230ff.; 13]. This method involves taking the slope of the concentration profile and calculating certain integrals; with the usual errors in measurements of diffusion profiles, this method is often subject to large errors in the diffusion coefficient calculated. The method frequently yields concentration dependencies of diffusion coefficients that are highly misleading, especially at small concentrations of diffusing material. A much better procedure is to assume a concentration dependence of D values, calculate a profile, and compare it with experimental data. The latter method is used in this book.

REFERENCES

1. J. Crank, *The Mathematics of Diffusion*, 2nd ed., Oxford University Press, Oxford, 1975.
2. H. C. Carslaw and J. C. Jaeger, *Conduction of Heat in Solids*, 2nd ed., Oxford University Press, Oxford, 1959.
3. M. D. Mikhailov and M. N. Özizik, *Unified Analysis and Solutions of Heat and Mass Diffusion*, Dover, New York, 1994.
4. S. C. Jain, *Proc. R. Soc.* **A243**, 359 (1958).
5. R. H. Doremus, *Glass Science*, 2nd ed., Wiley, New York, 1994, Chap. 8.
6. W. A. Rogers, R. S. Burlitz, and D. Alpert, *J. Appl. Phys.* **25**, 86 (1954).
7. R. M. Barrer, *Diffusion in and through Solids*, Cambridge University Press, Cambridge, p. 117ff.
8. R. W. Barrer and D. E. W. Vaughn, *Trans. Faraday Soc.* **63**, 2275 (1967).
9. K. N. Woods and R. H. Doremus, *Phys. Chem. Glasses* **12**, 69 (1971).
10. C. Wagner, *J. Chem. Phys.* **18**, 1227 (1950).
11. J. E. Wilkins, *J. Soc. Ind. Appl. Math.* **11**, 632 (1963).
12. W. D. Barber, *General Electric Corporate Research and Development Report 70-C-390* (1970).
13. C. Matano, *Jpn. J. Phys.* **8**, 109 (1932–1933).

3

DIFFUSION AND REACTION WITH LOCAL EQUILIBRIUM

3.1 INTRODUCTION

Chemical reactions of substances diffusing in solids or liquids are common. The rate of diffusion is influenced by removal or supply of reacting species; alternatively, a chemical reaction is slowed or speeded up by removal or supply of reacting species. There are many examples of these diffusion–reaction processes in physiological systems, gels, and polymers. In this book I focus on diffusion and reaction of dissolved molecules in solids or viscous melts. An example is the molecular diffusion of water in an oxide such as silica glass (SiO_2) and its reaction with the silicon–oxygen network:

$$\text{Si—O—Si} + H_2O = \text{SiOH} + \text{HOSi} \tag{3.1}$$

Another example is the reaction of molecular hydrogen dissolved in silicon:

$$H_2 + 2\text{Si} = 2\text{SiH} \tag{3.2}$$

In this chapter it is assumed that these reactions go to equilibrium at each place throughout the solid; this situation is called *local equilibrium*. Local nonequilibrium conditions are considered in Chapter 4.

In this chapter two distinctly different kinds of diffusion experiments are treated. In the first, called *reaction experiments*, a substance diffuses into a solid material that has no reacted groups in it. An example is the dissolution and diffusion of water vapor into silica glass containing no SiOH groups. The dissolved molecular water reacts with the silica network to form SiOH groups by Eq. 3.1. A related kind of experiment starts with silica glass with a certain uniform concentration of SiOH already in it. Then these groups combine by Eq. 3.1 to form molecular water in the glass; the water then diffuses out of the glass. In these reaction experiments the concentrations of the diffusing and reacting groups can be followed by analyzing chemically for the concentration as a function of distance and time at a constant temperature. For example, water can be analyzed from infrared spectra or with radioactive or isotopic tracers, such as deuterium, tritium, and oxygen-18. In these experiments the tracers are used simply as a way to measure the total chemical concentration of reacted SiOH groups plus dissolved molecular water.

In a second class of experiments, called *exchange experiments*, the starting material initially contains a uniform concentration of reacted groups: for example, SiOH groups. These reacted groups are at local equilibrium (by Eq. 3.1) with untagged dissolved molecular water ($H_2{}^{16}O$). Then an external source of tagged water (e.g., $H_2{}^{18}O$) is introduced and dissolves and diffuses in the material. The total water concentration in the material ($H_2{}^{16}O + H_2{}^{18}O$) is constant in time and space. The ^{18}O oxygen in the dissolved molecular water exchanges with ^{16}O in the material. If the tracer is deuterium or tritium, the exchange is only with the hydrogen in the SiOH groups, but if the tracer is oxygen-18, it can exchange both with oxygen in the SiOH groups and in the remaining silicon–oxygen network.

3.2 REACTION EXPERIMENTS

To derive a differential equation for diffusion plus reaction, one can consider the differential volume in Fig. 2.1. In addition to fluxes of material entering and leaving the volume, it is now possible to remove or generate diffusing material in the differential volume. The rate of removal of reacting material from the dissolved to the reacted state is $(\partial S/\partial t)_x$ at each point x corresponding to the differential volume; $(\partial S/\partial t)_x$ is also the rate of increase of reacted material in the volume. S is therefore the concentration of reacted groups (SiOH groups in the example above). The quantity $(\partial S/\partial t)_x$ must be subtracted from the rate of change of flux in the volume, because $(\partial S/\partial t)_x$ represents a reduction in the rate of dissolved diffusing material. The concentration of this dissolved material is designated as C; thus from Eq. 2.14 the usual diffusion equation without reaction must be modified to

$$\left(\frac{\partial C}{\partial t}\right)_x = D\left(\frac{\partial^2 C}{\partial x^2}\right)_t - \left(\frac{\partial S}{\partial t}\right)_x \tag{3.3}$$

with the assumption that the diffusion coefficient D of diffusing substance is not a function of x, t, or C (see Ref. 1, p. 326ff.).

To solve Eq. 3.3 for diffusion profiles for different geometries and boundary conditions, one must first have a relation between the dissolved and reacted concentrations S and C. The simplest situation is a linear relation between S and C

at a particular temperature:

$$S = K_1 C \tag{3.4}$$

where K_1 is not a function of concentrations. Then

$$\frac{\partial S}{\partial t} = K_1 \frac{\partial C}{\partial t} \tag{3.5}$$

and from Eq. 3.3,

$$\frac{\partial C}{\partial t} = \frac{D}{1 + K_1} \frac{\partial^2 C}{\partial x^2} \tag{3.6}$$

This equation is just the usual continuity equation (2.14) with a constant diffusion coefficient $D/(1+K_1)$. If the concentration C of dissolved material is much greater than that of the reacted concentration S, the diffusion of dissolved material is not much influenced by the reaction. If $S \gg C$, the effective diffusion coefficient is much less than that of the dissolved material alone. A large flux of dissolved material must flow to provide material for the reacted groups. Solutions of the diffusion equations valid for a constant diffusion coefficient are applicable for all geometries and boundary conditions, with the effective diffusion coefficient $D/(1+K_1)$.

In many reactions the simple relation of Eq. 3.4 does not hold. A more general relation is

$$S^n = K_n C \tag{3.7}$$

where K_n and n are constants and n is usually a small integer. From Eq. 3.7,

$$\left(\frac{\partial C}{\partial x}\right)_t = \left(\frac{nS^{n-1}}{K_n} \frac{\partial S}{\partial x}\right)_t \tag{3.8}$$

Equation 3.3 can be written

$$\frac{\partial C}{\partial t} = \frac{\partial}{\partial x}\left(D \frac{\partial C}{\partial x}\right) - \frac{\partial S}{\partial t} \tag{3.9}$$

Then from Eq. 3.8,

$$\left(\frac{\partial S}{\partial t}\right)_x = \left(1 + \frac{S}{nC}\right) \frac{\partial}{\partial x}\left(\frac{DnS^{n-1}}{k_n} \frac{\partial S}{\partial x}\right)_t \tag{3.10}$$

If $S \gg C$, which is often valid,

$$\frac{\partial S}{\partial t} = \frac{\partial}{\partial x}\left(\frac{DnS^{n-1}}{K_n} \frac{\partial S}{\partial x}\right)_t \tag{3.11}$$

This equation is just that for an effective diffusion coefficient (DnS^{n-1}/K_n) that is a function of concentration. Equation 3.11 is a formal description of the concentration S of reacted groups; in reality, these groups are immobile compared to diffusion of the dissolved component (concentration C). Even if the relation between S and C is different from Eq. 3.7, one can find an effective concentration-dependent diffusion coefficient for profiles of S if $\partial S/\partial t \gg \partial C/\partial t$.

In many reactions of interest, the exponent n in Eq. 3.7 is equal to 2: for example, in both Eqs. 3.1 and 3.2. Then the effective diffusion coefficient D_e for Eq. 3.11 is

$$D_e = \frac{2SD}{K_2} \tag{3.12}$$

so that D_e is directly proportional to the concentration S. Solutions of Eq. 3.11 for this functionality are discussed in Chapter 2. Other values of n different from 1 lead to a more complicated concentration dependence of the effective diffusion coefficient; the concentration profiles can be determined by a numerical integration of Eq. 3.11. I have not found solutions of Eq. 3.11 for n values other than 1 or 2.

The solutions of the diffusion equations for the diffusion coefficient of Eq. 3.12, in which D_e is directly proportional to the concentration, give an advancing front at which the concentration of diffusing material is zero in semi-infinite geometry. This front moves with a distance proportional to the square root of time (see Fig. 3.1). Another important geometry is cylindrical, because of the

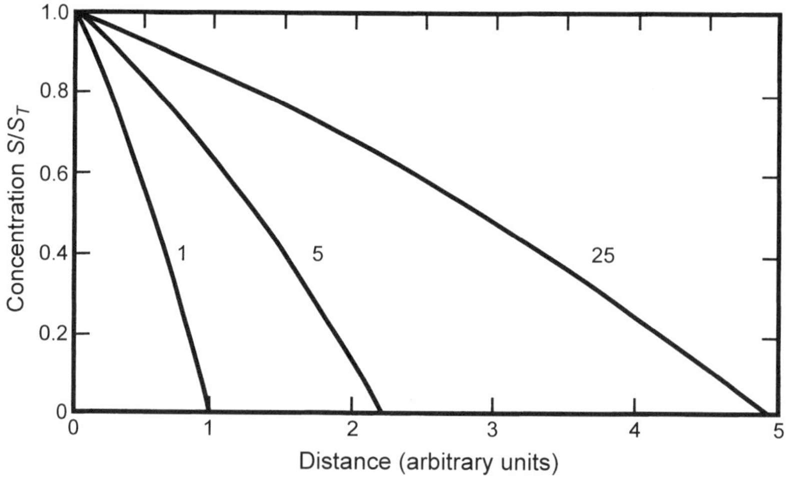

Figure 3.1 Profiles of reacted concentration S/S_T in a solid as a function of normalized time $D_e t/a^2$ (numbers on curves), where D_e is the effective diffusion coefficient at distance zero, t the time, and a the distance at which the concentration is zero. Diffusion coefficient proportional to concentration (Eq. 3.12).

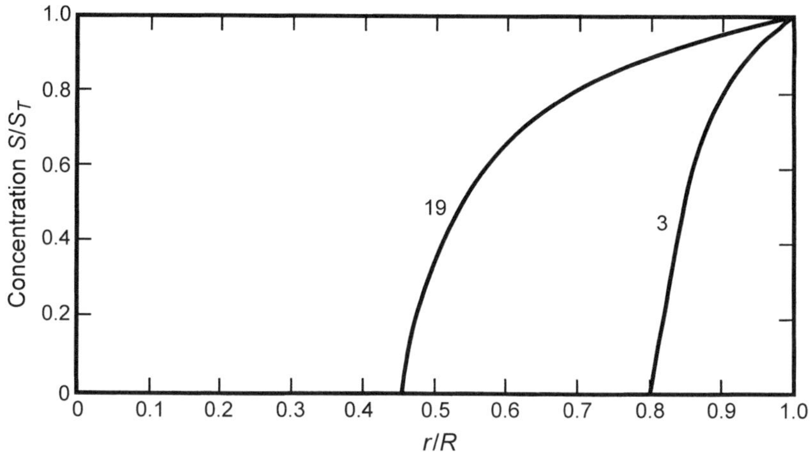

Figure 3.2 Profiles of normalized reacted concentration S/S_T in a cylinder as a function of normalized distance r/R, in which r is the distance from the cylinder surface and R the cylinder radius. The numbers on the curves are normalized times $D_e t/R^2$, where D_e is the effective diffusion coefficient and t the time. Diffusion coefficient proportional to concentration (Eq. 3.12).

diffusion of water and hydrogen into fiber optic cables of silica glass; subsequent reactions can degrade the quality of the fiber. Wiesenfeld *et al.* [3] have solved the diffusion equations numerically for this geometry, including reaction and an equilibrium constant of the form $K=S^2/C(Y-S)$, in which Y is the total number of reactive sites. Some calculated profiles of S and C from their paper are given in Fig. 3.2. Again in the reaction case the diffusant concentration goes to zero at a finite distance from the fiber surface, and this distance moves into the fiber nearly proportional to the square root of time, at least not too deep into the fiber.

3.3 EXCHANGE EXPERIMENTS

In these experiments there is no net chemical change in the solid or liquid material. There are two species of diffusing and reacting material, tagged and untagged, and there is no net chemical change. As an example, consider a silicate that contains a concentration S of SiOH groups in equilibrium with dissolved molecular water (Eq. 3.1) of concentration C. The ratio of reacted to dissolved groups is given by the equilibrium constant for Eq. 3.1 (see Eq. 3.7):

$$K_2 = \frac{S^2}{C} \tag{3.13}$$

Now tagged water is diffused into the material, for example, $H_2{}^{18}O$. The oxygen exchange between oxygen in the dissolved water and in the silicate network can be written

$$\mathrm{Si{-}^{16}O{-}Si} + \mathrm{H_2{}^{18}O} = \mathrm{Si{-}^{18}O{-}Si} + \mathrm{H_2{}^{16}O} \tag{3.14}$$

An asterisk designates concentrations of tagged atoms. Thus C^* is the concentration of dissolved $H_2{}^{18}O$ and C is the concentration of dissolved $H_2{}^{16}O$. S^* is the concentration of ^{18}O in the silicon–oxygen network plus that in ^{18}OH groups, and S is the network ^{16}O plus ^{16}OH. Usually, the concentration of oxygen in OH groups is much smaller than oxygen in the network, so for exchange the OH concentration will be neglected for simplicity. It is possible that the exchange rate between oxygen in dissolved water and in OH groups is faster than for oxygen in the network, so if the OH concentration is not small compared to network oxygen concentration, a more complicated treatment is needed. However, in a $Na_2O{\cdot}2SiO_2$ glass, Zotov and Keppler showed that the reaction of dissolved molecular water prefers Si–O–Si groups, in which each silicon is bonded to four other silicon–oxygen tetrahedra [2] (Q_4 bonding). Thus Si–O–OH bonds with nonbridging oxygens or OH groups on one silicon atom (Q_3 bonding) are apparently less reactive than Q_4 bonded groups.

The total concentration S_T of network oxygen atoms is the sum of ^{16}O and ^{18}O and is constant with distance and time:

$$S_T = S + S^* \tag{3.15}$$

The total concentration of dissolved molecules C_T is also composed of tagged C^* and untagged species C:

$$C_T = C + C^* \tag{3.16}$$

For C_T to be constant with time and uniform in space, the conditions at the sample surface must be such that the initial uniform concentration of C_T is maintained.

An equilibrium constant K for Eq. 3.14 can be written

$$K = \frac{CS^*}{C^*S} \tag{3.17}$$

The isotopic ratio at equilibrium should be the same in the dissolved and reacted species, because the forward and backward reactions in Eq. 3.14 have the same reaction coefficients k; at equilibrium:

$$kCS^* = kC^*S \tag{3.18}$$

so the equilibrium constant K for the reaction $K{=}1$. Then from Eqs. 3.15, 3.16, and 3.17,

$$(C_T - C^*)S^* = (S_T - S^*)C^* \tag{3.19}$$

and

$$\frac{S^*}{S_T} = \frac{C^*}{C_T} \tag{3.20}$$

for all x and t.

The quantity actually measured in an isotope exchange reaction when $S^* \gg C^*$ is S^*, the concentration of exchanged isotope (Si–^{18}O–Si in the example above). To derive equations for profiles of S^*, the differentials from Eq. 3.20 are needed, which are

$$\left(\frac{\partial C^*}{\partial t}\right)_x = \left(\frac{C_T}{S_T}\frac{\partial S^*}{\partial t}\right)_x \qquad \left(\frac{\partial^2 C^*}{\partial x^2}\right)_t = \left(\frac{C_T}{S_T}\frac{\partial^2 S^*}{\partial x^2}\right)_t \tag{3.21}$$

if C_T and S_T are uniform in time and space. Substituting Eq. 3.21 into Eq. 3.3 for C^* and S^* yields

$$\left(\frac{\partial S^*}{\partial t}\right)_x \left(\frac{C_T}{S_T} + 1\right) = \frac{DC_T}{S_T}\frac{\partial^2 S^*}{\partial x^2} \tag{3.22}$$

Usually, $S_T \gg C_T$ (the concentration of reacted or network atoms is much greater than the concentration of dissolved molecules), so Eq. 3.22 becomes the continuity equation for diffusion with a constant effective diffusion coefficient D_e:

$$D_e = \frac{DC_T}{S_T} \tag{3.23}$$

Thus the solution for diffusion with this effective diffusion coefficient is just that for a constant diffusion coefficient (Chapter 2).

3.4 PRESSURE EFFECTS

The influence of pressure on the effective diffusion coefficients can be deduced from Eqs. 3.12 and 3.23. In the reaction experiments (Eq. 3.12) the effective diffusion coefficient at the sample surface ($x = 0$) is

$$D_e(0) = \frac{2S_T D}{K_2} = \frac{2C_T D}{S_T} \tag{3.24}$$

because at x=0, $K_2 = {S_T}^2/C_T$. Thus the effective diffusion coefficient $D_e(0)$ should be proportional to the pressure p of reactant molecules in the gas phase as long as Henry's law is valid for the molecular dissolution of reactant molecules. Then

$$C_T = Hp \tag{3.25}$$

in which H is the Henry's law constant, which does not depend on C_T or p. Fugacity is substituted for pressure for nonideal gases. From Eqs. 3.23, 3.24, and 3.25 the effective diffusion coefficient is directly proportional to the pressure of reactant molecules in the gas phase. This result is true even if the concentration of reacted atoms (e.g., of SiOH groups) is proportional to the square root of the pressure, because

$$S = \sqrt{K_2 C} = \sqrt{K_2 H p} \tag{3.26}$$

The diffusing species are the water molecules, and their surface concentration is directly proportional to the pressure.

3.5 MOVING BOUNDARY-TARNISHING REACTIONS

Some reactions of gases in solids or melts lead to changes in optical absorption in the condensed phase. Examples are formation of metal particles (noble metals, arsenic), changes of oxidation state of a transition metal ion (iron, manganese), and formation and removal of lattice defects such as color centers. In certain special cases the color change takes place sharply at a moving boundary. If the distance the boundary moves is proportional to the square root of time, its motion is probably controlled by the diffusion of dissolved gas to a fixed reaction site. This process is mathematically similar to the diffusion-controlled growth of a reaction layer (oxide, sulfide) on the surface of a metal, which is called a *tarnishing reaction*. The mathematics of this process are described in Ref. 1, p. 305 ff., and Ref. 4 and are summarized below, also giving the limiting conditions for the validity of this treatment.

First the limiting case of a sharp diffusing boundary is described. Consider the diffusion of a reacting gas into a solid containing a fixed concentration R of reactive sites. As an example, the reduction of silver ions to atomic silver in a silicate glass by hydrogen can be written

$$2SiO^{-}Ag^{+} + H_2 = 2SiOH + 2Ag \tag{3.27}$$

The silver atoms agglomerate to form silver particles, which color the glass yellow. Under some conditions a sharp yellow front moves into the glass proportional to the square root of time. The various concentrations are shown in Fig. 3.3. The flux J of hydrogen required to move the front a velocity dL/dt is given by a mass balance:

$$J = R\frac{dL}{dt} \tag{3.28}$$

in which R is the concentration of reactive sites (silver ions). The diffusion of hydrogen through the reacted portion of the glass is assumed to be slow enough so that it reaches a steady-state gradient that is constant throughout the reacted layer,

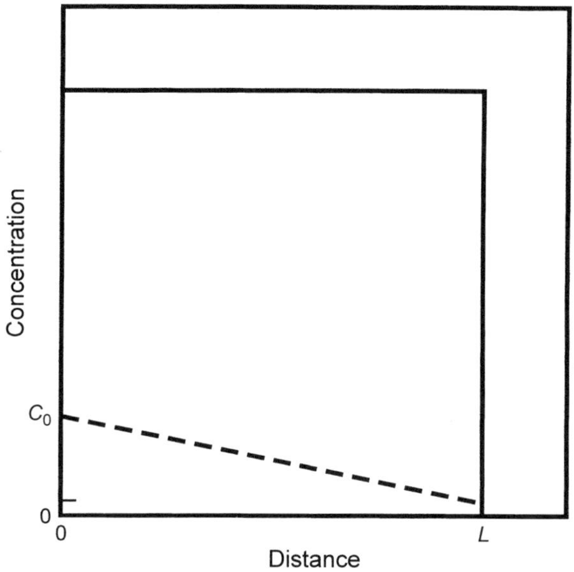

Figure 3.3 Sharp boundary moving into a solid as a result of a tarnishing reaction (e.g., Eq. 3.27). Solid line, concentration of reacted material; dashed line, concentration of dissolved and diffusing reactant, such as hydrogen.

as shown in Fig. 3.3. The flux J of hydrogen that reaches $x=L$, where L is the thickness of the layer, is given by

$$J = \frac{-D\,dC}{dx} = \frac{D(C_0 - C_L)}{L} \tag{3.29}$$

in which D is the diffusion coefficient of reacting gas (hydrogen) in the glass, assumed independent of concentration, C_0 is the concentration of gas dissolved in the glass at its surface ($x=0$) and C_L is the concentration at $x=L$. Equating Eqs. 3.28 and 3.29 with the assumption that $C_0 \gg C_L$,

$$\frac{dL}{dt} = \frac{DC_0}{RL} \tag{3.30}$$

Integrating this equation from zero time and zero layer thickness gives

$$L^2 = \frac{2DC_0 t}{R} \tag{3.31}$$

so the layer thickness moves proportional to the square root of time. For a reaction such as that in Eq. 3.27, $2C_0$ must be substituted for C_0 in Eq. 3.31, because each mole of H_2 gives 2 mol of product Ag^0.

One might have an uncomfortable feeling at this point because it seems that this problem is the same as the diffusion of a reactive gas into a solid treated in

Section 3.2, but the result is much different. Thus we must carefully define the conditions under which these two different treatments are valid. One condition for the formation of a sharp reaction front is that all the reactive sites at $x=L$ are used up rapidly, so that at $x=L$,

$$\frac{R - S_L}{R} \ll 1 \tag{3.32}$$

in which S_L is the number of sites reacted. This condition is now written in terms of reaction rates. The net rate of reaction is given by

$$\left(\frac{\partial S}{\partial t}\right)_x = k_f(R - S)C - k_r S^2 \tag{3.33}$$

in which k_f and k_r are forward and reverse reaction rates, S the concentration of reacted sites (silver atoms), and C the concentration of dissolved gas (hydrogen). The condition of Eq. 3.32 means that the rate of the forward reaction must be much greater than the rate of the reverse reaction, or at $x=L$,

$$k_f(R - S_L)C_L \gg k_r {S_L}^2 \tag{3.34}$$

when S approaches R (nearly complete reaction). In at least some reactions these conditions require that the dissolved concentration of gas at the surface (C_0) is about the same as the concentration R of reactive sites, as shown in later chapters. If Eqs. 3.33 and 3.34 are not fulfilled, the sharp reaction front becomes spread as the reverse reaction becomes appreciable, and when the concentration of reactant sites becomes large compared to the dissolved molecular concentration, the treatments of Section 3.2 are valid. There is no analytical treatment of diffusion and reaction profiles for the intermediate case that I know of. When $R \gg S$, K_2 is given by Eq. 3.7. Thus in the intermediate situation in which C_0 is too small for Eqs. 3.33 and 3.34 to be valid, and R is not much larger than S, a numerical solution of the diffusion equation can provide concentration profiles. These profiles should change from the sharp front of Eq. 3.31 to the profile of Fig. 2.3, for a diffusion coefficient proportional to concentration, valid with Eq. 3.12 and the treatment in Section 3.2, as shown schematically in Fig. 3.4.

Another requirement for the validity of Eq. 3.31 is that the layer thickness L be proportional to the square root of time. Booth [4] found that this condition was always valid, but for a slow surface reaction the condition is not valid, as shown next. The flux J taken up by a reaction, assumed to be first-order for simplicity, at the reaction front $x=L$ is

$$J = \alpha(C_L - C_e) \tag{3.35}$$

The parameter α is in cm/s, and C_e is the equilibrium concentration of dissolved molecules at $x=L$. Equating Eqs. 3.29 and 3.35 and solving for C_L gives

$$C_L = \frac{C_0 - C_e}{1 + \alpha L/D} + C_e \tag{3.36}$$

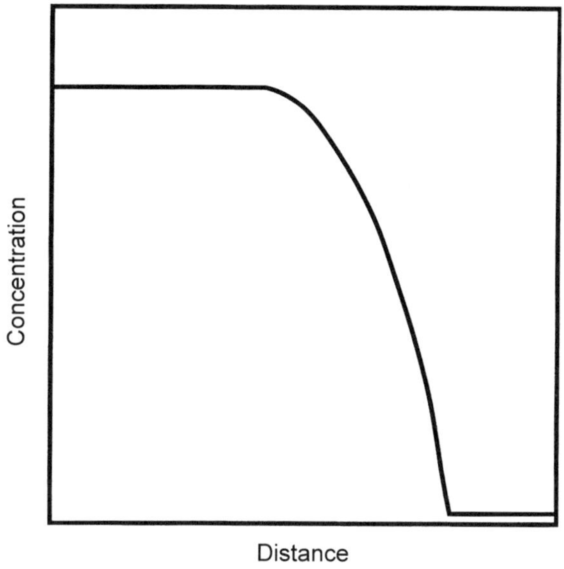

Figure 3.4 Change of concentration profile from a sharp front (Fig. 3.3) to a profile for $D_e = 2SDK_2$ (Eq. 3.12). Solid line, intermediate profile of reacted material.

When $\alpha L/D \gg 1$, $C_L \rightarrow C_e$, and the process is controlled by diffusion. If this condition is not met, the concentration C_L depends on the thickness L and the time. The flux from Eq. 3.29 or 3.34 is

$$J = \frac{\alpha(C_0 - C_e)}{1 + \alpha L/D} = R\frac{dL}{dt} \quad (3.37)$$

Thus if the condition

$$\frac{\alpha L}{D} \gg 1 \quad (3.38)$$

is not met, the thickness L is given by

$$L^2 + \frac{2D}{\alpha}L = \frac{2D(C_0 - C_e)t}{R} \quad (3.39)$$

so that the layer thickness is not proportional to the square root of time.

Yet another requirement for the validity of Eq. 3.31 is that the gradient of reactant concentration in the reacted region be linear with distance, as shown in Fig. 3.3. Booth [4] discussed this requirement and showed that for it to be valid, the dissolved concentration C_0 at $x=0$ must be much smaller than the concentration of reactive sites, or

$$C_0 \ll R \quad (3.40)$$

If this condition is not valid, the reaction front still moves proportional to the square root of time [4], but the proportionality in Eq. 3.31 is modified by a factor B, calculated by Booth [4]:

$$L^2 = \frac{2BDC_0t}{R} \tag{3.41}$$

The factor B is found from a solution of the equation

$$\left(\operatorname{erf}\sqrt{\frac{C_0B}{2R}}\right)^2 = \frac{2C_0}{\pi BR}e^{-BC_0/R} \tag{3.42}$$

If $u \ll 1$, $\operatorname{erf}(u) \to 2u/\sqrt{\pi}$, so if $C_0 \ll R$, the solution of Eq. 3.42 is $B=1$, since the exponential term is unity. Booth calculated values of B as a function of C_0/R, as shown in Fig. 3.5. From this graph the value of B to correct Eq. 3.31 when condition 3.39 does not hold can be found. When $B < 1$, the profile of dissolved gas molecules is no longer linear with distance and becomes increasingly more concave upward as B decreases [4].

This section shows that Eq. 3.31 for the tarnishing reaction is valid for a reaction front in a solid or melt only for a rather narrow range of dissolved and reactive concentrations, as defined by Eqs. 3.34, 3.37, and 3.39. As long as Eqs. 3.32, 3.34, and 3.37 are valid, there is still a reaction front moving with the square root of time. If Eqs. 3.32 and 3.34 are not met, the front spreads, and if Eq. 3.37 is not met, the front does not move proportional to the square root of time.

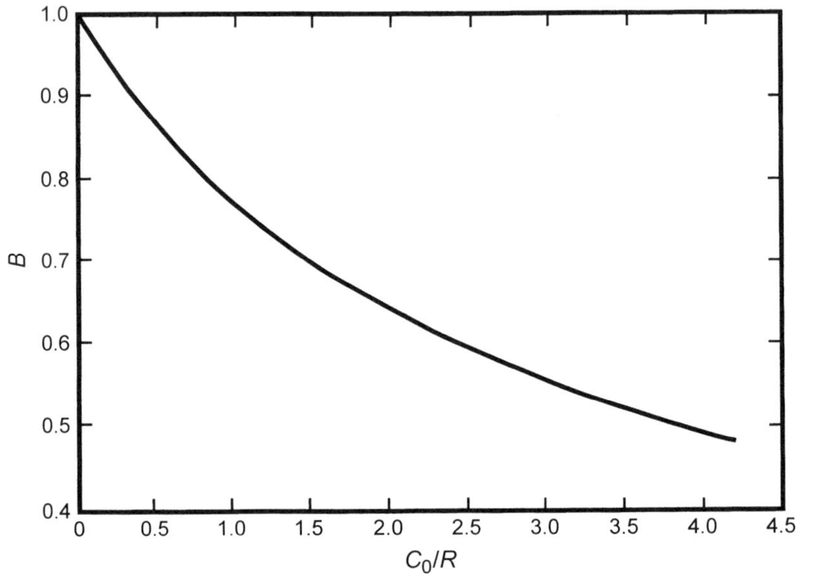

Figure 3.5 Parameter B (Eqs. 3.41 and 3.42) as a function of C_0/R (see the text).

REFERENCES

1. J. Crank, *The Mathematics of Diffusion*, 2nd ed., Oxford University Press, Oxford, 1975.
2. N. Zotov and H. Keppler, *Am. Mineral.* **83**, 823 (1998).
3. J. M. Wiesenfeld, J. Stone, D. Maruse, C. A. Burrus, and S. Yang, *J. Appl. Phys.* **61**, 5447 (1987).
4. F. Booth, *Trans. Faraday Soc.* **44**, 796 (1948).

4

DIFFUSION AND NONEQUILIBRIUM REACTIONS

4.1 INTRODUCTION

In Chapter 3, diffusion plus a chemical reaction was considered in which the reaction has reached local equilibrium. Local equilibrium means that at each point in the solid or melt there is equilibrium between dissolved molecules and the reaction groups in the solid or melt. Examples of reactions are

$$\mathrm{Si{-}O{-}Si} + \mathrm{H_2O} = \mathrm{SiOH} + \mathrm{HOSi} \tag{4.1}$$

and

$$2\mathrm{Si} + \mathrm{H_2} = 2\mathrm{SiH} \tag{4.2}$$

In this chapter, reactions not at local equilibrium are considered. A manifestation of no local equilibrium is an increase of concentration of reacted SiOH groups (Eq. 4.1) in a silicate glass at its surface. At local equilibrium the surface concentration of reacted SiOH groups rapidly reaches the value given by the equilibrium relation

$$K_2 = \frac{S^2}{C} \tag{4.3}$$

in which S is the concentration of the reacted SiOH groups in the glass, C the concentration of dissolved water molecules in the glass, and K_2 an equilibrium constant for Eq. 4.1; the subscript 2 refers to a binary reverse reaction. In an experiment in which a glass containing reacted OH groups and dissolved water molecules is held in a vacuum, one would expect zero concentration of both OH groups and dissolved water molecules in the glass at its surface if there is local

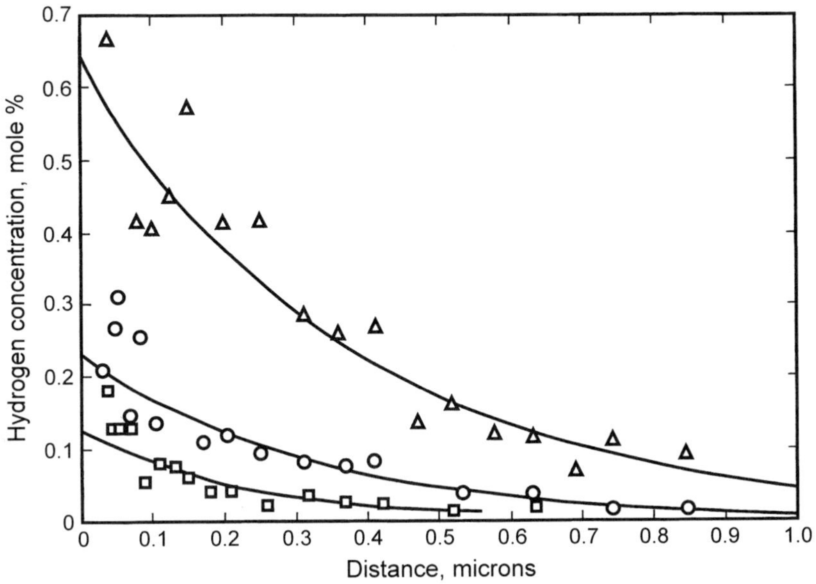

Figure 4.1 Profiles of hydrogen in crystalline quartz after diffusion of water at 200°C for times: □, one day; ○, three days; △, nine days. Lines drawn for an exponential function (Eq. 4.20) by linear regression, with correlation coefficients; one day, 0.914; three days, 0.969; nine days, 0.955; from Ref. 1, Fig. 2a.

equilibrium; if there is not local equilibrium, the surface concentration of dissolved water molecules is zero, but the surface concentration of reacted OH groups is not zero, and decreases to zero as the time at temperature increases. An example of the increase in concentration of reacted OH groups in quartz at 200°C in water vapor is shown in Fig. 4.1.

In this chapter the equations for no local equilibrium are first developed for a first-order forward reaction only, and then for first-order forward and reverse reactions. There is apparently no solution for a second-order reverse reaction, which is important in many practical problems. Then equations for exchange reactions, in which there is no net chemical reaction, but only exchange of isotopes in a uniform material, are presented. Finally, other possible systems are discussed.

4.2 NO LOCAL EQUILIBRIUM, FORWARD REACTION ONLY

The general differential equation for diffusion plus chemical reaction was derived in Section 3.2:

$$\frac{\partial C}{\partial t} = D\frac{\partial^2 C}{\partial x^2} - \frac{\partial S}{\partial t} \tag{4.4}$$

in which C is the concentration of dissolved molecules, S the concentration of reacted groups, and D the diffusion coefficient of dissolved molecules, assumed to be independent of distance x and time t. If there is a first-order forward reaction with negligible reverse reaction,

$$\frac{\partial C}{\partial t} = \lambda C \tag{4.5}$$

in which λ is a reaction rate coefficient with units of reciprocal time. The differential equation is then

$$\frac{\partial C}{\partial t} = D\frac{\partial^2 C}{\partial x^2} - \lambda C \tag{4.6}$$

In semi-infinite geometry and one kind of experiment, the initial concentration of dissolved molecules C_i is uniform, and the concentration at the surface is held constant at C_0. The boundary conditions are

$$\begin{aligned} C &= C_0 \quad \text{at } x = 0,\ t > 0 \\ C &= C_i \quad \text{at } t = 0, \text{ all } x \\ C &\to C_i \quad \text{as } x \to \infty \end{aligned} \tag{4.7}$$

The solution of Eq. 4.6 for these boundary conditions is [2, p. 334]

$$\begin{aligned} \frac{C - C_i}{C_0 - C_i} &= \frac{1}{2}\exp\left(-x\sqrt{\frac{\lambda}{D}}\right)\operatorname{erfc}\left(\frac{x}{2\sqrt{DT}} - \sqrt{\lambda t}\right) \\ &\quad + \frac{1}{2}\exp\left(x\sqrt{\frac{\lambda}{D}}\right)\operatorname{erfc}\left(\frac{x}{2\sqrt{Dt}} + \sqrt{\lambda t}\right) \end{aligned} \tag{4.8}$$

The total amount of material M (dissolved plus reacted) in the solid or melt as a function of time is

$$M = (C_0 - C_i)\sqrt{\frac{D}{\lambda}}\left[\left(\lambda t + \frac{1}{2}\right)\operatorname{erf}\left(\sqrt{\lambda t}\right) + \frac{\sqrt{\lambda t}}{\sqrt{\pi}}e^{-\lambda t}\right] \tag{4.9}$$

After an initial transient during which the reacted concentration is small compared to the dissolved concentration, the condition $\lambda t \gg 1$ is approached, for which there is a steady state. Then the first erfc term in Eq. 4.8 approaches 2, and the second term approaches zero, so

$$\frac{C - C_i}{C_0 - C_i} = \exp\left(-x\sqrt{\frac{\lambda}{D}}\right) \tag{4.10}$$

When $\lambda t \gg 1$, erf $\left(\sqrt{\lambda t}\right) \rightarrow 1$, and the last term in brackets in Eq. 4.9 is small, so

$$M = (C_0 - C_i)\sqrt{D\lambda}\left(t + \frac{1}{2\lambda}\right) \tag{4.11}$$

In this steady-state condition the concentration profile of dissolved molecules is constant with time and decreases exponentially with distance from the sample surface (Eq. 4.10). The total amount of molecular water M diffused into the solid becomes linear with time, by Eq. 4.11, and the intercept of the linear portion is $(C_0 - C_i)\sqrt{D/\lambda}/2$ on the M axis at $t = 0$.

To find the relationship between the concentration C of dissolved molecules and the concentration S of reacted groups, let us first assume that the amount of dissolved molecules initially in the solid is zero, so $C_i = 0$. Then the surface concentration S_0 of reacted groups at $x = 0$ can be found from integration of the equation

$$\frac{dS_0}{dt} = \lambda C_0 \tag{4.12}$$

from time zero to t:

$$S_0 = \lambda C_0 t \tag{4.13}$$

Thus the surface concentration of reacted groups increases proportional to time. The concentration S of reacted groups at any point x is found from

$$\frac{dS}{dt} = \lambda C \tag{4.14}$$

in which the small concentration of reacted groups formed during the transient $t\lambda < 0.2$ is neglected. Then, since C is not a function of time (Eq. 4.10),

$$S = \lambda C t \tag{4.15}$$

and

$$S = S_0 \exp\left(-x\sqrt{\frac{\lambda}{D}}\right) = \lambda C_0 t \exp\left(-x\sqrt{\frac{\lambda}{D}}\right) \tag{4.16}$$

Thus the concentration of reacted groups has an exponential profile in this regime rather than a error function or complementary error function profile. Furthermore, profiles of S do not superpose as a function of $x/\sqrt{t}$ during this regime. See Fig. 4.2 for profiles of reacted groups calculated from Eq. 4.16.

A comparison of Eqs. 4.16 and 4.10 shows that

$$\frac{S}{C} = \frac{S_0}{C_0} \tag{4.17}$$

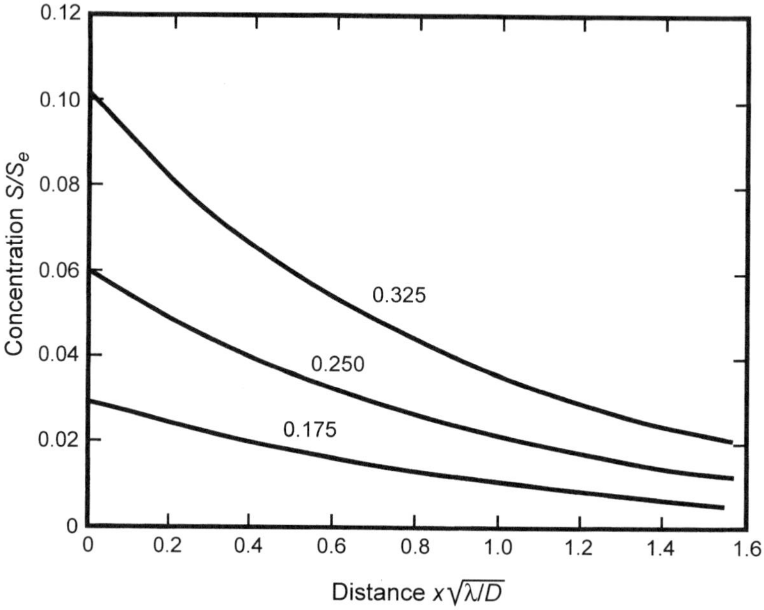

Figure 4.2 Calculated profiles of reacted groups S as a function of different diffusion times, during the time regime in which the forward reaction (Eq. 4.5) controls. Numbers on the curves are $\sqrt{\mu t}$ (see Fig. 4.5).

In this equation S and C are functions of x and t, S_0 is a function of t only, and C_0 is constant. At equilibrium, Eq. 4.3 shows that $S^2/C = K_2$; Eq. 4.17 is valid at early times, as defined in Section 4.3, and it shows that profiles of S and C have the same shape even though there is no local equilibrium.

If there are some dissolved reactive molecules in the solid or melt before the diffusion treatment, their influence depends on the thermal history of the solid. If this initial dissolved material ($C_i > 0$) is at equilibrium with reacted groups (Eq. 4.3 is valid throughout the silica),

$$\frac{dS}{dt} = \lambda(C - C_i) \tag{4.18}$$

and from integration of Eq. 4.18, and Eq. 4.10:

$$S - S_i = \lambda(C - C_i)t = \lambda(C_0 - C_i)t \exp\left(-x\sqrt{\frac{\lambda}{D}}\right) \tag{4.19}$$

or

$$S - S_i = (S_0 - S_i) \exp\left(-x\sqrt{\frac{\lambda}{D}}\right) \tag{4.20}$$

in which S_i is the initial concentration of reacted groups in equilibrium with C_i, and is uniform in the solid. However, often S_i and C_i are not at equilibrium, because this condition requires long preheating at or above the diffusion temperature; such heating is not always done in diffusion experiments. If S_i is different from the equilibrium amount, Eqs. 4.10, 4.11 and 4.18 will not hold exactly, and additional terms are needed to account for the reaction (either forward or background) of the residual concentrations S_i and C_i; these terms will depend on the treatment history of the material.

From a comparison of Eqs. 4.19 and 4.10,

$$\frac{S - S_i}{S_0 - S_i} = \frac{C - C_i}{C_0 - C_i} \tag{4.21}$$

which is analogous to Eq. 4.17 if the initial dissolved and reacted components are initially at equilibrium. In this case S_i and C_i are related by Eq. 4.3.

The treatment described above considered net diffusion into the solid or melt (sorption). It is also possible to diffuse the reactive molecules from the solid to the surroundings (desorption). This experiment starts with uniform concentrations S_i and C_i in the solid or melt. Usually, the surroundings have none of the diffusing molecules, so the surface concentration of dissolved molecules is equal to zero ($C_0 = 0$) for all times. However, the surface concentration of reacted molecules is initially not zero and decreases to zero with time. Very similar profiles to those above are found in this case:

$$\frac{C}{C_i} = 1 - \exp\left(-x\sqrt{\frac{\lambda}{D}}\right) \tag{4.22}$$

and

$$\frac{S_i - S}{S_i - S_0} = \exp\left(-x\sqrt{\frac{\lambda}{D}}\right) \tag{4.23}$$

Equation 4.23 results from Eqs. 4.21 and 4.22, with $C_0 = 0$ in Eq. 4.21. Equations 4.20 and 4.23 show that the concentration of reacted groups S has an exponential profile both for sorption and desorption when the forward first-order reaction (Eq. 4.5) controls. The time regime of validity of these exponential profiles is explored in the next section. Some experimental profiles of both S and C for the diffusion of water in obsidian are shown in Fig. 4.3 from Zhang *et al.* [3]; these profiles demonstrate the relation

$$\frac{S_i - S}{S_i - S_0} = \frac{C_i - C}{C_i} \tag{4.24}$$

from Eq. 4.21 with zero surface concentration of dissolved molecules.

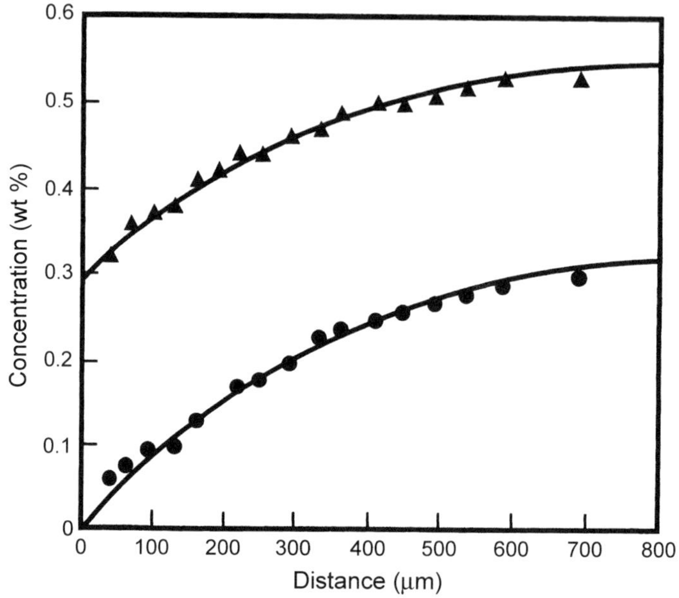

Figure 4.3 Profiles of molecular water (•) and OH groups (▲) in obsidian after 51.7 days at 450°C. The lines are drawn from Eqs. 4.22 and 4.23, and also fit Eq. 4.24. (From Ref. 3.)

4.3 NO LOCAL EQUILIBRIUM, FIRST-ORDER FORWARD AND REVERSE REACTIONS

At times longer than those during which the treatment of Section 4.2 is valid, both the forward and reverse reactions are significant. For simplicity, the case of a first-order reverse reaction will be treated first and then a second-order reverse reaction discussed. The reaction equation is

$$\frac{\partial S}{\partial t} = \lambda C - \mu S \tag{4.25}$$

in which the rate coefficient μ for the reverse reaction has units of reciprocal time. In the first type of experiment considered, the surface concentration of dissolved molecules C_0 is held constant by a source at the solid or melt surface, and the initial concentrations of dissolved and reacted molecules in the solid is zero:

$$\begin{aligned} C &= C_0 && \text{at } x = 0 \quad \text{for } t > 0 \\ C_i &= S_i = 0 && \text{at } t = 0 \quad \text{for all } x \end{aligned} \tag{4.26}$$

Then Eq. 4.25 can be integrated to find the surface concentration S_0 of reacted groups as a function of time:

$$S_0 = S_e(1 - e^{-\mu t}) \tag{4.27}$$

In this equation S_e is the equilibrium concentration of reacted groups after a long reaction time. At equilibrium,

$$S_e = C_0 K_1 \tag{4.28}$$

where K_1 is an equilibrium constant for a reaction that is first order in both the forward and reverse directions. As an example of such a reaction, if the OH or SiH groups in Eqs. 4.1 and 4.2 are completely immobile, they will react only with the same companion group as first formed, so the reverse reaction is then first order; the rate of this reaction is proportional to the first power of S, not the second power, which results only if more distant neighboring OH or SiH groups can react with a central group.

Equation 4.27 shows that the surface concentration of reacted groups increases exponentially with time, and finally reaches equilibrium at long times, when from Eq. 4.25,

$$S_e = \frac{\lambda C_0}{\mu} = K_1 C_0 \tag{4.29}$$

so $K_1 = \lambda/\mu$ when $\partial S/\partial t = 0$. The reaction coefficient μ can be calculated from Eq. 4.27 and surface concentrations S_0 as a function of time. A logarithmic plot of surface concentrations of reaction OH groups in silica shown in Fig. 4.4; a straight line results, from which $\mu = 5.6(10)^{-5}\ \mathrm{s}^{-1}$.

With Eq. 4.25 the diffusion–reaction Eq. 4.4 becomes

$$\frac{\partial C}{\partial t} = D\frac{\partial^2 C}{\lambda x^2} + \lambda C - \mu S \tag{4.30}$$

A solution for Eq. 4.30 for C for a plane sheet has been given by Crank [2, p. 338ff.]. As a boundary condition at the sheet surfaces, Crank used a stirred solution of limited extent. The resulting solution for C or S involves real and imaginary roots and two infinite series. Crank has given an extensive discussion on the evaluation of his solution, numerical solutions, and some limiting conditions. For small times the series converge slowly, so the semi-infinite solution cannot be easily derived from them. Furthermore, it appears that the factor suggested by Crank to obtain a solution for S from that for C is oversimplified, because it results in a constant surface concentration of reacted groups (S_0) for constant C_0 (infinite extent of stirred solution).

It is desirable to have an equation to calculate profiles in semi-infinite and infinite geometries. I proposed an approximate solution to Eq. 4.30 for semi-infinite geometry, with the following boundary conditions:

$$\begin{aligned} &S = C = 0 && t = 0,\ \text{all } x \\ &C = C_0\ (\text{constant}) && \text{at}\quad x = 0,\ t > 0 \\ &S, C \to O && \text{as}\quad x \to \infty \end{aligned} \tag{4.31}$$

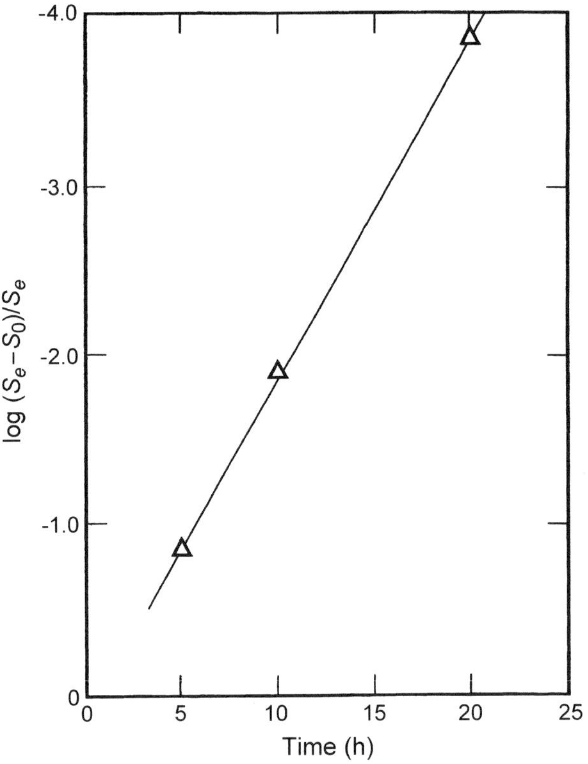

Figure 4.4 Log concentration of reacted SiOH groups at the surface of fused silica as a function of time (see Eq. 4.27). (From Ref. 4.)

The solution for $S(x, t)$ is [4]

$$S(x, t) = S_e[\mathrm{erfc}(r) - e^{-r^2}u(\sqrt{\mu t}, r)] \tag{4.32}$$

In this equation μt is a dimensionless time and

$$r = \frac{x}{2}\left[\frac{1 - \exp(-\mu t)}{D\mu t/\lambda}\right]^{1/2} \tag{4.33}$$

is a dimensionless distance. The function u arises from error functions with complex argument and is tabulated in Ref. 5, p. 486. These functions are discussed in an appendix to this chapter. At long times such that $\mu t \gg 1$, $r \to x/2\sqrt{D_e t}$), in which D_e is the effective diffusion coefficient of S at local equilibrium:

$$D_e = D\frac{\mu}{\lambda} \tag{4.34}$$

The effective diffusion coefficient D_e is time dependent, and only at long times $\mu t \gg 1$ does it approach the constant value of Eq. 4.34. Sometimes, diffusion coefficients are calculated with the assumption that the distance parameter is $x/2\sqrt{Dt}$, and it is found that the calculated D value is time dependent; examples are the diffusion of hydrogen and arsenic in silicon. These experiments are discussed in later chapters.

The solution of Eqs. 4.32 and 4.33 at long times ($\mu t \gg 1$) becomes that for local equilibrium, as discussed in Chapter 3. At intermediate times ($\mu t < 1$, $\lambda t \ll 1$) the solution tends to those of Eqs. 4.10, 4.11, 4.16, 4.22 and 4.23. Thus this solution has the correct limiting profiles for these time conditions. A summary of the ranges of validity of various solutions of the diffusion–reaction equations of nonequilibrium is given in Section 4.6.

Profiles of S/S_e as a function of dimensionless distance $x\sqrt{\lambda/D}$ and dimensionless time $\sqrt{\mu t}$ are given in Fig. 4.5. S_e is the value for S at equilibrium. At $\sqrt{\mu t} = 3$, the profile is close to that of the erfc solution for local equilibrium over the distance range in the figure but deviates progressively at larger distances

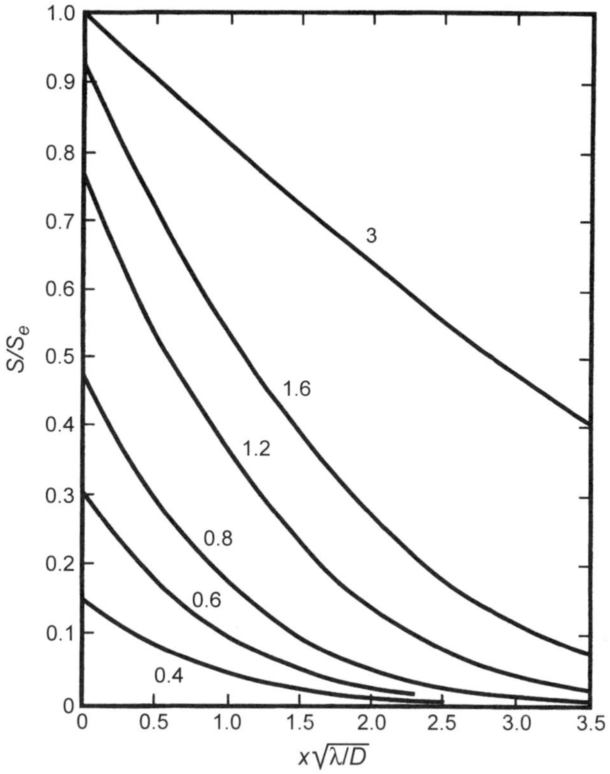

Figure 4.5 Profiles of normalized concentration S/S_e of reacted groups as a function of dimensionless distance $x\sqrt{\lambda/D}$, calculated from Eqs. 4.32 and 4.33. Different values of $\sqrt{\mu t}$ are shown on the curves.

(see below and Fig. 4.6). For $\sqrt{\mu t} < 1$ the profiles are close to exponential, as expected from Eq. 4.16. Thus the solution of Eqs. 4.32 and 4.33 has the correct values at these time limits; at times between these limits, the full solution of these equations must be used. Equation 4.30 has also been solved for nonequilibrium in a plane sheet by Rigo *et al.* [6]. Their profiles at small times are similar to those shown in Fig. 4.5.

To help in calculating diffusion coefficients from experimental data, a plot of $x\sqrt{\lambda/D}$ versus $\sqrt{\mu t}$ at $S/S_0 = 0.5$ is given in Fig. 4.6. From experimental data on the change of surface concentration with time, one can calculate μ from Eq. 4.27. If only one surface concentration has been measured, the value of μ can be calculated from Eq. 4.27 if the equilibrium value S_e of the concentration of reacted groups at the surface at long times is known. Then the value of $x\sqrt{\lambda/D} = p$ corresponding to $\sqrt{\mu t}$, where t is the experimental time, is found from Fig. 4.6. Then from the experimentally measured penetration depth $x(0.5)$ at half the surface concentration and this value of p, D/λ can be calculated:

$$\frac{D}{\lambda} = \left(\frac{p}{x}\right)^2 \tag{4.35}$$

The value of λ can be found from equilibrium value of the reacted concentration (S_e) and the concentration of dissolved moleculary water (C_e), since

$$\lambda = \mu \frac{S_e}{C_e} \tag{4.36}$$

from Eq. 4.29.

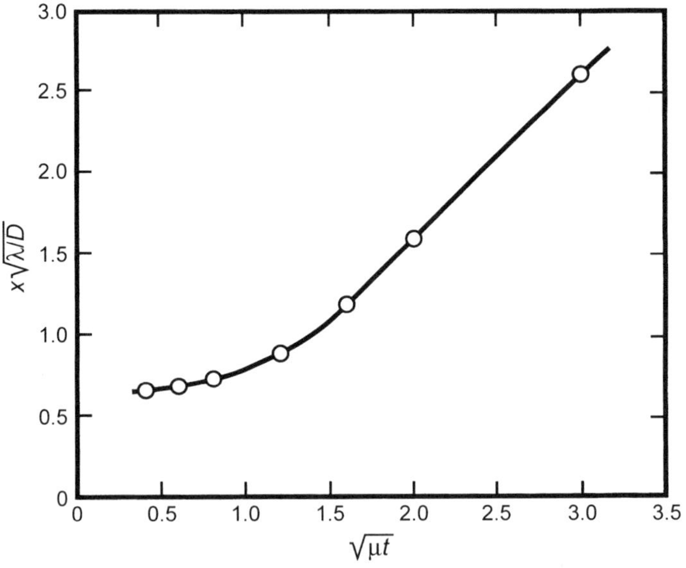

Figure 4.6 Plot of $x\sqrt{\lambda/D}$ versus $\sqrt{\mu t}$ from Fig. 4.5 and Eq. 4.32 at $S/S_0 = 0.5$.

Profiles of diffusing reactive molecules have been measured over a wide range of concentration. With techniques such as secondary ion mass spectroscopy of hydrogen in silicon, a concentration range of three orders of magnitude can be measured. Thus it is desirable to examine the diffusion–reaction solution for a large range of concentration. In Fig. 4.7 the profile of S from Eqs. 4.32 and 4.33 is plotted as log S versus normalized distance for $\sqrt{\mu t} = 2$. The fit with an erfc function is good up to about $S = S_0/2$, but then the calculated concentration deviates progressively from the erfc (local equilibrium) profile. There are experimental examples of this deviation in diffusion of deuterium in hydrogenated silicon, as discussed in Section 4.4. Even though the surface concentration of reacted groups is close to the equilibrium value, there may not be local equilibrium deeper into the solid.

When the boundary conditions are initial concentrations S_i and C_i at local equilibrium, and a surface concentration $C = 0$ at $t > 0$, the solution above can be used directly. The surface concentration of $S = S_0$, from integration of Eq. 4.25, is

$$S_0 = S_i e^{-\mu t} \tag{4.37}$$

and the solution of Eq. 4.32 can be modified directly to

$$S = S_i[\text{erf}(r) + e^{-r^2} u(\sqrt{\mu t}, r)] \tag{4.38}$$

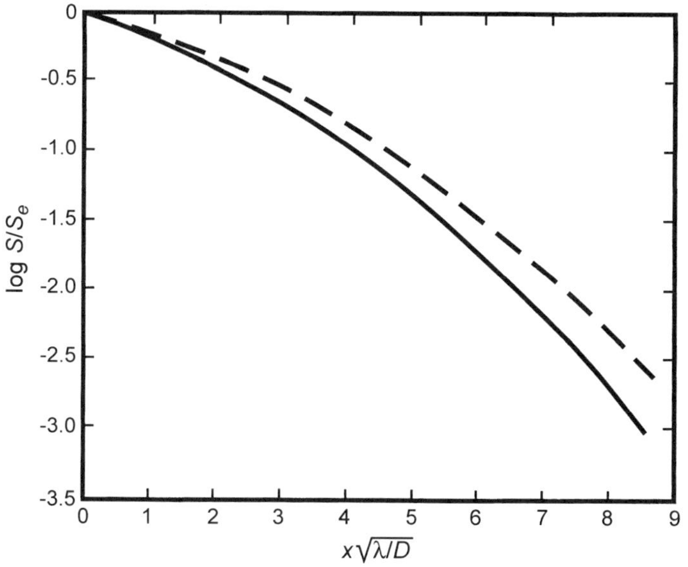

Figure 4.7 Solid line, profile of reacted concentration S for $\sqrt{\mu t} = 2$ from Eq. 4.32 (see Fig. 4.5) over a wide range of time; dashed line, erfc profile (Eq. 2.15).

The solution above for semi-infinite geometry can be modified to give a solution in infinite geometry. The initial boundary conditions are shown in Fig. 4.8. As diffusion progresses, the interfacial concentration of dissolved molecules becomes C_0 at $x=0$, where $C_0=C_1/2$, and C_1 is the initial concentration of dissolved molecules in region 1. There is a step in the concentration S_0 at the interface $x=0$. In region 1

$$S_0(1) = \frac{S_1}{2}(1 + e^{-\mu t}) \tag{4.39}$$

analogous to Eq. 4.37. In region 2,

$$S_0(2) = \frac{S_1}{2}(1 - e^{-\mu t}) \tag{4.40}$$

analogous to Eq. 4.27. At long time $\mu t \gg 1$, the interfacial concentrations in both regions become equal at $S_1/2$. The profiles in regions 1 and 2 result from equations

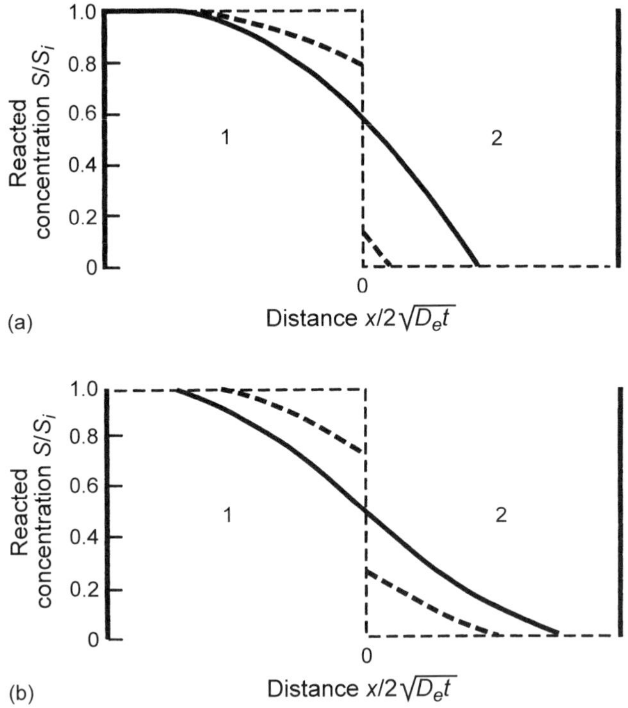

Figure 4.8 Schematic diagram of diffusion from region 1 into region 2 with first-order forward and reverse reactions. Light dashed line, initial concentration of both C and S; heavy solid line, profile with local equilibrium; heavy dashed line, profile of reacted groups after a time of diffusion, no local equilibrium. (*a*) Diffusion coefficient directly proportional to concentration; (*b*) constant diffusion coefficient.

similar to Eqs. 4.32 and 4.38:

$$S(1) = \frac{S_1}{2}\left[\operatorname{erf}(r) + e^{-r^2}u(\sqrt{\mu t}, r)\right] \tag{4.41}$$

and

$$S(2) = \frac{S_1}{2}\left[\operatorname{erfc}(r) + e^{-r^2}u(\sqrt{\mu t}, r)\right] \tag{4.42}$$

At long time $\mu t \gg 1$, the profile of S is that for a constant effective diffusion coefficient $D_e = \lambda D/\mu$, as shown by the solid line in Fig. 4.8.

4.4 NO EQUILIBRIUM, FIRST-ORDER FORWARD REACTION, SECOND-ORDER REVERSE REACTION

The condition in which there is no equilibrium, a first-order forward reaction, and a second-order reverse reaction is often met in experiments: for example, for Eqs. 4.1 and 4.2 for diffusion into a solid without reacted groups. The rate of reaction to form or remove reacted groups is

$$\frac{\partial S}{\partial t} = \lambda C - \frac{\mu S^2}{S_e} \tag{4.43}$$

in which λ and μ are reaction coefficients with units of reciprocal time. S_e is the final uniform equilibrium concentration of reacted groups and is included in the last term so that the reverse reaction coefficient has the same units as the forward coefficient. At local equilibrium $K_2 = S^2/C = S_e\lambda/\mu$, from Eq. 4.43.

Consider the removal of reacted groups of initial uniform concentration S_e by holding the surface concentration of dissolved molecules at zero, $C = 0$ at $x = 0$. Then the concentration S_0 of reacted groups at the solid surface ($x = 0$) is

$$\frac{dS_0}{dt} = -\frac{\mu {S_0}^2}{S_e} \tag{4.44}$$

and integrating this equation with limits $S_0 = S_e$ at $t = 0$ and S_0 at t gives

$$\frac{S_0}{S_e} = \frac{1}{\mu + t} \tag{4.45}$$

At long times $\mu t \gg 1$, S_0 goes to zero.

Now consider diffusion of dissolved molecules into the solid with fixed surface concentration C_0:

$$\frac{dS_0}{dt} = \lambda C_0 - \frac{\mu {S_0}^2}{S_e} \tag{4.46}$$

The integration of this equation from $S_0 = 0$ at $t = 0$ to S_0, t; C_0 constant, is

$$\frac{S_0}{S_e} = \frac{1 - e^{-2\mu t}}{1 + e^{-2\mu t}}. \tag{4.47}$$

Equations 4.39 and 4.40 give the interface concentrations in an infinite couple (see Fig. 4.8) for a first-order reverse reaction. If there is a second-order reverse reaction in infinite geometry with boundary conditions $S = S_i$ for $x < 0$ (region 1) and $S = 0$ for $x > 0$ (region 2) at $t = 0$, Eq. 4.46 for the interfacial concentration can be written

$$\frac{dS_0}{dt} = \lambda C_0 - \frac{\mu S_0{}^2}{S_1} \tag{4.48}$$

At equilibrium $S_0 = 0.59 S_1$, as shown in Chapter 2 and Ref. 2, Fig. 7.6, p. 122. For a diffusion couple, let us assume that the concentration C_0 of dissolved molecules at the interface $x = 0$ is $(0.59)^2 C_1 = 0.348 C_1$, which should be valid after a short initial transient. Then at $x = 0$, Eq. 4.48 becomes

$$\frac{dS_0}{dt} = 0.348 S_1 \mu - \frac{\mu S_0{}^2}{S_1} \tag{4.49}$$

since $C_1 = S_1 \mu / \lambda$. Integrating Eq. 4.49 with limits $S_0 = 0$ at $t = 0$ and S_0, t gives

$$\frac{S_0(2)}{0.59 S_1} = \frac{1 - e^{-1.18\mu t}}{1 + e^{-1.18\mu t}} \tag{4.50}$$

for the concentration $S_0(2)$ of reacted groups at $x = 0$ in region 2. The factor 1.180 arises in the integration and equals $2\sqrt{0.348}$. When $t = 0$, $S_0(2) = 0$, and as $1.18\mu t \gg 1$, $S_0(2) \to 0.59 S_i$ or equilibrium at the interface. For region 1, integration of Eq. 4.49 gives

$$\frac{S_0(1)}{0.59 S_1} = \frac{1 + 0.258 e^{-1.18\mu t}}{1 - 0.258 e^{-1.18\mu t}} \tag{4.51}$$

The factor $0.258 = (1 - 0.59)/(1 + 0.59)$. As $\mu t \gg 1$, $S_0(1) \to 0.59 S_1$, or equilibrium at the interface. The difference $S_0(1) - S_0(2)$ gives a "step" in the interface concentration of reacted groups and is found experimentally, for example, in the diffusion of hydrogen in silicon (see Chapter 15).

Various relations for the increase or decrease of concentrations of reacted groups at the solid surface are compared in Fig. 4.9. For sorption (curves 3 and 4), equilibrium is reached more rapidly with the second-order reverse reaction; for desorption, equilibrium is reached more rapidly for the first-order reverse reaction. For a first-order reverse reaction, the solutions for the profiles of S correspond, as shown by Eqs. 4.32 and 4.38; however, for the second-order reverse reaction, there is no direct correspondence between profiles for sorption and

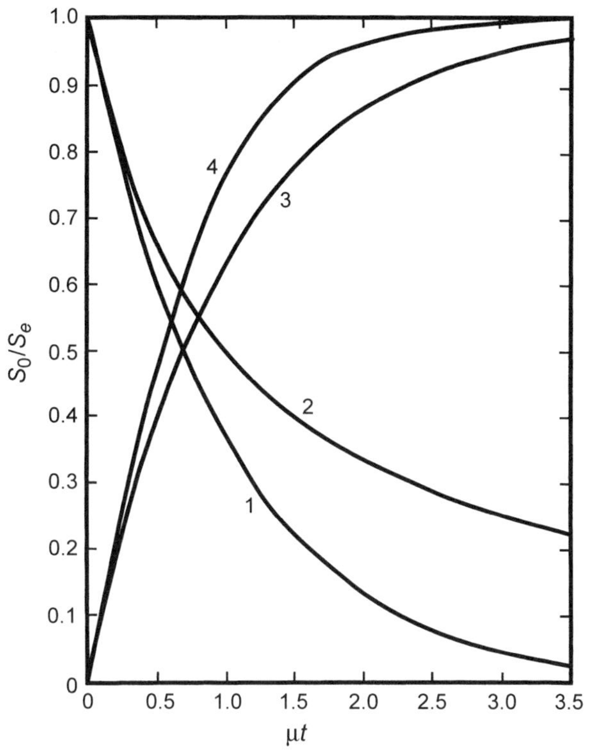

Figure 4.9 Comparison of the change of concentration of reacted groups S_0/S_e at the solid surface as a function of time. Curve 1, desorption, first-order reverse reaction (Eq. 4.37); curve 2, desorption, second-order reverse reaction (Eq. 4.45); curve 3, sorption, first-order reverse reaction (Eq. 4.27); curve 4, sorption, second-order reverse reaction (Eq. 4.47).

desorption. A similar result for a second-order reverse reaction was found in Chapters 2 and 3 for local equilibrium; the profiles for sorption (Fig. 2.3) and desorption (Fig. 2.4) are very different. Paradoxically, it is found that the amount of material M that diffuses through a unit area of solid surface is greater for sorption than for desorption for the same time of diffusion for local equilibrium. From Table 7.6 of Ref. 2, $M/C_0\sqrt{D_i t} = 1.255$ for sorption and 0.939 for desorption. D_i is the diffusion coefficient at the solid surface ($x = 0$), where the concentration of dissolved molecules is C_0; the effective diffusion coefficient $D_e = 2SD/K_2$ is directly proportional to the concentration S of reacted groups, so $D_i = 2S_eD/K_2$.

The reason that sorption is faster than desorption with a second-order reverse reaction is related to the effective diffusion coefficient D_e, which is proportional to the concentration S. During sorption the concentration S is high at the solid surface, so D is high there and the flux of material through the surface is high. During desorption the concentration of S is low at the surface, so D_e is low and the flux out of the solid is lower than for sorption. In contrast, the rate of reaction at the surface is given by Eq. 4.46, and is $\lambda C_0 - \mu {S_0}^2/S_e$. During desorption $C_0 = 0$, so the reverse reaction is unhindered, and equilibrium is rapidly reached. During sorption the forward reaction λC_0 is reduced by the reverse reaction $-\mu {S_0}^2/S_e$, so equilibrium is reached more slowly at the surface.

There is apparently no analytical solution for Eq. 4 with a second-order reverse reaction. Crank [2, p. 350] refers to some solutions with a second-order forward reaction only, as in Eq. 4.6. For the present purpose, the solutions for local equilibrium (Chapter 3) with the effective diffusion coefficient $D_e = 2SD/K_2$ should be approached as $\sqrt{\mu t}$ becomes larger than about 3. The profile of S goes to zero at a finite distance $x/2\sqrt{D_e t} = 0.81$. At times shorter than about $\sqrt{\mu t} = 1$, the forward reaction dominates, and the profiles of S are exponential in distance. Most experimental profiles are in either of these two time regimes and can be compared directly with the diffusion–reaction model. In the time range $1 < \sqrt{\mu t} < 3$, the profiles are uncertain. Of particular interest is at what time the profile goes to zero at a finite distance, but this time has not been determined.

4.5 NO LOCAL EQUILIBRIUM, TRACER DIFFUSION

In Section 4.4 the diffusing molecules moved into a solid or melt that contained no reacted groups. In addition, tracer experiments have been done in which isotopically labeled molecules have diffused into a solid or melt that is already at equilibrium between dissolved molecules and reacted ones. The isotopes exchange between the dissolved and reacted or matrix atoms. Some examples are the diffusion of tagged water molecules, with either deuterium or ^{18}O, into an oxide that contains reacted OH groups. The deuterium on the dissolved D_2O molecules exchanges with hydrogen on the OH groups:

$$D_2O + 2OH = H_2O + 2OD \tag{4.52}$$

The ^{18}O in the dissolved water molecules can exchange with either the oxygen 16 in the OH groups or in the oxide structure. Since the concentration of oxygen in the oxide network is usually much larger than that in the OH groups, it is this network exchange that is most significant. This exchange can be written

$$H_2^{18}O + {-}^{16}O{-} = H_2^{16}O + {-}^{18}O{-} \tag{4.53}$$

In these exchange reactions there is no net chemical reaction in the solid or melt.

In Section 3.4 these exchange reactions were discussed when they are at local equilibrium. It was found that diffusion plus exchange is governed by an effective diffusion coefficient D_e:

$$D_e = \frac{DC_T}{S_T} \tag{4.54}$$

in which D is the constant diffusion coefficient of dissolved molecules and C_T and S_T are the (uniform) total concentrations of dissolved and exchanged molecules, respectively:

$$C_T = C^* + C \quad S_T = S^* + S \tag{4.55}$$

where the asterisks denote tagged molecules.

For no equilibrium the rate of formation of tagged reacted groups is

$$\frac{\partial S^*}{\partial t} = k^*(C^*S - CS^*) \tag{4.56}$$

in which k^* is a reaction rate constant for exchange such as in Eq. 4.53 and has units of reciprocal concentration over time. From Eqs. 4.55,

$$\frac{\partial S^*}{\partial t} = k(C^*S_T - C^*S^* - S^*C_T + C^*S^*) = k(C^*S_T - C_TS^*) \tag{4.57}$$

For convenience it is assumed that all molecules supplied at the solid surface are tagged, so the surface concentration of dissolved molecules ${C_0}^* = C_T$. A lower ratio can readily be incorporated into the equations. Equation 4.57 then becomes

$$\frac{\partial {S_0}^*}{\partial t} = k^*C_T(S_T - {S_0}^*) = k(S_T - {S_0}^*) \tag{4.58}$$

at $x = 0$ with $k = k^*C_T$, with units of reciprocal time. Integration of Eq. 4.58 from ${S_i}^* = 0$ at $t = 0$ to ${S_0}^*$, t gives

$${S_0}^* = S_T[1 - \exp(-kt)] \tag{4.59}$$

Thus the surface concentration of tagged groups has a time dependence similar to that of Eq. 4.27 for first-order forward and reverse reactions. Thus the solutions valid for this condition are valid for the diffusion–exchange case. In particular, from Eq. 4.32,

$$S^*(x, t) = S_T\left[\operatorname{erfc}(r) - e^{-r^2}u(\sqrt{kt}), r\right] \tag{4.60}$$

with

$$r = \frac{x}{2}\left[\frac{1 - \exp(-kt)}{D_e t}\right]^{1/2} \tag{4.61}$$

and D_e from Eq. 4.54. At long times $kt \gg 1$, local equilibrium is approached and

$$S^* = S_T \operatorname{erf}\left(\frac{x}{2\sqrt{D_e t}}\right) \tag{4.62}$$

However, Fig. 4.7 shows that even though the profile may fit the erfc equation for some distance into the solid, at low S^* the diffusion–exchange profile shows lower S^* than the erfc profile of Eq. 4.62. Such deviations are found in, for example, the diffusion of deuterium into hydrogenated amorphous silicon [7].

At early times $kt < 1$, the forward exchange reaction dominates, and the differential equation becomes

$$\frac{\partial S^*}{\partial t} = D_e \frac{\partial^2 S^*}{\partial x^2} - kS^* \tag{4.63}$$

and the solution for semi-infinite geometry for constant surface concentration is (see Eqs. 4.19 and 4.20)

$$S^* = S_T kt \exp\left(-x\sqrt{\frac{k}{D_e}}\right) \tag{4.64}$$

Thus at times when $kt < 1$, one expects an exponential profile in tagged reacted groups.

4.6 TIME RANGES OF VALIDITY OF EQUATIONS

The first question is the value of time at which local equilibrium can be assumed. Equation 4.27 shows that when $\sqrt{\mu t} \succeq 1.8$, $S_0 > 0.95 S_e$; in other words, if $\sqrt{\mu t}$ is greater than 1.5, S_0 is within 5% of local equilibrium at the solid surface, as can be read from curve 4 in Fig. 4.9. However, deeper into the solid the concentration S of reacted groups is progressively farther from equilibrium. For example, to reach an S value of 93% of equilibrium at $x\sqrt{\lambda/D} = 8.0$ requires that $\sqrt{\mu t} = 4$; the actual value of S/S_e for these conditions is 0.144, and $2\sqrt{\mu t} > x\sqrt{\lambda/D}$. As another example, at $S/S_e = 9(10)^{-4}$ and $\sqrt{\mu t} = 5$, the value of S/S_e is within 20% of equilibrium. Thus to reach 80% local equilibrium at S/S_e equal to about 10^{-3}, diffusion must take place for about eight times longer than to reach 95% of surface equilibrium. The parameters involved in reaching local equilibrium are the diffusion coefficient of dissolved molecules D and the rate coefficients λ and μ.

In the time interval $\mu t < 1$, $\lambda t \gg 1$ the extent of the reverse reaction is small compared to the forward reaction, so the equations of Section 4.2 are valid. The profiles of reacted concentration S are exponential in distance for both first- and second- order reverse reactions, and the profiles of dissolved concentration are also exponential; the relations of Eq. 4.21 hold between dissolved and reacted concentrations. At earlier times there is a transient, but the reacted concentration is small and it has apparently not been studied experimentally. These time ranges are valid for both direct reaction and exchange experiments.

4.7 SURFACE EXCHANGE

In many studies of diffusion, deviations from equilibrium at the sample surface are found. In this chapter these deviations are attributed to nonequilibrium of the surface reaction or isotope exchange in the solid. In most previous work, such deviations have been attributed to a slow reaction or exchange right on the sample surface rather than in the interior of the material. In this section the mathematics of these two kinds of surface process are compared.

If there is a slow surface reaction the boundary condition at the surface is no longer a constant concentration but is given by the reaction flux J at the surface $x = 0$:

$$J = -D\left(\frac{\partial C}{\partial x}\right)_{x=0} = hD(C_0 - C_s) \tag{4.65}$$

in which h is a reaction parameter with units of reciprocal distance, C_0 the equilibrium concentration at the sample surface, and C_s the surface concentration at time t. Equation 4.65 is the boundary condition at $x = 0$. For a semi-infinite material additional boundary conditions are $C = 0$ for all x at $t = 0$ and $C \to 0$ as $x \to \infty$. The solution of the differential equation of diffusion without reaction and constant diffusion coefficient D (Eq. 4.4 with $\partial S/\partial t = 0$) with these boundary conditions is (Ref. 2, Eq. 3.35)

$$\frac{C}{C_0} = \operatorname{erfc}\left(\frac{x}{2\sqrt{Dt}}\right) - \exp(hx + h^2Dt)\operatorname{erfc}\left(\frac{x}{2\sqrt{Dt}} + h\sqrt{Dt}\right) \tag{4.66}$$

in which x is the distance from the sample surface. At the surface $x = 0$,

$$\frac{C}{C_0} = 1 - \exp(h^2Dt)\operatorname{erfc}(h\sqrt{Dt}) \tag{4.67}$$

The function $\exp(u^2)\operatorname{erfc}(u)$ is given in Table 2.1 in Ref. 2 (p. 375) and in Ref. 5 (p. 585) and is plotted in Figure 4.10. At long times the right-hand side of Eq. 4.67 goes to $1 - 0 = 1$, and $C = C_0$, the equilibrium concentration.

Also plotted in Fig. 4.10 is the surface concentration S_0/S_e of reacted material from the diffusion–reaction model with constant surface concentration of dissolved molecules (Eq. 4.27). The figure shows that the surface concentrations of reacted material as a function of square root time are much different for the diffusion–reaction model and the surface exchange model. Thus it should be possible to

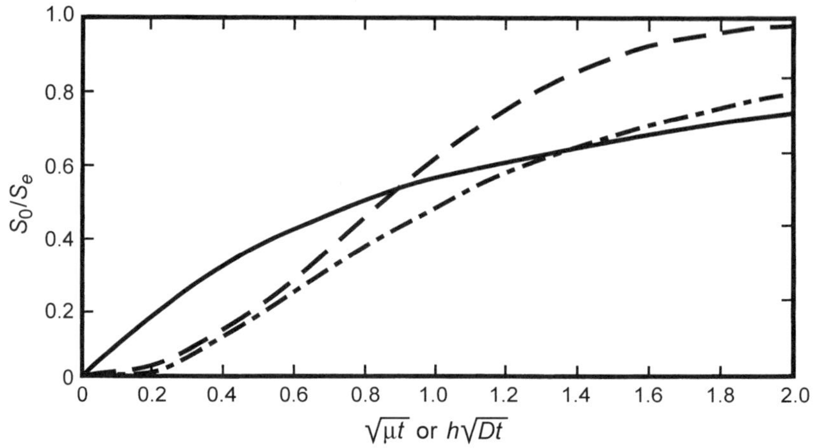

Figure 4.10 Concentration of reacted material at a sample surface S_0/S_e. Dashed line, diffusion–reaction model, Eq. 4.27, for first-order reverse reaction; solid line, Eq. 4.67, surface reaction model; dot-dashed line, Eq. 4.45, diffusion–reaction model, second-order reverse reaction.

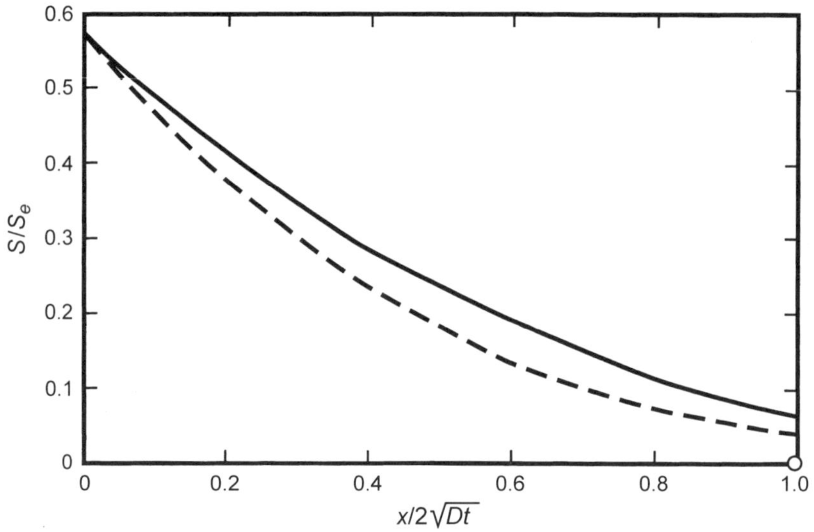

Figure 4.11 Profiles of reacted groups for the surface reaction model (solid curve) and diffusion–reaction model (dashed curve) at the same surface concentration of $0.57 = S_0/S_e$ ($h = 1$, $\mu t = 1$).

distinguish between these two models by measurements of surface concentrations of reacted material as a function of time.

The profiles of reacted groups as a function of distance from the surface are compared in Fig. 4.11 for the surface exchange and diffusion–reaction models with a surface concentration of 0.57 of the equilibrium value. The profiles are so similar that they cannot be distinguished by experimental measurements. Thus the only way I know of to distinguish between these two models is to measure the surface concentrations as a function of time, as described above.

REFERENCES

1. O. Dersch, A. Zouine, F. Rauch, and J. E. Ericson, *Fresenius J. Anal. Chem.* **358**, 217 (1997).
2. J. Crank, *Mathematics of Diffusion*, 2nd ed., Oxford University Press, Oxford, 1975.
3. Y. Zhang, E. M. Stolper, and G. J. Wasserburg, *Geochim. Cosmochim. Acta* **55**, 441 (1991).
4. R. H. Doremus, *J. Mater. Res.* **10**, 2379 (1995).
5. H. S. Carslaw and J. C. Jaeger, *Conduction of Heat in Solids*, 2nd ed., Oxford University Press, Oxford, 1959.
6. S. Rigo, F. Rochet, B. Agius, and A. Staboni, *J. Electrochem. Soc.* **129**, 867 (1982).
7. W. B. Jackson, C. C. Tsai, and P. V. Santos, *Mater. Res. Soc. Symp. Proc.* **258**, 319 (1992).

APPENDIX: ERROR FUNCTIONS OF COMPLEX ARGUMENT

Error functions of complex argument have been discussed by Carslaw and Jaeger [5, p. 484], and the functions u and v are tabulated by them (pp. 486 and 487). They also give references to the work from which this discussion is derived, especially Russian authors Faddeeva, Terentev, and Karpov, and other tabulations of these functions. Faddeeva and Terentev define the following functions:

$$\begin{aligned} w(z) &= u(x, y) + iv(x, g) \\ &= e^{-z^2}\left(1 + \frac{2i}{\sqrt{\pi}}\int_0^z e^{t^2}\,dt\right) \end{aligned} \qquad \text{(A4.1)}$$

$$z = x + iy \qquad \text{(A4.2)}$$

$$w(iz) = e^{z^2}\operatorname{erfc}(z) \qquad \text{(A4.3)}$$

With these definitions the series formulas for erf(z) and erfc(z) are the same as for real arguments.

Some additional properties of these functions are

$$\begin{aligned}
u(x,y) &= u(-x,y) \quad v(x,y) = -v(-x,y) \\
u(x,-y) &= -u(x,y) \quad v(x,-y) = -v(x,y) \\
\frac{\partial u(x,y)}{\partial y} &= 2\left(xv + yu - \frac{1}{\sqrt{\pi}}\right) \\
\frac{\partial v(x,y)}{\partial x} &= 2\left(\frac{1}{\sqrt{\pi}} - yu - xv\right) \\
\frac{\partial v(x,y)}{\partial y} &= 2(yv - xu)
\end{aligned} \tag{A4.4}$$

5

STRUCTURE AND SIZES OF ATOMS AND MOLECULES

5.1 INTRODUCTION

To understand the diffusion of atoms and molecules in solids and melts, one must determine the effective sizes of the dissolved constituents and the packing of atoms in the structures of the solids and melts. Gaseous and structural atoms are usually considered to be hard spheres. A *hard sphere* is one that has a specific radius; the atom does not extend beyond this radius, and other atoms and molecules cannot penetrate within it. This picture has been highly valuable in understanding gas kinetics and solid structures. This success is somewhat surprising, because quantum mechanics shows that the electrons around an atomic nucleus have a distribution and are not fixed at a particular radius, as described in more detail in the next section. Thus atoms are actually rather soft and squishy (deformable). Nevertheless, it is possible for many purposes to consider them to be hard spheres, which is a helpful simplification. However, we should always keep in the back of our minds that atoms are not truly hard spheres; we will find that some structures can only be understood by taking electronic distributions into account.

Sizes of dissolved atoms and molecules are described in Section 5.3. The dissolved atoms and molecules of interest here are usually gases (inert gases, water vapor, hydrogen, oxygen, nitrogen, and monoxides). The sizes of atoms in structures of solids and melts are listed and discussed in Section 5.4, and these sizes are then used to calculate the relative openness of different structures. This openness is found to correlate directly with the activation energy of diffusion of particular molecules (e.g., water) in the solids and melts. In Section 5.5 monoxides are discussed; in later chapters I try to show that monoxides can be diffusing constituents, especially in silicon.

5.2 ELECTRON DISTRIBUTIONS AROUND ATOMS AND MOLECULES

The probability of the electron in atomic hydrogen around the nucleus (proton) can be rigorously calculated from quantum mechanics (the Schrödinger equation). The result for the ground state of lowest energy is shown in Fig. 5.1. The probability increases from zero at the nucleus to a maximum and then decreases asymptotically to zero at large distances from the nucleus. This distribution is spherically symmetrical around the nucleus. The distance of the maximum in the probability distribution function is taken as the radius for the hard-sphere assumption of the hydrogen atom. This assumption emphasizes the comments made in the introduction; atoms are not truly hard spheres, because the electron has a spread in its probability distribution.

It is a remarkable result of studies of electron distributions around atoms of higher atomic number than hydrogen (1) that this relatively simple distribution is also a good approximation for some electron distributions around these atomic nuclei with higher charge and more than one electron. For example, see the discussions in Chapter 4 of Ref. 1 and Chapter 5 of Ref. 3. These *s* electrons have spherical probability distribution functions. The naming of electron types (*s*, *p*, *d*, *f*, etc.) and their distribution functions are discussed in books on quantum mechanics (e.g., Ref. 1) and also in many chemistry books (e.g., Ref. 2). A detailed description of these distribution functions is beyond the scope of this book.

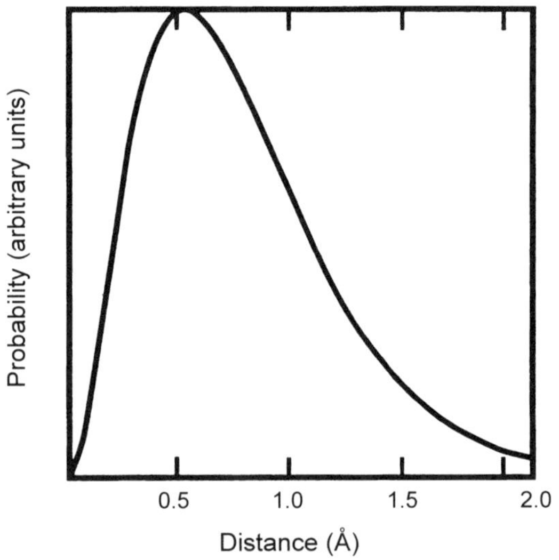

Figure 5.1 Probability distribution function of finding an electron at a distance *r* from hydrogen nucleus (proton). (From Ref. 1, Fig. 21-3, p. 143.)

The electrons designated as *p*, *d*, and *f* have probability distributions that are not spherically symmetrical superposed on a spherically symmetrical distribution. The nodes of maximum probability at particular angles around the nucleus lead to directional bonding, called *covalent bonding*. The assumption of a hard sphere for atoms bonded in this way is problematic. In a covalent bond, valence electrons from each of the two bonded atoms are shared by the atoms, so there is a distribution of electrons between the atoms as well as around their nuclei. Calculations of distributions of electrons in covalent bonds are described in many books and articles, for example, in Refs. 1, 3, and 4. A simple way to describe the "space" that two covalently bonded atoms occupy is to consider them as two spheres connected by a cylinder if they are the same size, or as a truncated circular pyramid if the atoms are of different sizes. Some depictions of covalently bonded atoms with this convention are given below and in Chapter 16.

Atom sizes are found by measuring interatomic distances in a number of different solids and molecules and assigning a consistent size to each atom [5]. This system works quite well for most metallic atoms. The bonding in a metal or alloy is nondirectional; it results from the sea of free electrons in the metal or alloy. These electrons can move freely throughout the entire sample and are not localized at a particular atom. Metallic compounds are often ionically bonded, for example, alkali halides; in these compounds each atom gives up one or more electrons to become ions. Again the bonding force or field is reasonably spherically symmetrical around each ion, so the assumption of a hard sphere for the ion is reasonable. Thus metal atoms in metals and alloys and ions in ionic compounds can be accurately portrayed as hard spheres. Examples of sizes of ions that are usually bonded in ionic compounds are given in Section 5.4.

It is for covalently bonded atoms in solids that the assumption of hard-sphere atoms becomes questionable. These atoms tend to be in, although not limited to, groups III, IV, and V of the periodic table. Their nominal sizes are often quite small and appear to vary strongly from one compound or solid to another. Examples are boron, aluminum, silicon, phosphorus, and arsenic. These atoms are of particular importance in many of the solid crystals and glasses of most interest in this book, such as silicates, minerals, and silicon and its III–V dopants. In strongly covalently bonded solids such as silicon, we need some way of designating atom sizes. An approximate compromise is to treat two atoms and their bond as a cylinder connecting two spheres rather than as touching spheres. Such a convention is described next for solid silicon. In many compounds, such as oxides, the atomic radius found for silicon is about 0.26 angstrom unit (Å). Angstrom units are used in this book for atomic sizes and volumes, even though they are not SI approved, because they are the most convenient units; atomic dimensions are from about 0.2 to 5 angstrom units. See the Appendix for a comparison of different length and volume units. Compared to this radius of 0.26 Å, the distance of closest approach of silicon atoms in solid silicon is about 2.35 Å [6], giving an effective radius for the silicon atom of 1.17 Å. The compromise is to consider the silicon atoms in solid silicon to be spheres of 1 Å radius joined by cylinders of 1 Å radius. There is a sketch of a (110) plane in solid silicon in Fig. 5.2, showing the atoms and bonds in

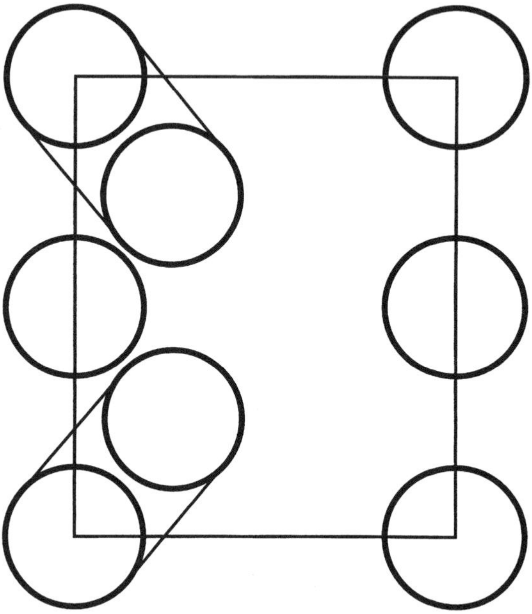

Figure 5.2 A (110) plane in the unit cell of silicon (diamond structure). The silicon atoms are shown as spheres of radius 1 Å and the bonds between silicon atoms as cylinders of this radius.

the plane in this way. This convention gives a fair approximation to the actual space occupied by the atoms and their bonds that are in this plane. This method emphasizes the openness of the silicon structure, giving about 47% open space when the space occupied by both the silicon atoms and their bonds is taken into account. The open space or *porosity* is comparable to that of the many multicomponent silicate glasses (see Section 5.4). If one uses the distance of closest approach of silicon atoms in silicon to estimate the atomic radius (1.17 Å; see above) and assumes the silicon atoms to be spheres of this size, the volume fraction of the silicon structure occupied by the silicon atoms is about 0.34, which is probably too low. This volume fraction gives an ionic porosity or open fraction of 0.66.

5.3 SIZES OF GASEOUS ATOMS AND MOLECULES DISSOLVED IN SOLIDS AND MELTS

A number of methods have been used to calculate the sizes of gaseous atoms and molecules. I believe the sizes calculated from the viscosities of gases and the kinetic theory are the best for atoms and molecules dissolved in solids. This view is supported by a comparison between these sizes and the activation energies of their

Table 5.1 Atomic and Molecular Radii from Gas Kinetic Viscosity (Å)

Gas	700°C [7]	0°C [8]
He	1.0	1.09
Ne	1.2	1.30
H_2	1.25	1.36
H_2O	(1.65)[a]	2.33
Ar	1.6	1.83
O_2	1.6	1.81
N_2	1.7	1.89
Kr	2.1	2.08
Xe	2.45	2.46

[a] A value of 1.38 Å is used for water instead of the gas kinetic values. See the text for an explanation.

diffusion in silica glass as described below. Sizes of a number of gaseous atoms and molecules are given in Table 5.1, where the values at 0°C are about 10% higher than those at 700°C, except for krypton and xenon, where the values for the two temperatures are about the same. I have chosen the values at 700°C for comparison with activation energies for diffusion because most diffusion experiments have been conducted at temperatures up to 1000°C or higher.

Zhang and Xu [25] have suggested an ingenious way to calculate sizes of inert gas atoms dissolved in solids. They plotted the radii of ions with the same number of electrons as the inert gas atom as a function of ion charge and interpolated to zero change to get the inert gas radii. For helium, neon, and argon atoms and ion coordination numbers of 6, this method gives atomic radii close to those in Table 5.1 at 700°C; however, the radii for krypton (1.78 and 1.87 Å) and xenon (1.96 and 2.03 Å) found by this method are smaller than the values in Table 5.1. Plots of ionic sizes for krypton (36 electrons) and xenon (54 electrons) are shown in Fig. 5.3. Zhang and Xu found their sizes by drawing a smooth curve through the size data, as shown in the figure; however, if the anion sizes are extrapolated to zero charge, larger radii of 1.94 Å for krypton and 2.19 Å for xenon are found, which are closer to the values in Table 5.1. The comparison below supports the values given in Table 5.1; nevertheless, the radii for krypton and xenon are not as certain as for the other atoms and molecules in the table.

To compare atomic and molecular sizes with activation energies for diffusion, a theoretical relation between these quantities is needed. It is well established that gas atoms and molecules dissolve in interstices or atomic-sized holes in silicate glasses [19, 20, 26]. Anderson and Stuart [27] assumed that the activation energy for diffusion of such dissolved entities is the energy required for the atoms or molecules to jump from one interstice to a neighboring one. The interstices are considered to be connected by doorways of average radius r_D; the activation energy

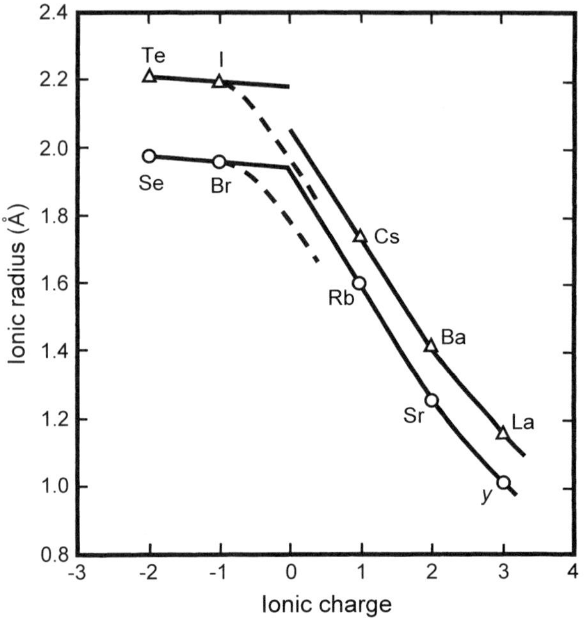

Figure 5.3 Sizes of ions isoelectronic with krypton (○) and xenon (△), as a function of the ionic charge from Ref. 5 for a coordination number of 8. The dashed curves are from Ref. 25.

is then the elastic energy required to enlarge the doorway to the radius r of the diffusing molecules, assuming them to be incompressible. This energy Q can be estimated from the equation

$$Q = 8\pi G r_D (r - r_D)^2 \tag{5.1}$$

derived by Frenkel [28] for the elastic energy to dilate a spherical cavity from the radius r_D to r. G is the elastic shear modulus for the glass. Anderson and Stuart [27] changed the factor of 8 in Eq. 5.1 to a factor of 4 to give better agreement with results on glasses. They argued that in the loose and more open structure of silica, the strain energy of dilation should be less than for a close-packed liquid, for which Eq. 5.1 was derived.

Diffusion coefficients of many atoms and molecules in silica glass have been measured and the activation energies for these diffusion coefficients are listed in Table 5.2. Shelby [9] has summarized many experimental results on diffusion of gases in silica glass. The activation energies Q in Table 5.2 were calculated from the Arrhenius relation:

$$D = D_0 \exp\left(-\frac{Q}{RT}\right) \tag{5.2}$$

Table 5.2 Activation Energies for Diffusion of Gaseous Atoms and Molecules in Silica Glass

Gas	Activation Energy (kJ/mol)	Refs.
Helium	25.3	10, 11
Hydrogen	43.4	12
Neon	44.3	13–17
Water	71.0	18
Argon	105	14, 15, 36
Nitrogen	109	22, 23
Oxygen	113	21
Krypton	173	14, 36
Xenon	297	24, 36

in which D_0 is a constant, R the gas constant, and T the absolute temperature. The data for helium and neon are extensive enough to show that they actually fit better an equation of the form

$$D = D_1 T \exp\left(-\frac{Q_1}{RT}\right) \tag{5.3}$$

with a preexponential temperature factor [26,29], but this factor is ignored here and only Eq. 5.2 used.

The activation energies are plotted as $\sqrt{Q}$ versus the atomic or molecular radius r in Fig. 5.4 to test Eq. 5.1. Data for all the gases except water fit very well on a line on this plot. Zhang and Xu [25] suggested that a smaller radius for water, instead of the gas kinetic values, is appropriate. A value of 1.38 Å for water is used in Fig. 5.4 instead of the radii in Table 5.1. This is exactly the radius of an oxygen atom in solid oxides (see Section 5.4), showing that when dissolved in a solid or melt, the water molecule behaves as if the hydrogen atoms are completely buried in the electron cloud of the oxygen atom.

The linear regression line in Fig. 5.4 shows an excellent fit to the experimental data, with a correlation coefficient of 0.993. This fit shows:

1. Equation 5.1 provides the correct functional dependence between activation energy for diffusion in silica and atomic and molecular size. A linear relation between Q and r gives a much poorer fit.
2. The correct sizes for atoms and diatomic molecules are those calculated from viscosity of the gases and the kinetic theory. The sizes and activation energies for krypton and xenon are subject to greater errors than those for other gases, but until better data are available, those in Table 5.2 are judged to be reasonable.
3. The correct size for water molecules dissolved and diffusing in silica glass is 1.38 Å rather than the gas kinetic value.

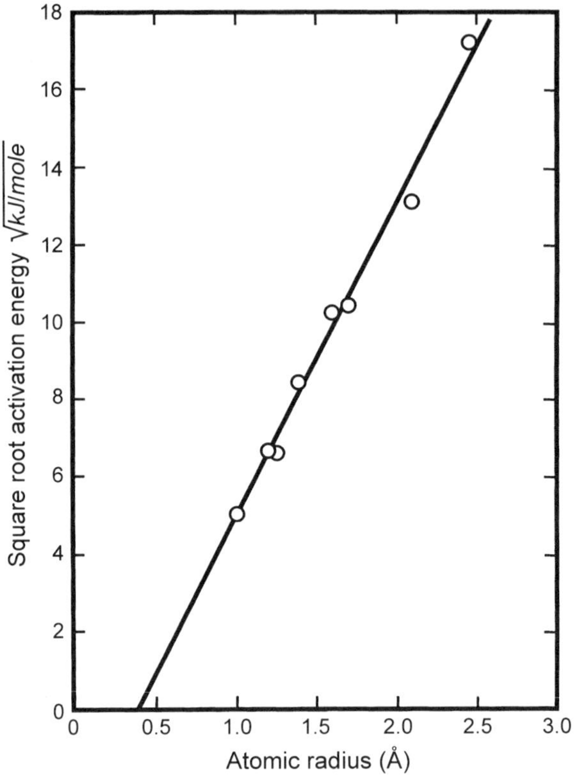

Figure 5.4 Square root of the activation energy for diffusion of various atoms and molecules in silica glass as a function of their radius (see Table 5.2).

The doorway radius r_D is the intercept of the line in Fig. 5.4 and the r axis, and equals 0.38 Å. This size fits well with models of the structure of silica glass [26, Chap. 3]. The shear modulus G calculated from the line in Fig. 5.4 is about $12(10)^9$ Pa or 12 GPa; this value is smaller than the shear modulus of silica glass, of about 30 GPa. This difference possibly results because the strain energy in the loose network structure of silica is less than in a close-packed liquid, for which Eq. 5.1 was derived, as suggested by Anderson and Stuart [27].

Silica glass is the only solid for which extensive diffusion measurements for many different atoms and molecules have been made. There are a few data for rhyolitic and albitic natural glasses. The compositions of these glasses are given in Table 5.3. Measurements for helium, water, and argon for these glasses appear to be reliable and are plotted in Fig. 5.5. Diffusion of neon in rhyolitic glasses was measured by outgassing powdered glass [30]; details of the calculations are not given. Apparently, uniform concentration of neon was assumed even after diffusion at a lower temperature; this assumption is unlikely to be correct. The resulting diffusion coefficients for three of four glasses tested in Ref. 30 showed curvature on

Table 5.3 Approximate Compositions of Glasses (wt % of Major Constituents)

Oxide	Rhyolite (Obsidian)	Albitic
SiO_2	77.3	69.3
Al_2O_3	13.0	20.3
Na_2O	4.1	10.5
K_2O	4.2	—
CaO	0.5	—
FeO	0.8	—

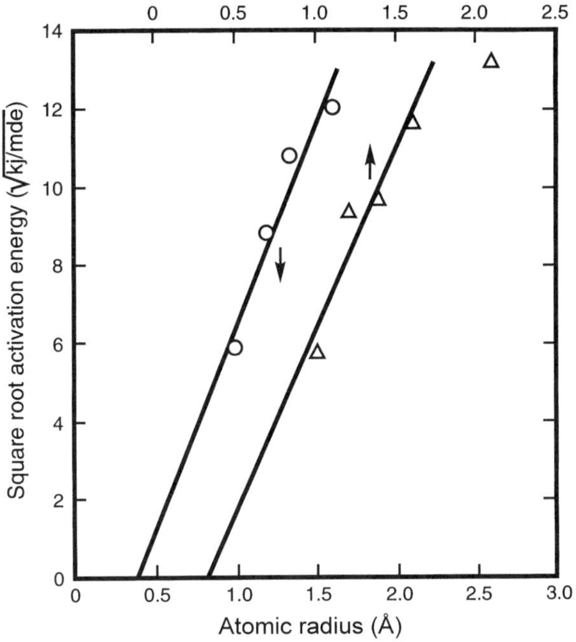

Figure 5.5 Square root of the activation energy for diffusion of atoms and molecules in rhyolite glass (○) and albitic glass (△), whose compositions are in Table 5.3. Refs.: rhyolite, He [31, 32], Ar [33], H_2O [34, 35]; albitic, He [31], Ne [36], Ar [33], Kr [39, 36], H_2O [37]. Regression line for albitic glass omits points for neon and krypton.

a log D versus $1/T$ plot, so these activation energies are uncertain and are not used here. The linear regression line in Fig. 5.5 for rhyolite glass had a correlation coefficient of 0.994 and a doorway radius (intercept with the r axis) of 0.43 Å. The data for albitic glasses are much more scattered than for the other two glasses, as shown in Fig. 5.5. The regression line for the data in Fig. 5.5 gave a correlation coefficient of 0.959 and a doorway radius of 0.13 Å, which is probably too small. If the data for neon and krypton in albitic glass are omitted, a doorway radius of about 0.34 Å and a correlation coefficient of 0.999 result as shown in Fig. 5.5. The shear moduli calculated from the lines and r_D values in Fig. 5.5 are 15 and 18 GPa for rhyolitic and albitic glasses, respectively.

The fraction of open space or structure porosity (see Section 5.4) of these glasses is 0.509, 0.488, and 0.441 for silica, rhyolitic, and albitic glasses, respectively. Thus the smaller doorway for the albitic glass is perhaps reasonable. These three glasses contain almost no nonbridging oxygen atoms; in the natural glasses four-coordinated aluminum ions substitute for silica in the glass network, and the extra negative charge on the aluminum ions is compensated by alkali ions. Therefore, these glasses should have similar network structures, giving similar doorway radii and shear moduli. The comparisons in this section show that the sizes of dissolved atoms and molecules in Table 5.1 are reasonable, and Eq. 5.1 fits diffusion data for them well, except for requiring a low shear modulus.

5.4 SIZES OF IONS IN SOLIDS AND MELTS, AND OPENNESS OF SOLID STRUCTURES

Sizes of some selected ions are listed in Table 5.4. These sizes were calculated mainly from oxide compounds, so they should be especially valid for oxides. For other solids and melts, these sizes may not be appropriate, as described in Section 5.2. For example, in strongly covalently bonded materials, the atom sizes can be much different from those in Table 5.4. Ionic volumes are also given in the table.

For understanding diffusion in solids and melts, it is useful to have some measure of their openness or ionic porosity Z. The openness can be calculated from the molecular volume and the volume of the ions assumed to be hard spheres, from Table 5.4. The molecular volume V in A^3 per molecule is

$$V = \frac{M}{\rho(0.602)} \tag{5.4}$$

in which M is the molecular weight in grams per mole and ρ is the density in grams per cubic centimeter. Values of V for some selected binary oxides are given in Table 5.5. The total volume of ions in the molecule V_i is the sum of their volumes from Table 5.4. Then the ionic porosity Z is

$$Z = 1 - \frac{V_i}{V} \tag{5.5}$$

Values of Z for some binary oxides are given in Table 5.5 and for many more oxide compounds are given in Chapter 9 (see also Ref. 38). This method is exactly the same as working with the crystalline unit cell and has the advantages that it does

Table 5.4 Radii and Volumes of Selected Ions

Ion	Charge	Coordination Numbers	Radius (Å)	Volume (Å^3)
Aluminum	+3	4	0.39	0.25
Antimony	+3	6	0.54	0.66
	+5	4, 6	0.76	1.84
Arsenic	+3	6	0.60	0.90
	+5	6	0.58	0.82
Calcium	+2	6	1.00	4.19
Carbon	+4	4	0.15	0.01
Chloride	−1	6	1.81	24.84
Chromium	+3	6	0.62	1.00
Copper	+1	6	0.77	1.91
	+2	6	0.73	1.63
Fluorine	−1	6	1.33	9.86
Gallium	+3	6	0.62	1.00
Germanium	+4	4	0.39	0.25
Iron	+2	6	0.78	1.99
	+3	6	0.64	1.10
Lead	+2	6	1.19	7.06
	+4	6	0.78	1.99
Lithium	+1	6	0.76	1.84
Magnesium	+2	6	0.72	1.56
Nitrogen	+5	6	0.13	0.01
Oxygen	−2	4, 6	1.38	11.01
Phosphorus	+3	6	0.44	0.36
	+5	6	0.38	0.23
Potassium	+1	6	1.38	11.01
Silicon	+4	4	0.26	0.07
Sodium	+1	6	1.02	4.46
Sulfur	−2	6	1.84	26.10
Titanium	+4	6	0.60	0.90
Yttrium	+3	6	0.90	3.05
Zinc	+2	6	0.74	1.70
Zirconium	+4	6	0.72	1.56
		7	0.78	1.99
		8	0.84	2.48

Source: Ref. 5.

not require knowledge of the crystal structure, is simpler, and can be used for glasses and melts.

The values in Table 5.5 vary from more than half to 20% openness. Some compounds, such as corundum (Al_2O_3), have less open space than for close packing of spheres of equal size. Other structures are surprisingly open, for example, wurtzite (ZnO) and quartz (SiO_2 and GeO_2). These structures are often thought of

Table 5.5 Porosity (Open Space) Z of Some Binary Oxides

Oxide	Structure	V (Å^3)	Z
Al_2O_3	Corundum	42.66	0.195
FeO	NaCl	20.04	0.351
Fe_2O_3	Corundum	50.48	0.302
GeO_2	Hexagonal (quartz)	41.09	0.458
	Tetragonal	27.84	0.200
MgO	NaCl	10.70	0.328
SiO_2	Amorphous	45.36	0.513
	α-Quartz	37.67	0.414
TiO_2	Rutile	31.05	0.262
	Anatase	33.48	0.315
	Brookite	32.19	0.288
ZnO	Wurtzite	23.66	0.463
ZrO_2	Monoclinic	35.13	0.317
Close packing of equal-sized spheres			0.26

as being close-packed, but this designation is misleading because it connotes limited space. For the present purpose of understanding diffusion, openness greater than close packing suggests that molecules can dissolve and diffuse in these open solids and melts. See also Chapter 9, where it is shown that the activation energy for diffusion of water in a wide variety of oxides, both crystalline and amorphous, correlates with their Z values.

5.5 MONOXIDES

At equilibrium with ambient air the surfaces of all metals except gold are covered with a stable oxide. The stabilities of metallic oxides can conveniently be compared with plots of their free energies of formation, called an *Ellingham diagram*. Such a plot is shown in Fig. 5.6. The diagram includes nomographs for calculation of the ratios H_2/H_2O and CO/CO_2 to maintain the metal surface clean of oxide and the oxygen pressure for this condition. The great stability of metal oxides shown by this diagram suggests to me that in the presence of even a small amount of oxygen, oxides are compounds likely to be present in solids. The easiest oxide to dissolve and diffuse in a solid or melt is a monoxide. I suggest that sometimes monoxides are the transporting species in diffusion in solids and melts. In Chapter 16 I argue that oxygen in silicon is transported by silicon monoxide (SiO). Most monoxides are about the same size as the oxygen atom (about 1.38 Å in radius) because of the small size of their atoms other than oxygen. In Section 5.3 it was shown that water has an effective radius about the same as that of the oxygen atom (1.38 Å); in Chapter 9 it will be shown that water dissolves and diffuses in a wide variety of oxides. Some atom–atom distances in group IV monoxides are C–O, 1.13 Å; SiO,

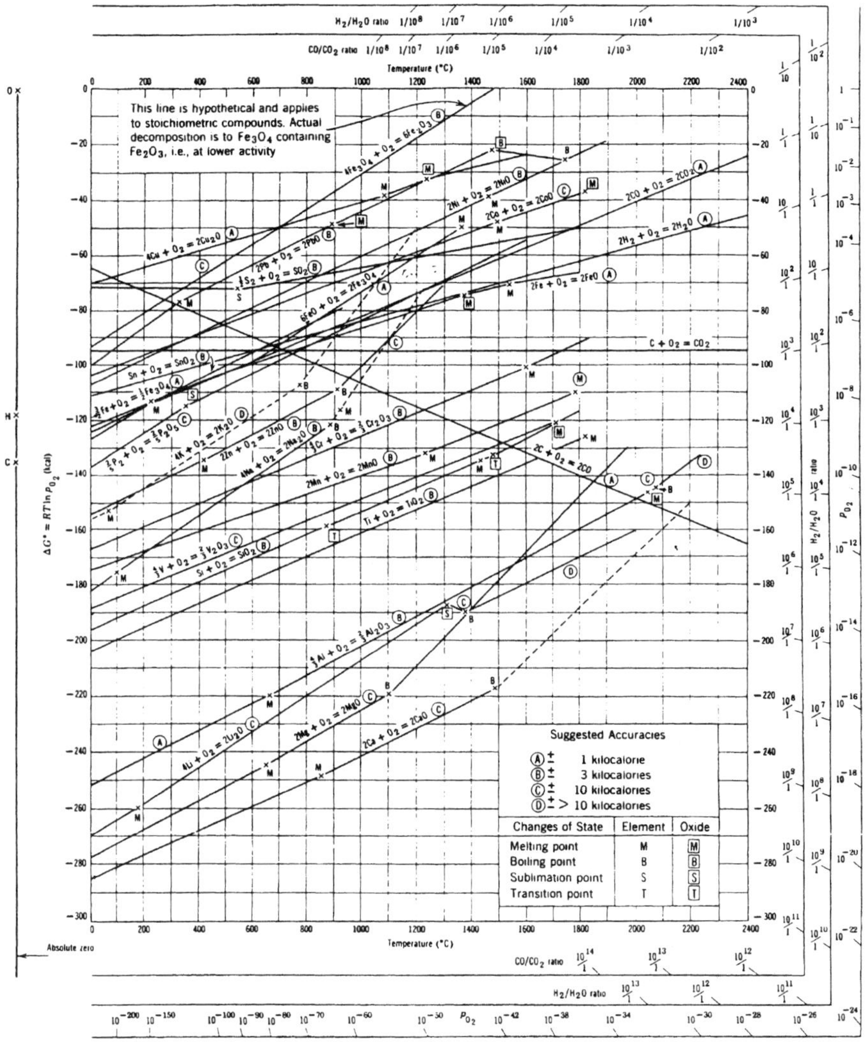

Figure 5.6 Gibbs free energy of formation of oxides as a function of temperature. (From Ref. 40, as modified in Ref. 41.)

1.4 Å, and GeO, 1.65 Å. These bond lengths suggest that these monoxides are small molecules, especially when they are dissolved in a solid or melt. Thus monoxides can be important diffusing species in solids and melts, and this section is devoted to a discussion of these compounds.

The Gibbs free energies of formation of a number of monoxides at 298°C are given in Table 5.6. The reaction is, for example, for SiO,

$$\mathrm{Si}(s) + \tfrac{1}{2}\mathrm{O}_2(g) = \mathrm{SiO}(g) \tag{5.6}$$

Table 5.6 Gibbs Free Energies of Formation of Selected Monoxides at 298°C (kJ/mol)

CO −137.2	BO −4.0	NO 86.6
SiO −126.4	AlO 65.3	PO −51.9
GeO −73.2	GaO 253.5	AsO 70.0 (ΔH)
SnO −8.4	YO −64.9	

Source: Ref. 42.

The entropy change for this reaction and for the other compounds in Table 5.6 is positive, so the Gibbs free energy of formation becomes more negative as the temperature increases. Thus for AlO the ΔG_f value becomes negative above about 500°C. Therefore, from the negative heats of formation in the table, the monoxides of carbon, silicon, germanium, tin, boron, yttrium, and phosphorus should be stable above room temperature, at 1 atm oxygen pressure. It is known that when SiO_2, GeO_2, SnO_2, and Al_2O_3 vaporize, the main constituent in the vapor phase is the monoxide [43]. All these conditions suggest that monoxides will be stable in solids under mildly oxidizing conditions and are candidates for transport in solids and melts.

REFERENCES

1. L. Pauling and E. B. Wilson, *Introduction to Quantum Mechanics*, McGraw-Hill, New York, 1935.
2. W. L. Masterton and C. N. Hurley, *Chemistry: Principles and Reactions*, Saunders, Philadelphia, 1993, Chaps. 6 and 7.
3. J. P. Lowe, *Quantum Chemistry*, Academic Press, San Diego, CA, 1978.
4. C. Kittel, *Quantum Theory of Solids*, 2nd ed., Wiley, New York, 1987.
5. R. D. Shannon, *Acta Crystallogr.* **A32**, 751 (1976).
6. B. D. Cullity, *Elements of X-ray Diffraction*, 2nd ed., Addison-Wesley, Reading, MA, 1978, p. 507.
7. Landolt-Börnstein, *Tabellen* **1**, 325 (1950).
8. J. Jeans, *An Introduction to the Kinetic Theory of Gases*, Cambridge University Press, Cambridge, 1948, p. 183.
9. J. E. Shelby, *Handbook of Gas Diffusion in Solids and Melts*, ASM International, Materials Park, OH, 1996.
10. D. E. Swets, R. W. Lee, and R. C. Frank, *J. Chem. Phys.* **34**, 17 (1961).
11. J. E. Shelby, *J. Am. Ceram. Soc.* **54**, 125 (1971).
12. R. W. Lee, *J. Chem. Phys.* **38**, 448 (1963).
13. R. C. Frank, D. E. Swets, and R. W. Lee, *J. Chem. Phys.* **35**, 1451 (1961).
14. W. G. Perkins and D. R. Begeal, *J. Chem. Phys.* **54**, 1683 (1971).
15. M. Abe, B. Rauch, and W. W. Brandt, *Z. Naturforsch.* **26**, 997 (1971).
16. J. E. Shelby, *Phys. Chem. Glasses* **13**, 167 (1972).

17. K. P. Srivastava and G. J. Roberts, *Phys. Chem. Glasses* **11**, 21 (1970).
18. R. H. Doremus, *J. Mater. Res.* **10**, 2379 (1995).
19. G. S. Nakayama and J. F. Shackelford, *J. Non-Cryst. Solids* **126**, 249 (1990).
20. M. R. Carroll and E. M. Stolper, *Geochim. Cosmochim. Acta* **55**, 211 (1991).
21. F. J. Norton, *Nature* **191**, 701 (1961).
22. R. M. Barrer, *Diffusion in and through Solids*, Cambridge University Press, Cambridge, 1941, p. 117ff.
23. J. Johnson and R. Burt, *J. Opt. Soc. Am.* **6**, 734 (1922).
24. H. Matzke, *Phys. Status. Solidi.* **18**, 285 (1966).
25. Y. Zhang and Z. Xu, *Am. Mineral.* **80**, 670 (1995).
26. R. H. Doremus, *Glass Science*, 2nd ed., Wiley, New York, 1994, Chap. 8.
27. O. L. Anderson and D. A. Stuart, *J. Am. Chem. Soc.* **37**, 573 (1954).
28. J. Frenkel, *Kinetic Theory of Liquids*, Oxford University Press, Oxford, 1946, pp. 10–11.
29. J. E. Shelby, in *Treatise on Materials Science and Technology*, Vol. 17, R. H. Doremus and M. Tomozawa, eds., Academic Press, San Diego, CA, 1979, p. 1.
30. J. J. Matsuda, K. Matsubara, H. Yajima, and K. Yamamoto, *Geochim. Cosmochim. Acta* **53**, 3025 (1989).
31. J. E. Shelby and R. J. Eagan, *J. Am. Ceram. Soc.* **59**, 420 (1976).
32. G. H. Frischat and H. J. Oel, *Phys. Chem. Glasses* **8**, 92 (1967).
33. M. R. Carroll, *Earth Planet. Sci. Lett.* **103**, 156 (1991).
34. Y. Zhang, E. M. Stolper, and G. J. Wasserburg, *Geochim. Cosmochim. Acta* **55**, 441 (1991).
35. R. H. Doremus, *J. Noncryst. Solids* **261**, 101 (2000).
36. K. Roselieb, W. Rammensee, H. Büttner, and M. Rosenhauer, *Chem. Geol.* **120** (1995).
37. B. J. Giletti, M. P. Semet, and R. A. Yund, *Geochim. Cosmochim. Acta* **42**, 45 (1978).
38. R. H. Doremus, *J. Mater. Res.* **14**, 3754 (1999).
39. M. R. Carroll, S. R. Sutton, M. L. Rivers, and D. S. Woolum, *Chem. Geol.* **109** (1993).
40. F. D. Richardson and J. H. E. Jeffer, *J. Iron Steel Inst.* **160**, 261 (1948).
41. L. S. Darken and R. W. Gurry, *Physical Chemistry of Metals*, McGraw-Hill, New York, 1953.
42. D. D. Wagman et al., *NBS Tables of Thermodynamic Properties*, National Bureau of Standards, Washington, DC, 1982.
43. J. Drowart, in *Condensation and Evaporation of Solids*, E. Rutner, P. Goldfinger, and J. P. Hirth, eds., Gordon & Breach, London, 1964, p. 255ff.

6

DIFFUSION AND REACTION OF WATER IN SILICA GLASS

6.1 INTRODUCTION

Silica glass (SiO_2) is one of the purest commercial materials, vying with silicon for this distinction, for which silica glass is used as a crucible material. Additionally, silica glass has excellent properties: high-temperature stability, low thermal expansion and consequently high thermal-shock resistance, excellent chemical durability, and good optical transmission in the ultraviolet. These properties have led to the use of silica glass for high-temperature vessels, lamp envelopes, optical components, and many specialized applications. The development of fiber optic cables of silica glass has revolutionized communications. The dissolution, diffusion, and reaction of hydrogen and water in these cables can degrade their performance and demonstrates a highly practical application of the diffusion–reaction models of this book.

Other names for *silica glass* are *vitreous silica, amorphous silica, fused silica, quartz glass*, and *fused quartz*. All but the last two terms are used synonymously in this book; inclusion of *quartz* in the name implies starting with quartz crystal, which is one way to make silica glass; it can also be made from liquids or gases such as $SiCl_4$. Use of the term *quartz* for silica glass is widespread and highly unfortunate, because it leads to confusion with crystalline quartz. I hope all readers will refrain from such use.

The diffusion, reaction, and exchange of water in silica have been studied more than in any other material and provide strong evidence for the diffusion–reaction model. These experimental results are examined in detail in this chapter. I suggested the application of the diffusion–reaction model to the diffusion of water in silica glass in 1969 [1]; at first there was strong resistance to the idea of

Table 6.1 Various Types of Fused Silica

		Maximum Impurity Concentration (ppm)									Manufacturers' Designations				
Type	Method of Manufacture	Al	Fe	Ca	Mg	K	Na	Li	Cl	OH	General Electric	Thermal Syndicate	Haeraeus	Corning	Quartz et Silice
I	Electrical fusion of quartz crystal	150	7	12	7	4	12	12	≈50	4	101	IR Vitreosil	Infrasil		Pursil
											204				
											125				
II	Flame fusion of quartz crystal	Uncertain								400		OG Vitreosil	Herosil		
													Homosil		
													Vittrasil		
III	Flame hydrolysis of $SiCl_4$[a]	10	6	4	3	2	2	1	60	1200	151	Spectrosil	Suprasil I	7940	Tetrasil
IV	Vapor-phase oxidation of $SiCl_4$	Similar to type III							500	Low		Spectrosil WF	Suprasil W	7943	

Source: Ref. 3, p. 7.

[a]Lower values of impurity concentrations are frequently quoted but are questionable.

solution and diffusion of water in a solid, but by now this process is widely accepted.

A variety of different types of silica glass are available commercially, as listed in Table 6.1. The differences involve impurity levels and methods of manufacture. Some silicas contain substantial amounts of water as reacted SiOH groups, which must be taken into account in interpreting diffusion data.

We begin the chapter with a discussion of the structure of silica glass to help in understanding how water dissolves and diffuses in it; then gas solubility in silica glass is considered, primarily to show that water molecules dissolve, as expected from their size, as compared to other gases. Next, experimental results are treated, starting with experiments at temperatures above about 650°C, where reaction equilibrium is reached rapidly. Then experiments at lower temperatures, at which reaction equilibrium is reached more slowly, are discussed. The conclusion of an examination of a wide variety of experimental results is that the diffusion–reaction model provides a complete, consistent, and accurate description of all these results.

6.2 STRUCTURE OF SILICA GLASS

The basic building block of all silicates is the silicon–oxygen tetrahedron, in which a central silicon atom is bonded to four oxygen atoms as its nearest neighbors, forming the corners of a tetrahedron. These tetrahedra have bond angles and distances that are uniform, especially in silica glass, but also in most other silicates. In silica glass these tetrahedra are connected at their corners into a three-dimensional network, shown schematically in Fig. 6.1 [2; 3, p. 27]. The x-ray experiments of Warren and his collaborators [4, 5] showed that this structure in silica glass can be described as a random network. This term does not mean that the atoms are distributed at random in the structure. The silicon–oxygen tetrahedra impose a rigid local order on the structure, so there is short-range order but no long–range order.

The pair distribution function calculated from the x-ray function calculated from the x-ray diffraction results is shown in Fig. 6.2. The method of calculating this function is described in Refs. 4 and 5. Interatomic distances are calculated directly from the peak positions on the distribution curve, and the areas under the peaks give the number of neighboring atoms. The first peak in Fig. 6.2 corresponds to the silicon–oxygen distance of 1.62 Å, and the second peak to an oxygen–oxygen distance of 2.65 Å. These distances are close to those found in crystalline silicates. The widths of these peaks are consistent with a tetrahedral distribution of oxygen atoms around each silicon atom, as shown by the calculated peaks shown in the figure. The third peak corresponds to the silicon–silicon distance of 3.12 Å and is much broader than the other two peaks because of the distribution in silicon–oxygen–silicon bond angles. The fourth peak at 4.15 Å is the silicon–second oxygen contribution, and the fifth peak at about

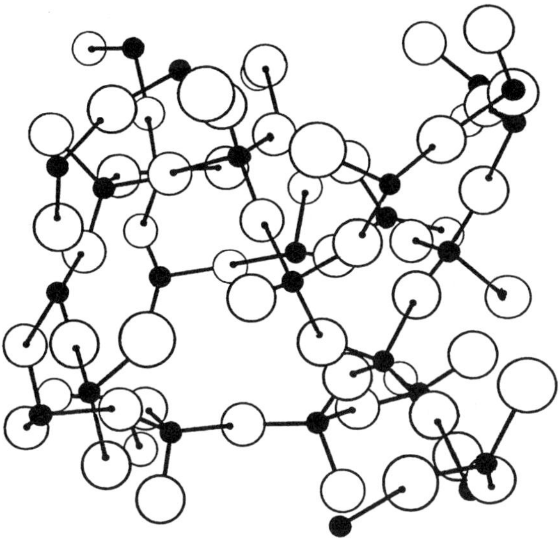

Figure 6.1 Schematic diagram of silica glass, showing silica tetrahedra linked at corners to form a random network with rings and cages. Silicon atoms are black, oxygen atoms are white. (From Refs. 2 and 3.)

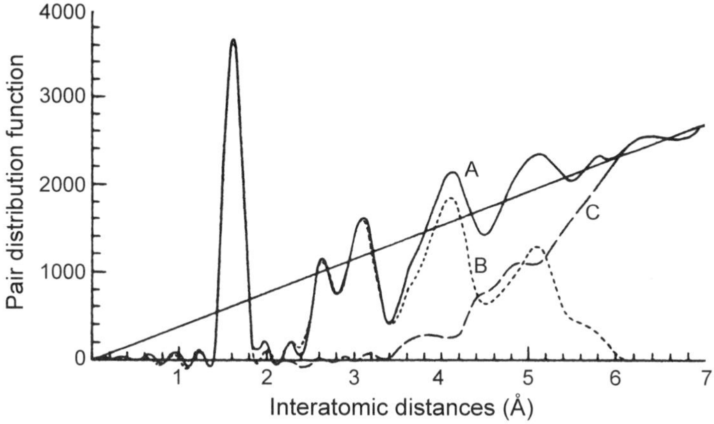

Figure 6.2 Pair distribution function for silica glass *A*, measured curve; *B*, sum of curves calculated from the first six contributions of atom pairs Si–O, O–O, Si–Si, Si–2nd O, O–2nd O, and Si–2nd O; *C*, difference between *A* and *B*. (From Refs. 3 and 5.)

5.1 Å is a combination of oxygen–second oxygen and silicon–third oxygen contributions. A sixth peak at about 6.4 Å could result from the silicon–third oxygen contribution. No more peaks are expected from a random network, and none were found experimentally.

A variety of other types of experimental results, especially spectroscopies, as well as physical models, are consistent with the random network model. Neutron diffraction results are also consistent with the random network model of silica glass [6]. Analysis by Warren and Briscoe [7] and low-angle scattering of x-rays show that the amount of crystalline regions in silica glass is small. Further references and discussion are provided in Ref. 3, p. 30ff.

For the present consideration of solubility and diffusion of water molecules in silica glass, the most important structural aspects are the interstices (cages) in the structure of silica glass and the openings between cages (see Fig. 6.1). Nakayama and Shackelford [8] estimated that there are about $2.2(10)^{22}$ interstitial sites per cubic centimeter in silica glass. Not all of these are accessible for gas solubility. For example, a maximum of about $2.3(10)^{21}$ helium atoms per cubic centimeter can dissolve in silica glass, $1.3(10)^{21}$ per cubic centimeter of neon atoms, and $1.1(10)^{21}$ per cubic centimeter of argon atoms, so that the accessibility to gas atoms decreases as the atom size increases. The effective size of water molecules is between that of neon and argon atoms from the analysis of diffusion activation energies (Section 5.3). Thus the number of interstitial sites accessible to water molecules is probably about $1.2(10)^{21}\,\text{cm}^{-3}$ (see also Section 6.3).

6.3 GAS SOLUBILITY IN SILICA GLASS

A variety of gases have been shown experimentally to dissolve and diffuse molecularly in silica glass [3, Chap. 8]. The activation energies for diffusion are directly related to atomic size, as shown in Section 5.3. The solubility is defined here as the ratio $\alpha = C_d/C_g$, where C_g is the concentration of gas dissolved in a solid or melt and C_g is the concentration of gas molecules in the gas phase. This solubility is unitless and is sometimes called the *coefficient of solubility* or the *Ostwald solubility.* It is used here because its temperature dependence gives directly the energy of interaction between the gas molecules and the solid or melt solvent, and this ratio does not depend on gas pressure.

The Ostwald solubilities of helium, neon, and argon in silica are close to constant over a wide range of temperatures (see Ref. 3, p. 123ff. for a summary). Average values for C_d/C_g are helium, 0.024; neon, 0.019, and argon, 0.010. Norton [9] found a solubility of about 0.01 for oxygen in silica glass; oxygen has about the same effective molecular diameter of 3.2 Å as argon. Because water has an effective diameter halfway between those of neon and argon (see Section 5.3), its solubility should be about $\alpha = 0.015$.

At low gas pressure (near atmospheric and below), the solubility C_d/C_g is constant with pressure change, which states that the concentration of dissolved gas is proportional to the gas pressure. At higher gas pressures, the fugacity of the gas rather than its pressure (concentration) must be used to provide a constant solubility. At pressures above about 100 atm, the number of dissolved molecules is an appreciable fraction of the interstices available to them. The solubility can then be described by a Langmuir adsorption isotherm [12–14] to account for competition for the sites. In this model the fraction of adsorption sites θ is given by (see also Chapter 10)

$$\theta = \frac{kf}{1 + kf} \tag{6.1}$$

where f is the fugacity of the gas and k is a constant. This equation can be derived from statistical mechanics [11, p. 427]. To apply this model to gas solubility, θ is given by C_d/N_d, where N is the number of interstices per unit volume for gas dissolution. Then

$$C_d = \frac{Nkf}{1 + kf} \tag{6.2}$$

Data on helium and neon solubilities in cristobalite and tridymite up to 5000 atm [12] and in vitreous silica [13] as well as hydrogen and deuterium in silica [15, 16] fit Eq. 6.2. The number N of solubility sites in vitreous silica is found to be one to two times $(10)^{21}$ per cubic centimeter in this model, close to the number accessible to these gases in Shackelford's calculations and those from the free-volume model (see below).

The lack of temperature dependence of solubility leads to a free-volume model of gas solubility [10] in which the gas molecules are considered to dissolve in an inert material with a certain amount of open space or free volume available for dissolution of the molecules. This volume is effectively constant over the range of temperatures over which the structure of the solid or melt does not change; the glass transition temperature for silica glass is about 1200°C, so its structure is frozen in for the temperatures of interest here.

The free volume is given directly by the ratio C_d/C_g. This relation can also be derived from statistical equations, such as those of Fowler and Guggenheim [11] for the chemical potential of an ideal gas μ_g and of this gas dissolved in a condensed phase μ_i:

$$\mu_g = RT \ln \phi + RT \ln C_g$$
$$\mu_i = -E + PV + RT \ln \phi + RT \ln C_d - RTv_f$$

where ϕ is a function involving the translational and internal degrees of freedom of the molecules, $-E$ a potential energy resulting from interactions between the dissolved gas and the solvent liquid, P the total pressure, R the gas constant, T the absolute temperature, V the partial molar volume of the solute gas, and v_f the fractional free volume available to the dissolving gas. At equilibrium these two chemical potentials must be equal in order to give a relation for C_d/C_g in terms of the free volume. In deriving this relation, it is assumed that the function ϕ is the same in the gaseous and dissolved states.

This is equivalent to assuming that the dissolved molecules preserve three degrees of translational freedom. Then

$$\ln\frac{C_d}{C_g} = \frac{+E - PV}{RT} + \ln v_f \tag{6.3}$$

The total volume of the glass does not change appreciably as gas is dissolved in it; therefore, $V = 0$, and the small temperature dependence of C_d/C_g, shows that $E \approx 0$ consequently, the free volume is equal to C_d/C_g, and is about 3% for silica glass. The ionic porosity Z of a solid is equal to 1-V_i/V, where V_i is the sum of the volumes of atoms per mole in the solid and V is its molar volume (see Chapter 5). For silica glass, $Z = 0.509$. Thus the free volume available for gas solubility is much smaller than this measure of open space in silica glass.

6.4 EXPERIMENTS OF ROBERTS AND CO-WORKERS AT TEMPERATURES OF 700 TO 1200°C

The diffusion–reaction mechanism as applied to silica glass involves the dissolution and diffusion of water molecules into the glass and their reaction with the silicon–oxygen network of the glass:

$$\text{Si—O—Si} + H_2O = \text{SiOH} + \text{HOSi} \tag{6.4}$$

In this mechanism the molecular water is the carrier of H_2O molecules into and out of the glass, and the OH groups formed are relatively immobile. If Eq. 6.4 reaches equilibrium, it can be described by the equilibrium constant K_2:

$$K_2 = \frac{S^2}{C} \tag{6.5}$$

in which S is the concentration of reacted SiOH groups as H_2O in the glass and C is the concentration of dissolved molecular water. The units of S are often referred to water molecules and are therefore one-half of the actual concentration of SiOH groups. Since the concentration of Si–O–Si groups is nearly constant, it is incorporated into K_2. The subscript 2 on K_2 indicates that the reverse reaction is bimolecular.

Roberts and his co-workers did diffusion experiments of water in silica of a variety of types [17–21]. Most of their experiments involved diffusion of water into unhydrated silica glass. In these experiments water molecules dissolve in the silica and react with the network by Eq. 6.4. Roberts *et al.* measured the profiles of hydrogen in the silica after diffusion with infra-absorption [17] and radioactive tritium (T) of atomic number 3 [18–20]. Under these conditions of diffusion of water in unhydrated glass, the effective diffusion coefficient D_e for OH groups is given by

$$D_e = \frac{2SD}{K_2} \tag{6.6}$$

in which D is the actual diffusion coefficient of molecular water. Thus D_e is directly proportional to the concentration S. The derivation of this equation is given in Section 3.2. The OH groups are actually immobile, because only the molecular water diffuses.

A profile of hydrogen (tritium) in silica glass is compared to the profile calculated from Eq. 6.6 in Fig. 6.3. The agreement between the calculated and experimental curves is excellent, and the profile is much different than expected for a diffusion coefficient that is not a function of concentration (erfc curve; see Chapter 2). Most of the profiles measured for diffusion of water into unhydrated silica glass in Refs. 17 to 20 had the same shape as that in Fig. 6.3.

The diffusion coefficient D of molecular water in the glass is calculated from Eq. 6.6 and the measured profile. From a comparison of the theoretical profile for a diffusion coefficient proportional to concentration and the measured profile in Fig. 6.3, one can calculate the effective diffusion coefficient at the sample surface $(x=0)$, $D_e(0)$. The concentration S_0 of SiOH groups at the sample surface is found from the surface concentrations measured by Burn and Roberts [20] and the estimates of S values from the infrared data of Moulson and Roberts [17]. In this calculation it was assumed that all the water had the same specific activity of tritium tracer. The resulting S_0 values are given in Table 6.2. Then $D=D_e(0)$ $S_0/2C_0$ from Eqs. 6.5 and 6.6.

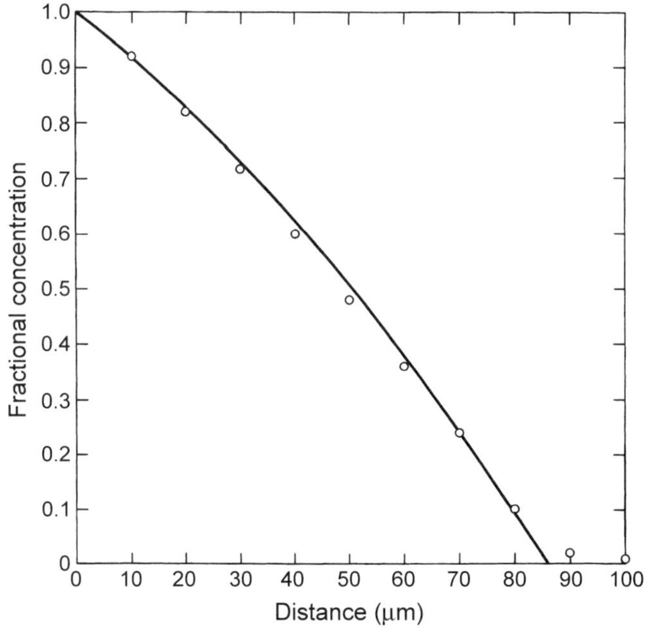

Figure 6.3 Profile of hydrogen (tritium) in silica glass after diffusion of tritiated water into the glass. Points from Fig. 1f of Ref. 18; line calculated from the diffusion coefficient of Eq. 6.6 (see Sections 2.6 and 3.2).

Table 6.2 Diffusion Coefficient D of Molecular Water[a]

T (°C)	S_0 (mol $H_2O/cm^3 \times 10^5$)	D ($cm^2/s \times 10^8$)	Ref.	Method[b]
1200	5.73	61.1	18	1
1100	5.73	39.3	18	1
1000	5.73	21.3	18	1
900	6.74	11.03	18	1
900	6.12	12.08	18	1
810	8.37	7.67	18	1
750	9.80	5.77	18	1
700	11.02	4.37	18	1
810	6.36	9.05	18	2
750	6.36	4.38	18	2
700	6.44	4.47	18	2
1293	5.00	74.9	19	1
1195	4.81	48.2	19	1
1095	5.04	25.2	19	1
895	6.44	8.39	19	1
794	9.38	5.88	19	1
1200	4.65	82.1	20	3
1100	4.57	46.2	20	3
1000	4.53	22.9	20	3
1160	5.78	61.1	17	4
1100	5.80	25.9	17	4

[a]Calculated from the diffusion–reaction model and the data of Roberts and co-workers [17–20].
[b]1, Diffusion of tritiated water into unhydrated glass, as-received samples (from Eqs. 6.7 and 6.8); 2, diffusion of tritiated water into unhydrated glass, samples heated for 48 h at 1100°C before diffusion (from Eqs. 6.24 and 6.25); 3, diffusion of tritiated water into hydrated glass (from Eqs. 6.24 and 6.25); 4, diffusion of untagged water into unhydrated glass.

The values of C_0 for the calculation of D were found as follows. Equilibrium between dissolved molecular water at the glass surface and the water vapor pressure of 700 mm Hg, the pressure in all the experiments of Roberts *et al.*, was assumed. The concentration C_g of water in the gas phase was calculated from the ideal gas law. It was assumed that C_d/C_g was the same at all temperatures and equal to 0.015 (see Section 6.3).

A sample calculation of D_e and D for diffusion of water in silica glass at 1000°C and 700 nm Hg water vapor pressure is now given. First the concentration C_g of water in the gas phase is calculated from the ideal gas law:

$$C_g = \frac{P}{RT} = \frac{700}{760} \frac{1.0(10)^5}{8.314(1273)} = 8.79 \text{ mol/m}^3 = 8.79(10)^{-6} \text{ mol/cm}^3 \tag{6.7}$$

From the factor $\alpha = C_d/C_g = 0.015$, C_d is calculated:

$$C_d = 0.015 C_g = 1.318(10)^{-7} \text{ mol } H_2O/\text{cm}^3 \tag{6.8}$$

Now the value of S_0, the concentration of reacted OH groups as H_2O is calculated. The optical absorption coefficient A (cm^{-1}) is related to the concentration S of reacted OH groups as H_2O by the formula

$$S = \frac{A}{E}$$

in which E is the extinction coefficient. From a study of optical absorption in different silica glasses, and by comparison with the results of other workers, Shelby *et al.* [22] found a value of E of $1.8(10)^5$ cm^2/mol. In Fig. 7 of Ref. 17 the absorption coefficient $A = 10.5\,cm^{-1}$ at the silica surface. Then $S_T = 10.5/1.81$ $(10)^5 = 5.80(10)^{-4}$ mol H_2O/cm^3. The diffusion profile for the diffusion coefficient proportional to concentration intersects the distance (x) axis at $x_0 = 0.81(2)$ $\sqrt{D_e(0)t}$, where $D_e(0)$ is the effective diffusion coefficient at $x = 0$ (the glass surface) [23, Fig. 7.5, p. 122]. This intersection is a good place to measure an accurate penetration depth (x_0) from which the effective diffusion coefficient can be calculated. From Fig. 7 of Ref. 17 this intersection for the 1000°C profile is at 200 μm or 0.0200 cm, and the diffusion time is 36 h. Then

$$D_e(0) = \left(\frac{x_0}{1.62}\right)^2 \frac{1}{t} = \left(\frac{0.02}{1.62}\right)^2 \frac{1}{36(3600)} = 1.176(10)^{-4} cm^2/s \qquad (6.9)$$

Then from Eq. 6.6 $D = S_T D_E / 2C_T$, so

$$D = \frac{(5.80)(10)^{-4}(1.176)(10)^{-4}}{2(1.318)(10)^{-7}} = 2.59(10)^{-7} cm^2/s \qquad (6.10)$$

Other values of D are listed in Table 6.2.

The form of Eq. 6.5 results from reaction 6.4 with a first-order forward reaction and a second-order reverse reaction. The rate of formation of OH groups can be written

$$\frac{\partial S}{\partial t} = \lambda C - \frac{\mu S^2}{S_e} \qquad (6.11)$$

in which λ and μ are reaction rate constants with units of reciprocal time, and S_e is the final uniform equilibrium concentration of reacted OH groups (see Eq. 4.43). At equilibrium,

$$K_2 = \frac{S^2}{C} = \frac{S_e \lambda}{\mu} \qquad (6.12)$$

This expression assumes that an individual OH group has enough mobility to interact with neighboring OH groups. If the pairs of OH groups are immobile, the reverse reaction becomes effectively of first order, so the equilibrium expression is

$$K_1 = \frac{S}{C} \qquad (6.13)$$

For the condition of Eq. 6.13, the effective diffusion coefficient D_e is (see Eq. 3.6)

$$D_e = \frac{D}{1 + K_1} \approx \frac{D}{K_1} \tag{6.14}$$

so it is not a function of concentration, in contrast with the effective diffusion coefficient with a second-order reverse reaction (Eq. 6.6).

When the silica glass was annealed at 1100°C for 48 h before the diffusion experiments, the same profiles as in as-received glass were found at temperatures of 900°C and above [17]. However, at temperatures of 850 to 700°C, the as-received samples had higher surface concentration S_0 of reacted water (SiOH) groups (see Table 6.2). The heated samples had values about the same as those measured at higher diffusion temperatures (900 to 1300°C); however, the profiles for these heated samples at 850 to 700°C followed an erfc function (Eq. 2.20) instead of the profile for concentration-dependent D_e found for the as-received samples. These differences can perhaps be understood as follows. The as-received silica samples contain defects and/or strains. These defects or strains are rapidly annealed out of the glass during diffusion at 900°C or higher temperatures, so the results for as-received and heated samples are the same for these diffusion temperatures. At lower temperatures the defects or strains lead to easier reaction of water with the glass, increasing the forward reaction rate constant λ, and consequently, the equilibrium constant K_2, giving higher S_0 values. After the defects or strains are removed by heating, the surface concentration S_0 is about the same as at higher temperatures. The defects and strains can also increase the rate of diffusion of OH groups in the glass. Thus, when they are present, the diffusion distances of these groups are high enough to maintain bimolecular collisions for the reverse reaction. Removal of these groups reduces the diffusion distances of the OH groups, so that they can react only with their neighboring group, giving a first-order reverse reaction, and the profile of Eq. 2.20 for a concentration-independent diffusion coefficient.

If water diffuses into the hydrated glass, there is no net reaction 6.4, which is at equilibrium. Then the tritium on dissolved water molecules exchanges with hydrogen atoms on OH groups already in the glass. In this case of exchange, the effective diffusion coefficient is (Section 3.3)

$$D_e = \frac{S_T D}{K_2} = \frac{D C_T}{S_T} \tag{6.15}$$

in which S_T is the concentration of reacted OH groups in the glass and C_T is the concentration of dissolved molecular water in the glass. Both S_T and C_T are uniform and constant with time in the glass. Thus D_e is not a function of concentration in these exchange experiments.

Burn and Roberts [20] diffused tritiated water into hydrated glass. In this case the diffusion coefficient D of molecular water was calculated from D_e and Eq. 6.15. D values calculated in this way agree well with D values calculated from the experiments diffusing into unhydrated glass, as shown in Table 6.2 and Fig. 6.4.

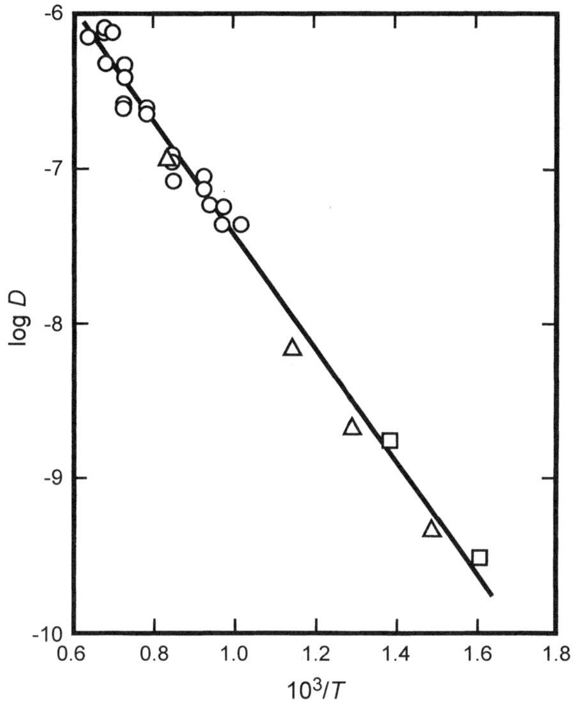

Figure 6.4 Log diffusion coefficients of molecular water in silica glass as a function of $10^3/T$. Points calculated from experimental results in Refs. 17 to 22 (○), 25 and 26 (△), and 29 (□); line, linear regression with an activation energy of 70 kJ/mol.

The diffusion coefficients calculated from the diffusion–reaction model all agree well, within experimental error.

Roberts and Roberts [21] found that water tagged with ^{18}O and diffused into unhydrated silica glass reacted in a narrow region at the glass surface, with much less apparent penetration of OH. This puzzling result can be explained with the diffusion–reaction model. It occurs because the total concentration of available exchange sites S_T is equal to all the available network oxygen atoms, which is much larger than the number of exchange sites for hydrogen within OH groups. The correct diffusion coefficient for this situation is that of Eq. 6.15. Even though the glass is unhydrated, the exchange of ^{18}O between dissolved water and network oxygen leads to a profile of ^{18}O in the glass that fits an erfc function, and the effective diffusion coefficient is not a function of concentration. This contention implies that the concentration of reacted OH groups is small compared to the total concentration of network oxygen, so Eq. 6.4 contributes a negligible amount of ^{18}O exchange. The concentration S_T for Eq. 6.15 is 0.0732 mol O/cm^3 of network oxygen. The histograms in Fig. 7 of Ref. 17 give an approximate value of $D_e = 5(10)^{-13}$ cm^2/s at 1100°C. The value of $C_T = 1.2(10)^{-7}$ mol dissolved H_2O per cubic centimeter, so from Eq. 6.15, $D = S_T/C_e/C_T = 3(10)^{-7}$ cm^2/s. This value is close to those in Fig. 6.4 and Table 6.2 at 1100°C, showing that the apparent small diffusion coefficient for ^{18}O exchange results because there are many more exchange sites than for hydrogen isotopes.

6.5 EXPERIMENTS OF SPITZER AND LIGENZA AT 650°C, AND RIGO ET AL. FROM 400 TO 900°C

Spitzer, Ligenza, and Rigo *et al.* studied the exchange of water and network oxygen in a silicon dioxide layer on silicon. Spitzer and Ligenza [24] followed the exchange with the shift of the Si–O vibrational absorption band at 9.1 μm in the infrared as ^{18}O substituted for ^{16}O. They interpreted their results in terms of the exchange reaction only; however, the results of Roberts *et al.* show that at 700°C the exchange reaction is fast enough to reach local equilibrium. Thus at 650°C both diffusion and the exchange reaction are involved in the kinetics.

The spectral shift measured by Spitzer and Ligenza measures the total amount M of ^{18}O that has exchanged in the silicon dioxide layer of thickness 0.60 μm. The profile of dissolved water in this layer is shown schematically in Fig. 6.5. At the water vapor–silica interface water vapor dissolves molecularly in the silicon dioxide to concentration $C(0)$. At the silica–silicon interface the dissolved water reacts rapidly with the silicon to form silicon dioxide:

$$Si + 2H_2O = SiO_2 + 2H_2 \tag{6.16}$$

Thus the boundary conditions are $C = C(0)$ at $x = 0$ and $t > 0$ and $C = 0$ at $x = L$, where C is the concentration of dissolved water and L is the film thickness. For purposes of studying the exchange reaction, the silicon dioxide layer is first grown

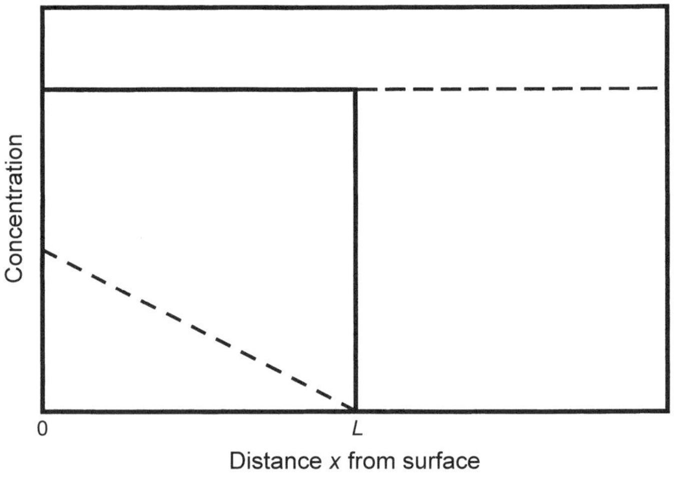

Figure 6.5 Profile of water in a silicon dioxide layer on silicon.

to thickness L in $H_2{}^{16}O$, and then the water vapor is changed to $H_2{}^{18}O$. It is assumed that the growth of the film thickness L is negligible during the exchange.

The effective diffusion coefficient is given by Eq. 6.15 and can be found by comparing the time-dependent increase of the total amount M of ^{18}O in the film with Eq. 2.25 (see also Fig. 4.6 of Ref. 23). This comparison gives an effective diffusion coefficient of about $7.7(10)^{-13}\,cm^2/s$. Then from Eq. 6.15 with $S_T = 0.0732$ mol of oxygen atoms per cubic centimeter and $C_T = 6.8(10)^{-6}$ mol/cm^3 for a water vapor pressure of 35 atm, $D = 8.3(10)^{-9}\,cm^2/s$ at 650°C, which fits reasonably well with the other values in Fig. 6.4.

Rigo and his collaborators [25, 26] measured the profiles of ^{18}O in silicon dioxide layers on silicon with resonant nuclear reactions, after diffusion and exchange of $H_2{}^{18}O$ in a layer already grown containing ^{16}O. In most of these experiments the penetration of ^{18}O was much smaller than the thickness of the layer, so equations for semi-infinite geometry were applicable (see Sections 3.3, 3.4, and 4.5). They calculated values of the effective diffusion coefficient of ^{18}O as a function of temperatures from 400 to 930°C and water vapor pressures at 930°C from $0.31(10)^3$ to $17.3(10)^3$ Pa. At 930°C and above $16.6(10)^3$ Pa pressure (0.017 atm), the exchange reaction was at local equilibrium, but at lower pressures and at temperatures from 400 to 600°C the exchange reaction was not at equilibrium, as shown by the lower surface concentration of ^{18}O than expected for complete exchange (equilibrium).

Some equations from Sections 4.4, and 4.5 help in understanding the results of Rigo *et al.* and show that for nonequilibrium exchange the surface concentration of ^{18}O is a function of water vapor pressure, whereas for a nonequilibrium reaction (Eq. 6.4), the surface concentration of reacted OH groups is not a function of water vapor pressure. In the latter case the surface concentration of OH groups S_0 is given by Eq. 4.47:

$$\frac{S_0}{S_e} = \frac{1 - e^{-2\mu t}}{1 + e^{-2\mu t}} \tag{6.17}$$

where S_e is the equilibrium concentration of OH groups. Thus in this case the surface concentration of OH depends only on the back-reaction constant μ and the time, not on the concentration C of dissolved molecular water. However, for exchange the surface concentration is given by Eq. 4.59:

$$\frac{S_0{}^*}{S_t} = 1 - e^{-k^* C_T t} \tag{6.18}$$

in which k^* is a reaction rate constant. Thus in this case the surface concentration is a function of the total dissolved water concentration C_T, which is directly proportional to the water vapor pressure. Rochet and Rigo [26] found the dependence of Eq. 6.18 for the concentration of ^{18}O as a function of water vapor pressure below a pressure of $16.6(10)^3$ Pa, from which a value of about $7(10)^{-7}$ $(s\text{–}Pa)^{-1}$ is calculated for $k^* C_T/P$.

Values of k^*C_T calculated by Rigo *et al.* [25] for exchange of ^{18}O between dissolved water and network oxygen at different temperatures are given in Table 6.3 and plotted in Fig. 6.6. The results fit reasonably well on a line with an activation energy of 85 kJ/mol. Also included in the table and figure is a measurement by Helmich and Rauch [27] at 200°C, which is discussed further below, where it is suggested that the fit may actually be better than is shown in Fig. 6.6.

Values of k^*C_T for a variety of samples treated at 600°C and 13 torr [0.017 atm or $1.73(10)^3$ Pa] are given in Table 6.4. The parameter γ_{nc} of Rigo *et al.* [25] is equal to k^*C_T in the present notation. There is a large difference, by more than an order of magnitude, in the exchange parameters in the table for different samples. Thus the exchange rate of oxygen atoms between dissolved and network atoms in silica glass

Table 6.3 Reaction Parameter k^*C_T from Eq. 6.18 for Exchange of ^{18}O between Dissolved and Network Oxygen in Silica Glass at Different Temperatures

T (°C)	k^*C_T (h^{-1})	Ref.
923	4.1	26
600	1.2	25
500	0.019	25
400	0.0030	25
200	$4.2(10)^{-5}$	27

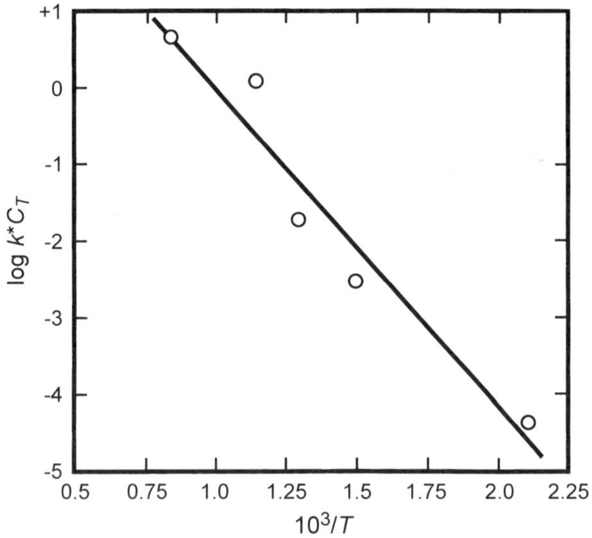

Figure 6.6 Log reaction rate coefficients k^*C_T from Table 6.3 as a function of $10^3/T$. Points, data; line, drawn with an activation energy of 78 kJ/mol from least squares (correlation 0.967).

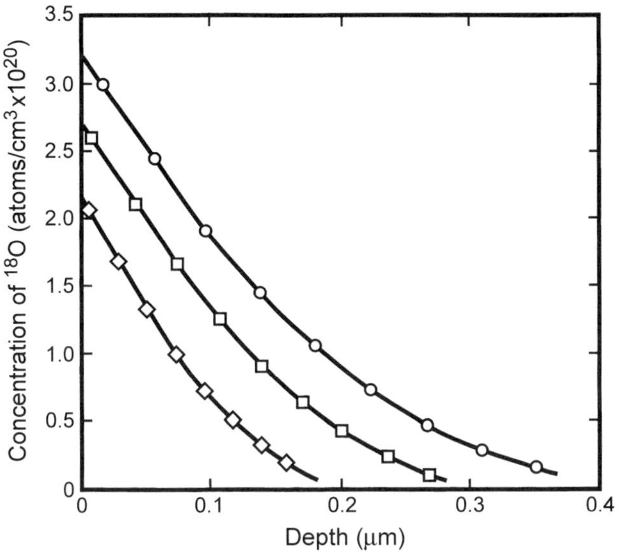

Figure 6.7 Profiles of ^{18}O in silica glass after different times at 200°C in 15.3 atm water vapor. 5 h (◇), 10 h (□), 20 h (○). (From Fig. 2 of Ref. 27.)

Table 6.4 Reaction Parameter k^*C_T from Eq. 6.18[a]

Batch	$K^*C_T(h^{-1})$	Treatment History
1	0.20	Anodic oxidation in diethylene glycol containing 0.4 wt % KNO_3 and H_2O
2	0.073	Thermal oxidation at 1000 and 1100°C, then annealed 10 min in dry N_2
3	1.2	Thermal oxidation at 1000°C

[a] Calculated by Rigo *et al.* [24, Table 2] from profiles of water in silica glass after treatment at 600°C for 1 or 3 h.

depends strongly on sample history. The reasons that the different treatments in Table 6.4 lead to these large differences in reaction rate are not clear; they probably result from unknown and uncontrolled factors.

Rochet and Rigo [26] also found an effective diffusion coefficient directly proportional to water vapor pressure at 930°C, as expected from Eq. 6.15. The values of D, the actual diffusion coefficient of molecular water, were calculated from this equation and are given in Table 6.5. These values are compared with an extrapolation of the line in Fig. 6.4 from the data of Roberts *et al.* and the agreement is good. Pfeffer and Ohring [28] also examined the diffusion and reaction of water in thin silica films on silicon. They implanted ^{18}O in the films and measured the change in the ^{18}O distribution in the film with a resonant nuclear reaction. They

Table 6.5 Diffusion Coefficients of Molecular Water in Silica Glass at Different Temperatures and Pressures[a]

T(°C)	Water Vapor Pressure (Pa × 10^3)	D (cm^3/s × 10^8)
400	17.3	0.046
500	17.3	0.21
600	17.3	0.68
930	17.3	14.5
	16.6	11.5
	8.7	9.3
	3.8	11.6
	1.5	10.9
	0.53	10.7
	0.31	12.2

[a]Data of Rigo *et al.* [24] and Rochet *et al.* [25, 26], calculated from the diffusion–reaction model.

measured the pressure dependence of the effective diffusivity at pressures from 0.06 to 1 atm and found it was proportional to the pressure, as expected from the diffusion–reaction model (Eq. 6.6). In a vacuum the ^{18}O was immobile, showing that network diffusion of oxygen was negligible. The temperature dependence of the effective diffusion coefficients was about 70 kJ/mol, close to that found for Fig. 6.4.

6.6 EXPERIMENTS OF WAKABAYASHI AND TOMOZAWA AT 350 AND 450°C

Diffusion coefficients of molecular water calculated from profiles of OH in silica glass measured with infrared absorption by Wakabayashi and Tomozawa [29] and the present diffusion–reaction model are listed in Table 6.6 and plotted in Fig. 6.4. The diffusion coefficients D calculated for profiles from heating at 350°C for 440 and 16 h are the same within experimental error, even though the profiles in Fig. 3 of Ref. 26 do not superpose when plotted as a function of $x/\sqrt{t}$, where x is the

Table 6.6 Diffusion Coefficients of Molecular Water in Silica Glass[a]

T(°C)	Diffusion Time (h)	D (cm^2/s × 10^{10})
350	0.52	2.8
	1.05	3.3
450	121	1.7
	245	1.7

[a]Calculated from the data of Wakabayashi and Tomozawa [29].

penetration distance and t the diffusion time. As the authors noted, this lack of superposition means that the apparent diffusion coefficient D_e of OH groups is a function of time. Nevertheless, the actual diffusion coefficient D of molecular water, the only diffusing component, is not a function of time. The time dependence of D_e arises from the diffusion and reaction mechanism.

At 450°C profiles from heating for 121 and 245 h superimposed, suggesting equilibrium between dissolved molecular water and reacted SiOH groups. The diffusion coefficients in Table 6.6 were calculated from the long-time portion of the profiles. At short times there was a more rapid reaction superposed on the slower one. I propose that the long-time profile results from the first-order reaction (Eqs. 6.13 and 6.15) and the short-time curve from the second-order reaction of Eqs. 6.5 and 6.6.

The diffusion coefficients D of molecular water calculated from the profiles of Wakabayashi and Tomozawa at 450 and 350°C fit very well with the extrapolation of these diffusion coefficients calculated from the data of Roberts *et al.* at higher temperatures, as shown in Fig. 6.4. Thus, the finding of unexpectedly high and time-dependent apparent diffusion coefficients by Wakabayashi and Tomozawa results from the diffusion–reaction mechanism, not from any other change of mechanism between high and low temperatures. The experimental results of Davis and Tomozawa [30, Fig. 5] on the diffusion coefficients of water in silica glass calculated from an erfc solution at 350 and 650°C as a function of time are quite similar to those calculated from the equations in section 4.4 (R. H. Doremus, unpublished).

6.7 EXPERIMENTS OF RAUCH ET AL. AND LANFORD ET AL. AT TEMPERATURES OF 200 TO 90°C

These authors measured profiles of hydrogen and ^{18}O in bulk silica glass after diffusion in water vapor. The penetration depths of hydrogen were about 0.1 μm for 20 h at 200°C and 0.4 μm for 16 days at 90°C. Profiles of ^{18}O at 200°C are shown in Fig. 6.7. The concentration of ^{18}O was just one-half that of hydrogen, showing that there was negligible exchange of ^{18}O between dissolved water molecules and network oxygen up to 20 h at 200°C. Thus the ^{18}O profiles measure just the concentration of OH groups formed by Eq. 6.4. The surface concentration of both hydrogen and ^{18}O increased with time, showing that the reaction was not at equilibrium. Equation 4.27 was followed by surface concentrations at 160 and 200°C as a function of time; a value of μ of about 0.20 h^{-1} was calculated at both temperatures.

At long times (6 to 20 days at 200°C) the hydrogen concentrations at the silica surface were all about $7.8(10)^{20}$ H atoms/cm^3 [$S_0 = 3.9(10)^{20}$ molecules H_2O/cm^3] for a water vapor pressure of 15.3 atm, showing that at these times reaction equilibrium was reached. However, at these times the ^{18}O increased substantially, as shown in Fig. 6 of Ref. 27. Thus at these times oxygen exchange between ^{18}O in the silica network was becoming appreciable. A value of μ for this exchange reaction at

200°C was calculated to be about $4.2(20)^{-5}\,h^{-1}$, a factor of about $4.8(10)^3$ smaller than the μ value for the reaction 4 to form OH groups. Thus exchange of ^{18}O between dissolved water and the network is much slower at 200°C than is the bond-breaking reaction to form OH groups. Limited results at higher temperatures (about 600°C) suggest that at that temperature there is not nearly so much difference between these two rates, because equilibrium for the two kinds of reactions seems to be reached at similar temperatures. These results suggest that the activation energy for the exchange reaction is much greater than that for the reaction to form OH groups.

Diffusion coefficients for molecular water from the profiles of Helmich and Rauch [27] were calculated from the equations and methods described in Sections 4.3 and 4.5 and are listed in Table 6.7. For both 200 and 160°C the value of μ was $0.20\,h^{-1}$. Values of λ were about 18 and $29\,h^{-1}$ at 200 and 160°C, respectively; vapor pressures were 15.3 and 6.1 atm at these temperatures. The mean diffusion coefficient of $3.9(10)^{-12}\,cm^2/s$ at 200°C from Table 6.7 is close to the value of $2.7(10)^{-12}\,cm^2/s$ extrapolated from the high temperature line at Fig. 6.4. The mean diffusion coefficient of $2.5(10)^{-12}$ at 160°C is somewhat higher than the extrapolated value of $5.4(10)^{-13}\,cm^2/s$ at 160°C. Thus from about 200°C to 1200°C diffusion coefficients of water molecules in silica glass, as calculated from experimental data and the diffusion–reaction model, fit on the same Arrhenius line.

Values of k^*C_T, calculated from Eq. 6.18 for surface concentrations of ^{18}O, are given in Table 6.3 and plotted in Fig. 6.6. The value at 200°C was taken from data of Helmich and Rauch [27, Fig. 6] on ^{18}O exchange between network oxygen and dissolved water in silica glass. The value fits fairly well with the others in the table and figure. However, the fit may be better than shown in the figure. If the average of k^*C_T values in Table 6.4 is taken instead of the value corresponding to the other samples in Table 6.3, the k^*C_T value at 600°C is about 0.5, which also fits better with values at other temperatures in Fig. 6.6. The important point is that the diffusion-exchange model that results in Eq. 6.18 provides a good explanation for the exchange rates measured between ^{18}O in dissolved water and network oxygen.

Lanford *et al.* [31] measured the profiles of water diffusing into optosil silica glass, which has a low water content. The water vapor came from liquid water at 90°C, so the water vapor pressure was 0.69 atm. After holding the silica for 16 days

Table 6.7 Diffusion Coefficients of Molecular Water in Silica Glass[a]

T(°C)	Diffusion Time (h)	$S_0 - S_i / S_e - S_i$	$D_1(cm^2/s \times 10^{12})$
160	5	0.50	2.8
	10	0.82	2.8
	20	0.96	2.0
200	5	0.68	3.0
	10	0.83	4.0
	20	0.97	4.6

[a]Calculated from the data of Helmich and Rauch [27].

at 90°C, the surface concentration appeared to be the equilibrium value and was $4.0(10)^{20}$ H atoms/cm^3; thus $S_e = 2.0(10)^{20}\,\text{cm}^{-3}$. The concentration of water molecules in the gas phase was $1.38(10)^{19}$ H_2O/cm^3, so if $C_d/C_g = 0.015$, $C_d = 2.07(10)^{17}\,\text{cm}^{-3}$. The penetration depth of hydrogen at $S_0/S_e = 1/2$ was about 0.48 μm after 16 days, so $D_e = x^2/4t = 4.17(10)^{-16}\,\text{cm}^2/\text{s}$. Then $D = D_e S_e / C_e$, so $D = 4.0(10)^{-13}\,\text{cm}^2/\text{s}$. This value is higher than that expected from an extrapolation of the data at higher temperatures.

If the diffusion coefficients of water in silica glass at 200°C, 160°C, and 90°C are plotted on a log D versus $1/T$ plot, a straight line results with a lower activation energy of about 30 kJ/mole, as compared with the activation energy of 70 kJ/mole at higher temperatures. The reason for this lower activation energy below about 200°C is uncertain. It is possible that the solubility C_e/C_g of molecular water decreases at temperatures below 200°C, giving a lower C_e value and a higher D value, perhaps because of interaction of water molecules with the glass network. Another possibility is that K_1, the equilibrium constant of Eq. 6.13, increases below 200°C, giving an apparent increase in D. The values of S_0 found at 160°C and 90°C are higher than at higher temperatures, supporting this possibility.

6.8 DISCUSSION

The diffusion and reaction mechanism and its comparison with experimental data provide a unified description of the diffusion of water into silica glass over the temperature range from 90 to 1200°C. At temperatures from 900 to 1200°C, the molecular water diffuses into the glass and reacts to equilibrium with the silicon oxygen lattice (Eq. 6.1). The OH groups are nearly immobile but do diffuse far enough to sample neighboring SiOH groups, making the reaction second-order (bimolecular). From 700 to 900°C this mechanism holds if there are quenched-in defects or strains in the glass, but if these defects or strains are annealed out of the glass, the reaction becomes first order, because the SiOH pairs cannot diffuse far enough to sample other pairs, and react only with themselves to form water.

At temperatures below about 500°C the first-order reverse reaction is dominant. However, it does not go to equilibrium at short diffusion times. Under these nonequilibrium conditions the surface concentration of reacted SiOH groups increases with time until a constant surface concentration is reached at long times. These conditions also cause the apparent diffusion coefficient of OH groups to be time dependent; however, the diffusion coefficient D of the only actual diffusing species, molecular water, is not time dependent.

Experiments with diffusing $H_2{}^{18}O$ give much smaller effective diffusion coefficients than other experiments, because the entire silicon–oxygen network acts as a source of oxygen for exchange with the oxygen in the dissolved water molecules. Nevertheless, the same diffusion coefficient of molecular water is calculated from these experiments as from those measuring reacted SiOH groups.

A variety of phenomena such as strain relaxation, long-range diffusion of OH groups, and surface reactions have been invoked to explain the time dependence of surface OH concentrations and apparent OH diffusion coefficients. The present work shows that these explanations are unnecessary. There is evidence of strain relaxation in the silica glass, but I feel that it has only a small influence on diffusion and reaction rates. The present work suggests that the diffusion distances of network OH groups are very small even at high temperatures, and negligible at low temperatures.

The values of the calculated diffusion coefficients of molecular water are plotted as log D versus $1/T$ in Fig. 6.4. The linear regression line for temperatures above 200°C gives an activation energy of 70 kJ/mol. The line follows the equation

$$D = 1.3(10)^{-4} \exp\left(\frac{-8360}{T}\right) \qquad \mathrm{cm}^2/\mathrm{s} \tag{6.19}$$

The correlation coefficient for the fit is 0.994, showing an excellent fit from 200 to 1200°C. In this temperature range the results of four different independent investigators for D agree well, even though the values of D_e do not fit on a single log D_e versus $10^3/T$ curve. These results include conditions of local equilibrium and nonequilibrium, diffusion in and out of hydrated glass with both hydrogen and oxygen isotopes, and time dependence of measured effective diffusion coefficients. The good agreement of diffusion coefficients of molecular water calculated from these diverse experiments with the diffusion–reaction mechanism is strong evidence that the details of this mechanism as expounded in this book are valid and that diffusion and reaction of water in silica glass follow this mechanism.

A theoretical equation for the D_0 factor of $1.3(10)^{-4}$ cm^2/s in Eq. 6.19 is [32, 33]

$$D_0 = \frac{\lambda^2 \nu}{6} \exp\left(\frac{\Delta S}{R}\right) \tag{6.20}$$

in which ΔS is the entropy of activation for diffusion, λ is about 1 nm, and ΔS is about −40 J/mol-deg. Zener [34] has shown that interstitial diffusion that involves only straining the network should be proportional to the negative of the temperature coefficient of the elastic modulus. Since silica glass has the unusual property of an increase of modulus with temperature increase, this negative entropy of activation is plausible.

REFERENCES

1. R. H. Doremus, in *Reactivity of Solids*, J. W. Mitchell, R. C. Devries, R. W. Roberts, and P. Cannon, eds., Wiley, New York, 1969, p. 667.
2. T. F. Soules, *J. Non-Cryst. Solids* **315** (1985).
3. R. H. Doremus, *Glass Science*, 2nd ed., Wiley, New York, 1994.
4. B. E. Warren, H. Krutter, and O. Morninstar, *J. Am. Ceram. Soc.* **19**, 202 (1936).
5. R. L. Mozzi and B. E. Warren, *J. Appl. Crystallogr.* **2**, 164 (1969).
6. D. J. Grimley and A. C. Wright, *J. Appl. Crystallogr.* **119**, 49 (1990).

7. B. E. Warren and J. Briscoe, *J. Am. Ceram. Soc.* **21**, 49 (1938).
8. D. J. Nakayama and J. F. Shackelford, *J. Non-Cryst. Solids* **126**, 249 (1990).
9. F. J. Norton, *Nature* **191**, 701 (1961).
10. R. H. Doremus, *J. Am. Ceram. Soc.* **49**, 461 (1966).
11. R. H. Fowler and E. A. Guggenheim, *Statistical Thermodynamics*, Cambridge University Press, Cambridge, 1939, pp. 81, 373.
12. R. W. Barrer and D. E. W. Vaughn, *J. Appl. Phys.* **63**, 2275 (1967).
13. J. E. Shelby, *J. Appl. Phys.* **47**, 135 (1976).
14. J. F. Shackelford and J. S. Masaryk, *J. Non-Cryst. Solids* **30**, 127 (1978).
15. C. M. Hartwig, *J. Appl. Phys.* **47**, 956 (1976).
16. J. E. Shelby, *J. Appl. Phys.* **48**, 338 (1977).
17. A. J. Moulson and J. P. Roberts, *Trans. Faraday Soc.* **57**, 1208 (1961).
18. G. J. Roberts and J. P. Roberts, *Phys. Chem. Glasses* **5**, 26 (1964).
19. J. Drury, G. J. Roberts, and J. P. Roberts, in *Advances in Glass Technology*, Plenum Press, New York, 1965, p. 249.
20. J. Burn and J. P. Roberts, *Phys. Chem. Glasses* **11**, 106 (1970).
21. G. J. Roberts and J. P. Roberts, *Phys. Chem. Glasses* **7**, 82 (1966).
22. J. E. Shelby, J. Vitko, and R. E. Benner, *J. Am. Ceram. Soc.* **65**, C-59 (1982).
23. J. Crank, *Mathematics of Diffusion*, 2nd ed., Oxford University Press, Oxford, 1975.
24. G. W. Spitzer and J. R. Ligenza, *J. Phys. Chem. Solids* **17**, 196 (1961).
25. S. Rigo, F. Rochet, B. Agius, and A. Straboni, *J. Electrochem. Soc.* **129**, 867 (1982).
26. F. Rochet and S. Rigo, *Philos. Mag.* **747** (1987).
27. M. Helmich and F. Rauch, *Glastech. Ber.* **66**, 195 (1993).
28. R. P. Pfeffer and M. Ohring, *J. Appl. Phys.* **52**, 777 (1981).
29. H. Wakabayashi and M. Tomozawa, *J. Am. Ceram. Soc.* **12**, 1850 (1989).
30. K. M. Davis and M. Tomozawa, in *Diffusion in Amorphous Materials*, H. Jain, and D. Gupta, eds., The Metallurgical Society, Warrendalc, PA, 1994, p. 119.
31. W. A. Lanford, C. Burman, and R. H. Doremus, in *Mater. Sci. Res.* **1**, 19 (1985); in *Advances in Materials Characterization II*, R. L. Snyder, R. A. Condrate, and P. F. Johnson, eds., Plenum Press, New York, 1985, p. 203.
32. P. Shewmon, *Diffusion in Solids*, 2nd ed., Minerals, Metals and Materials Society, Warrendale, PA, 1989, Chap. 2.
33. M. E. Glicksman, *Diffusion in Solids*, Wiley, New York, 2000.
34. C. Zener, in *Imperfections in Nearly Perfect Crystals*, Wiley, New York, 1952, p. 289.

7

DIFFUSION, REACTION, AND EXCHANGE OF WATER IN QUARTZ

7.1 INTRODUCTION

In Chapter 6 it was shown that experimental results from many different investigators on the diffusion and reaction of water with vitreous silica are explained by the diffusion–reaction model. Effects of temperature, pressure, and diffusion time on diffusion coefficients of water are all consistent with this model. Peculiarities such as changing profile shapes, lack of surface equilibrium, and different diffusion coefficients found for different experimental conditions and by different investigators can all be explained by the model. Originally, experimenters tried to explain the diffusion of water in silica glass in terms of diffusion of network hydrogen and oxygen, but many features of the diffusion results cannot be explained in this way.

In the diffusion–reaction model, molecular water dissolves and diffuses in a solid or melt and then reacts with the oxide to form immobile OH groups:

$$\text{Si—OH—Si} + H_2O = \text{SiOH} + \text{HOSi} \tag{7.1}$$

The idea of molecular solution and diffusion of molecules in silica glass was supported by experiments with inert gases and other diatomic gases such as hydrogen and oxygen. As shown in Chapter 5, the activation energies for diffusion of all these molecules in silica glass are a function of their size, consistent with their solution and diffusion as molecules.

The idea that molecules such as water can dissolve and diffuse in a "dense" crystalline solid such as quartz has been less accepted than such diffusion in "open" silica glass. As described in Chapter 5, the "openness" of a solid or melt can be measured from the ionic porosity Z, given by

$$Z = 1 - \frac{V}{V_c} \tag{7.2}$$

in which V is the volume of atoms in a unit cell and V_c is the volume of the cell. V can also be considered to be the volume of atoms in a molecule and V_c the molecular volume calculated from the bulk density and the molecular weight. See Chapter 5 for a more complete discussion. The Z value for silica glass was found to be 0.509, for β-quartz (see Section 7.2) to be 0.443, and for α-quartz to be 0.414, corresponding to the densities 2.20, 2.41, and 2.64 g/cm^3, respectively. Thus both α- and β-quartz actually have rather open structures compared to close-packed atoms all of the same size, which have a Z value of 0.26.

To show that the diffusion–reaction model applies to a crystalline solid as well as silica glass, experiments on diffusion and reaction of water in quartz are examined in this chapter. Although there are not as many experimental results as for silica glass, the results strongly support the diffusion–reaction model for water in quartz. Thus I conclude that water dissolves and diffuses molecularly in crystalline quartz just as it does in silica glass.

In the next section we describe briefly the crystal structure and allotropic forms of silicas, especially quartz. Then experimental results on diffusion of water in quartz are compared to the diffusion–reaction model and discussed.

7.2 STRUCTURE OF QUARTZ

Silica (SiO_2) can have a large number of different crystal structures. At ambient pressures at least seven structures are possible, as shown by this scheme, taken from Kats [1] (temperatures in degrees C)

$$\begin{array}{ccccccc}
 & 573^\circ & & 870^\circ & & 1470^\circ & \\
\alpha\text{-quartz} & = & \beta\text{-quartz} & = & \gamma\text{-tridymite} & = & \beta\text{-cristobalite} \\
 & & & & 163^\circ \ \| & & 270^\circ \ \| \\
 & & & & \beta\text{-tridymite} & & \alpha\text{-cristobalite} \\
 & & & & 117^\circ \ \| & & \\
 & & & & \alpha\text{-tridymite} & &
\end{array}$$

In recent work it has been found that trdymite is not stable in "pure" silica; some low concentration of impurities, usually alkalies, is needed to stabilize tridymite.

The various forms designated by Greek letters have structures that are quite similar, so they can transform easily and rapidly from one structure to another at the temperatures given. However, the forms quartz, tridymite, and cristobalite have quite different structures, so transformations among them are quite slow. For example, β-cristobalite formed on pure silica glass at high temperatures can easily be cooled to room temperature without transformation, although its equillibrium transformation temperature is 870°C to quartz.

The crystal structures of α- and β-quartz are quite complicated. An excellent description of the structure of α-quartz is given by Kats [1]. α-Quartz belongs to the trigonal (rhombohedral) crystal system, which has no plane of symmetry. The vertical axis of a crystal (called the *c-axis*, *z-axis*, or *optical axis*) is an axis of trigonal symmetry. There are also three horizontal axes of binary symmetry, perpendicular to the *x*-axis, and making angles of 120°C with one another. In the unit cell there are three silicon atoms and six oxygen atoms. In the vertical direction the silicon and oxygen atoms are spirally arranged. Open tunnels are formed between the spirals in a vertical direction parallel to the *c*-axis. Thus one would expect diffusion of molecules to be faster parallel to the *c*-axis, as is found experimentally for water. There are also tunnels in the quartz structure parallel to the horizontal axes of symmetry. Thus one would also expect that diffusion should be relatively rapid parallel to the prism faces of the crystal compared to other directions perpendicular to the *c*-axis. These expectations are also borne out by experiments [2–4].

The structure of β-quartz has hexagonal symmetry and the crystals are less dense than α-quartz crystals (α-quartz, $2.64\,g/cm^3$; β-quartz, $2.41\,g/cm^3$), but the crystal structure of β-quartz is quite similar to that of α-quartz. The same tunnels parallel to the *c*-axis and the horizontal axes are present, so that one would expect similar diffusion behavior in the two phases.

Impurities can substitute for silicon in the quartz structure but not for oxygen. Aluminum substitutes for four-coordinated silicon and thus requires a compensating univalent cation to maintain charge balance; these cations can be hydrogen or alkali ions. Germanium easily substitutes for silicon; tetravalent tin and lead substitute much less easily because of their large size. Increasing pressure increases the α→β transition temperatures because it favors the more dense α-phase. Yoder found an increase of about 1°C for every 40 atm pressure increase [5].

7.3 EXPERIMENTS OF KATS ON HYDROGEN–DEUTERIUM EXCHANGE

In an excellent study, Kats [1] examined the positions of hydrogen in α-quartz with infrared spectroscopy. He also followed the exchange of deuterium for hydrogen in hydrated quartz, from which he calculated diffusion coefficients. To facilitate understanding of Kats's experiments, some equations from Chapters 3 and 4 are repeated here. The exchange of deuterium (D) in dissolved water molecules with hydrogen in OH groups (see Eq. 7.1) can be written

$$D_2O + 2OH = H_2O + 2OD \tag{7.3}$$

since pairs of OH groups result from Eq. 7.1. An equilibrium constant K for this reaction can be written

$$K = \frac{[H_2O][2OD]}{[D_2O][2OH]} \tag{7.4}$$

in which the brackets designate thermodynamic activities. Concentrations can be used instead of activities because there are no chemical gradients in the system. The expressions [2OD] and [2OH] designate pairs of deuterated or hydrated groups. It is assumed that the pairs exchange as units. In this book the actual concentrations of hydrogen or deuterium reacted in the solid are designated as S and S^*, respectively, so the number of pairs is half of S or S^*. The concentration of dissolved D_2O is C^* and H_2O is C. Since the forward and backward reactions are the same, $K = 1$ and from Eq. 7.4,

$$\frac{S^*}{S} = \frac{C^*}{C} \tag{7.5}$$

At a particular location in the quartz crystal the rate of change of the concentration S^* of tagged groups is

$$\left(\frac{\partial S^*}{\partial t}\right)_x = k^*(C^*S - CS^*) \tag{7.6}$$

in which k^* is a reaction rate coefficient with units of reciprocal time and concentration. The total concentrations S_T and C_T at a particular point in the quartz are

$$S_T = S^* + S \qquad C_T = C^* + C \tag{7.7}$$

S_T and C_T are constant in time and uniform in space. Then from Eqs. 7.5 and 7.7,

$$\frac{S^*}{S_T} = \frac{C^*}{C_T} \tag{7.8}$$

when there is local equilibrium. In Chapter 3 it is shown that the effective diffusion coefficient D_E is given by

$$D_E = \frac{DC_T}{S_T} \tag{7.9}$$

in which D is the actual diffusion coefficient of molecular water in the quartz. Thus D_E is independent of tracer concentrations, so the profile of deuterium is given by

$$S^* = S_0{}^* \operatorname{erfc}\left(\frac{x}{2\sqrt{D_E t}}\right) \tag{7.10}$$

in which $S_0{}^*$ is the surface concentration of deuterium when there is local equilibrium. The total amount M of network hydrogen exchanged in time t is

$$M = \frac{2S_T\sqrt{D_E t}}{\pi} \tag{7.11}$$

in which it is assumed that the water vapor at the quartz surface is 100% deuterium, so $S_0{}^* = S_T$ at local equilibrium.

At shorter times ($kC_T\,t$ less than about 4) the local equilibrium assumed above is not valid. Then the surface concentration of deuterium is less than S_T, because exchange is not complete at the surface. The dissolved surface concentration $C_0{}^* = C_T$ because it is controlled by the water vapor. If D_E is calculated from profiles (Eq. 7.10) or total amount (Eq. 7.11), it becomes time dependent, as described in Chapter 4. At short times the values of D_E calculated from Eq. 7.10 or 7.11 become larger than those at local equilibrium (long times), as shown in Section 7.4.

The experimental effective diffusion coefficients calculated from the experimental data of Kats [1] and Eq. 7.11 are given in Table 7.1. Some of the values in Table 7.1 are different from those in Ref. 1, as noted in the table. These D_E values are plotted in Fig. 7.1. The least-squares line in the figure, excluding the value at 400°C, gives

$$D_E = 0.27 \exp\left(-\frac{123,000}{RT}\right) \tag{7.12}$$

with the D_E values in cm^2/s and the activation energy 123 kJ/mol. Experiments at 1000, 900, 820, 700, and 475°C were preceded by heating for 16 h at 1000°C. This treatment appears to reduce the values of D_E at 475°C, contrary to the statement in

Table 7.1 Diffusion Coefficients for Water in Quartz[a]

T(°C)	M/M_D	t (h)	Sample Thickness (cm)	D_T (cm^2/s)
1000	Various	—	—	$1.79(10)^{-7b}$
900	0.533	5	0.172	$9.16(10)^{-8}$
820	0.521	16.5	0.141	$1.76(10)^{-8}$
700	0.653^b	400	0.140	$1.14(10)^{-9b}$
620	0.730	161	0.085	$1.23(10)^{-9b}$
550	0.557	420	0.086	$1.98(10)^{-10}$
500	0.423^b	280	0.084	$2.46(10)^{-10}$
475	0.168	424	0.085	$2.62(10)^{-11b}$
400	0.093	425	0.139	$2.14(10)^{-11b}$

[a]Calculated from the experimental results of Kats [1].
[b]Different values from those in Ref. 1.

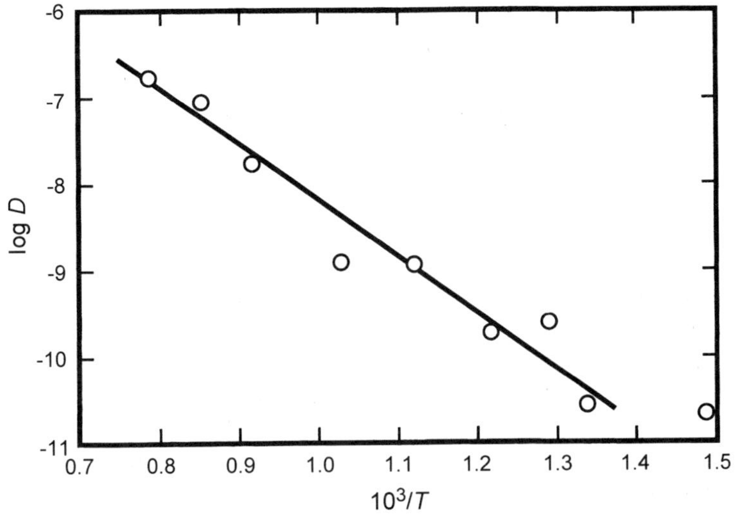

Figure 7.1 Log D_E as a function of reciprocal temperature. Points from Ref. 1 and Table 7.1; line from least squares, excluding point at 400°C. $Q = 123$ kJ/mol; correlation coefficient 0.974.

Ref. 1, but not at higher temperatures. The proximity of these temperatures to the $\alpha \rightarrow \beta$ quartz transformation temperature of 573°C appears to be a coincidence.

The measured D_E values begin to show a deviation from the high-temperature line in Fig. 7.1 at about 450°C. In silica glass, apparent diffusion coefficients of water deviated from the extrapolated coefficients at temperatures below about 550°C [6,7]. The higher D_E values calculated from Eq. 7.11 for temperatures below about 450°C show that local equilibrium is no longer occurring at these lower temperatures. This apparent increase can be understood physically as follows. As deuterated water molecules diffuse into the quartz, they do not exchange completely with the hydrogen on OH groups already in the quartz. Thus some deuterated molecules diffuse deeper into the quartz than they would if there were local equilibrium. This deeper penetration of deuterated water into regions of the quartz with lower deuterium concentration means that the back reaction of Eq. 7.3 is smaller, so the total amount M^* of reacted deuterium is greater than for local equilibrium. Thus the diffusion exchange model provides an explanation of the higher effective diffusion coefficients at lower temperatures. These D_E values should also be time dependent, decreasing to those expected from an extrapolation from high temperatures at long times of diffusion.

7.4 EXPERIMENTS ON ^{18}O–^{16}O EXCHANGE

There have been several studies of the diffusion of water tagged with ^{18}O into quartz [2–4,8]. The profile of ^{18}O exchanged into the silicon–oxygen network was analyzed by SIMS [2–4] or a nuclear reaction [8]. At temperatures from 697 to 849°C the results of three studies [2–4] at 100 MPa water pressure and transport parallel to the c-axis of the quartz agreed reasonably well and give the temperature dependence of the effective diffusion coefficient D_E in cm^2/s (least squares):

$$D_E = 7.1(10)^3 \exp\left(-\frac{129,000}{RT}\right) \tag{7.13}$$

with an activation energy of 129 kJ/mol, close to that for the data of Kats. Normal to the rhombohedral plane $(10\bar{1}1)$, the diffusion coefficient is about a factor of 2 smaller than that parallel to the c-axis at temperatures from 700 to 800°C [3]. Perpendicular to the c-axis the apparent diffusion coefficient of tagged water is about a factor of 100 smaller than for diffusion parallel to the c-axis; the activation energy of combined data from Refs. 1 and 2 between 700 and 850°C give a value of about 228 kJ/mol for diffusion perpendicular to the c-axis. These differences are just what one would expect from the structural differences in different directions in quartz crystals and their influence on diffusion of molecules in these directions. These directional effects would not be expected for diffusion of network oxygen ions.

The effective diffusion coefficients in these studies with ^{18}O isotopes were calculated from profiles and Eq. 7.10. These effective diffusion coefficients are related to the actual diffusion coefficients D of molecular water by Eq. 7.9, as long as there is local equilibrium. The total network oxygen concentration S_T is the same under all conditions at local equilibrium. However, C_T is directly proportional to the water vapor pressure and also depends on the solubility factor α, which is the ratio between the concentration of water dissolved molecularly in the quartz C_T and the concentration C_g of water in the vapor. The ratio α is about constant with temperature for silicate glasses, above about 200°C, but it could be temperature dependent for quartz. Thus comparisons of D_E values must be made with care, and further studies are needed to calculate actual D values of molecular water in quartz.

At temperatures below about 600°C, the diffusion coefficients measured for $H_2{}^{18}O$ in quartz by different investigators differ widely, by up to an order of magnitude or more at 500°C. I suggest that the reason for this difference is that the exchange rate between ^{18}O in molecularly dissolved and ^{16}O in the quartz network becomes so slow below 600°C that the effective diffusion coefficient D_E becomes time dependent. In Table 7.2 some selected data for apparent diffusion coefficients of water in quartz show that when the time of diffusion is similar, results of different investigators are similar, but when diffusion times are different, the diffusion coefficients at shorter times are higher, just as predicted. This difference is more marked as the temperature becomes lower; even at 700°C there seems to be a small difference. Especially striking is the difference of more than an order of magnitude in measured diffusion coefficients at 600°C for times of 6.37 and 120 h.

It is possible that some of the time dependence of the calculated diffusion coefficients arises because of a reaction layer on the surface of the quartz. At the temperatures and pressures of the experiments there is some solubility of quartz in

Table 7.2 Dependence of Diffusion Coefficients of Water in Quartz on the Time of Diffusion below 700°C

T (°C)	*D* (cm^2/s)	*t* (h)	Orientation with Respect to *c*-Axis	Ref.
500	$1.3(10)^{-17}$	600	Parallel	2
515	$3.2(10)^{-16}$	287	Parallel	4
595	$2.2(10)^{-15}$	169	Parallel	4
600	$1.1(10)^{-15}$	144	Parallel	2
591	$9.6(10)^{-18}$	135.5	Perpendicular	4
600	$1.2(10)^{-18}$	672	Perpendicular	2
	$3.6(10)^{-14}$	6.37	Parallel	[8]
700	$1.0(10)^{-15}$	120	Parallel	[8]
	$9.6(10)^{-15}$	20	Perpendicular	2
	$7.0(10)^{-15}$	114	Perpendicular	4
	$4.7(10)^{-17}$	43	Perpendicular	4
	$2.1(10)^{-17}$	337	Perpendicular	2

water, so dissolution and reprecipitation on the crystal surface is possible. However, there is no evidence for such a layer in the published tracer profiles, so it is unlikely that it contributes significantly to the diffusion process. Examination of profiles at lower temperatures should clarify this point.

In a definitive study, Farver and Yund [4] showed that the measured apparent diffusion coefficient of water in quartz is directly proportional to the fugacity of water vapor in the gas phase; this result is just what is expected from the diffusion and exchange mechanism, as shown in Eq. 7.9. The effective diffusion coefficient is directly proportional to C_T, the concentration of water dissolved molecularly in the quartz at its surface, which at equilibrium should be proportional to the fugacity of water vapor.

The proposal here is that both the exchange of deuterium for hydrogen [1] and ^{18}O for ^{16}O are controlled by the diffusion of molecularly dissolved water. The apparent diffusion coefficient D_E at 800°C from Eq. 7.13 is $3.4(10)^{-14}\ cm^2/s$, whereas for the data of Kats (Eq. 7.12) it is $1.6(10)^{-8}\ cm^2/s$ at 800°C, giving a ratio of $4.6(10)^5$. From Eq. 7.9 the difference between these two apparent diffusion coefficients results from two different factors, the different water vapor pressures and the different exchange capacities (S_T values). In Ref. 1 the water vapor pressure was 25 atm, whereas in the data for Eq. 7.13 it was 1000 atm. Kats reported a concentration of $5.0(10)^{18}$ hydrogen atoms for his natural quartz, and the calculated value of oxygen concentration in the quartz is $5.3(10)^{22}\ atoms/cm^3$. Thus the expected ratio between the two kinds of diffusion coefficients is $5.3(10)^{22}(40)/5(10)^{18} = 4.2(10)^5$, in excellent agreement with the measured value of $4.6(10)^5$. This agreement is evidence that the diffusion of molecular water is the controlling mechanism in both kinds of experiments.

Two sets of investigators measured the diffusion of "dry" tagged oxygen $^{18}O_2$ in quartz [9,10], and agreed rather well. The apparent diffusion coefficient of oxygen extrapolated to 800°C from Ref. 9 was $3.6(10)^{-20}$ cm^2/s at oxygen pressure of 90 torr or 0.118 atm. It is possible that even in these experiments the diffusing species is molecular water. The ratio of apparent diffusion coefficients from Eq. 7.13 and Ref. 9 is $8.8(10)^5$ at 800°C. The ratio of gas pressures in these experiments was $1000/0.118 = 8.5(10)^3$; thus the fraction of water vapor needed in the gas phase to give the diffusion coefficient found in the $^{18}O_2$ experiments was $8.5(10)^3/8.84(10)^5 = 0.0096$, or about $1.1(10)^{-3}$ atm of water vapor pressure. In an analysis of experiments of diffusion of nominally dry oxygen in silica glass [11] (see Section 12.2 and Table 12.1), it was found that when no special precautions were taken to absorb water, the gas contained about 0.0059 molar fraction of water vapor, or about $5.9(10)^{-3}$ atm of water vapor pressure. Thus it seems possible that the oxygen of Refs. 9 and 10 contained this much residual water.

7.5 EXPERIMENTS OF DERSCH AND RAUCH AT TEMPERATURES OF 200°C AND LOWER

These experiments on the diffusion and reaction of water with quartz can be compared with similar results on silica glass (see Section 6.7) by the same group of research workers. They measured the profiles of hydrogen in unhydrated quartz after treatments in water vapor from a nuclear reaction [12,13]. The surface concentration of reacted hydrogen S_0 increased with time at 200°C, showing that Eq. 7.1 had not reached equilibrium, as expected for this low temperature. To analyze the data I assumed that the profiles were exponential, as expected from the diffusion–reaction model at low nonequilibrium surface concentration (Eqs. 4.19 and 4.20):

$$S = S_0 \exp\left(-x\sqrt{\frac{\lambda}{D}}\right) \qquad (7.14)$$

in which S is the concentration of hydrogen, S_0 this concentration at the sample surface ($x=0$), x the distance from the surface, λ the forward reaction rate coefficient (with units of reciprocal time), and D is the diffusion coefficient of molecular water in the quartz, assumed to be independent of time, distance, and concentrations. See Fig. 4.1 for some experimental data.

The experimental data at times of 1, 3, and 9 days heating at 200°C in water vapor in a closed vessel over liquid water were plotted as log S versus distance. The profile after 9 days should actually be closer to a complementary error function (erfc) but can be approximated to be exponential. Linear regression fits were good, with correlation coefficients of 0.91, 0.97, and 0.96 for 1, 3, and 9 days, respectively. The resulting values of S_0/S_e, X at $S/S_0 = \frac{1}{2}$, and D_E values are given in Table 7.3. The ratio of surface concentration S_0 to the equilibrium surface concentration is S_0/S_e. The effective diffusion coefficient D_E was calculated in two different ways. First it was calculated with the equations of the diffusion–reaction model (Eq. 7.12; see also Chapter 4), and then with the assumption of an erfc profile. In the latter case the effective diffusion coefficient depended on the time, as

Table 7.3 Effective Diffusion Coefficients of Water in Quartz at 200°C[a]

t (days)	X at $S/S_0 = 1/2$ (nm)	S_0/S_e	$D_E \times 10^{15}$ Diffusion–Reaction	$D_E \times 10^{15}$ erfc
1	158	0.16	1.0	3.3
3	210	0.2	1.6	1.9
9	270	0.80	1.7	1.0

Source: Refs. 12 and 13.

[a]See the text for methods of calculation.

shown in Table 7.3, as expected when local equilibrium was not reached. At 1 day the D_E value was more than three times that at 9 days; D_E values calculated from the diffusion–reaction equations were more nearly constant, in light of the considerable experimental spread.

The effective diffusion coefficients are larger than expected from extrapolation of values at higher temperatures, as found for silica glass at 90°C (Section 6.7). Again the reason may be that the solubility α of water in the solid (quartz) decreases at lower temperatures. The effective diffusion coefficients in quartz are roughly three orders of magnitude lower than in silica glass, as one might expect from the more dense quartz structure. These comparisons are not exact, because the effective diffusion coefficients depend on the time of diffusion and on the equations used to calculate them, but the trends should be approximately correct.

Dersch and Rauch also measured profiles of ^{18}O resulting from exchange of dissolved $H_2{}^{18}O$ and the ^{16}O in the silicon–oxygen network of the quartz, and also from the reaction of Eq. 7.1 of dissolved molecular water to form OH groups. In silica glass (Chapter 6) these two processes of reaction and exchange were clearly separated at 200°C, because the reaction of Eq. 7.1 was considerably faster than the ^{18}O exchange with the network. In quartz these two reactions are closer in rate, so it is difficult to separate the two contributions to the ^{18}O profile [12, Fig. 2b].

The rate of formation of reacted OH groups is

$$\frac{\partial S}{\partial t} = \lambda C - \mu S \tag{7.15}$$

in which λ and μ are reaction rate coefficients with units of reciprocal time. The reverse reaction is first order, because at these low temperatures the OH groups are completely immobile, so neighboring groups do not interact; this interaction is required for a second-order reverse reaction. To find the time dependence of the surface concentration S_0 of reacted OH groups from Eq. 7.15, it is assumed that the surface concentration of dissolved water C_0 is constant and equal to the final total water concentration C_T throughout the quartz. Then an integration of Eq. 7.15 at the surface ($x = 0$) gives

$$S_0 = S_e(1 - e^{-\mu t}) \tag{7.16}$$

in which S_e in the equilibrium value of reacted OH concentration at long times. A value of μ for the surface concentrations given in Table 7.3 is about $0.065\,h^{-1}$ at 200°C. This value can be compared with a μ value of about $0.20\,h^{-1}$ at 200°C in silica glass, as given in Section 6.6. Thus the reaction of Eq. 7.1 to form hydroxyl groups is faster in silica glass than in quartz at 200°C. These low-temperature results on quartz are consistent with the diffusion–reaction model and support its validity over a wide range of temperature.

7.6 DISCUSSION

The experiments provide three factors of strong evidence for the molecular diffusion and exchange mechanism of water in quartz: (1) The apparent diffusion coefficient is proportional to the fugacity of the water vapor; (2) the ratio of diffusion coefficients of water in experiments on deuterium–hydrogen and ^{18}O–^{16}O exchange experiments is explained by Eq. 7.9; and (3) below 700°C the measured diffusion coefficients become time dependent. No other mechanism has been put forward that explains these factors.

These conclusions suggest that in any study of diffusion of gaseous isotopes that can exchange with a solid network (e.g., water and oxygen in oxides), the time dependence of the diffusion coefficient should be carefully tested over a wide range of diffusion times, especially below 700°C, and diffusion profiles examined carefully for deviations near the sample surface. Another way to detect the diffusion and exchange mechanism is by a reduction of the expected exchange ratio at the sample surface, which should occur along with the time dependence of the diffusion coefficient. Only a few profiles of ^{18}O are shown in the papers [2–4] on water in quartz, mostly at temperatures at or above 700°C, so it is not feasible to examine this possibility in these studies. I urge other workers to examine their profiles, especially at 500°C, for a lower surface exchange fraction. Furthermore, possible surface reaction layers (see above) should be examined.

Sharp *et al.* [14] measured ^{18}O profiles in quartz in contact with $C^{18}O_2$ vapor. They found diffusion coefficients between those in water and in nominally dry oxygen. Experiments of these authors on ^{18}O profiles in the quartz provide evidence that ^{18}O is not diffusing in the oxygen network. They formed an ^{18}O profile by diffusion from $C^{18}O_2$ and then annealed the sample in vacuum. The profile changed very little during this annealing step, with a nominal diffusion coefficient at least a factor of 20 smaller than in the tagged CO_2 vapor. This experiment shows that a constituent of the gas phase, probably water, is responsible for the apparent transport of oxygen in the quartz. Sharp *et al.* concluded that "there is probably a diffusing species, other than oxygen, that enhances the oxygen diffusion rate in these quartz–CO_2 systems... ." These comparisons between exchange experiments in quartz provide strong evidence that this model applies to crystalline quartz, and that molecular water dissolves, diffuses, and reacts and exchanges in quartz.

REFERENCES

1. A. Kats, *Philips Res. Rep.* **17**, 133–195, 201–279 (1962).
2. P. F. Dennis, *J. Geophys. Res.* **89**, 4047 (1984).
3. B. J. Giletti and R. A. Yund, *J. Geophys. Res.* **89B**, 4039 (1984).
4. J. R. Farver and R. A. Yund, *Chem. Geol.* **90**, 55, (1991).
5. H. S. Yoder, *Am. Geophys. Union Trans.* **31**, 827 (1950).
6. R. H. Doremus, *J. Mater. Res.* **10**, 2379 (1995).
7. H. Wakabayashi and M. Tomozawa, *J. Am. Ceram. Soc.* **72**, 1850 (1989).
8. A. Choudhury, D. W. Palmer, G. Amsel, H. Curien, and P. Baruch, *Solid State Commun.* **3**, 119 (1965).
9. R. Haul and G. Dümbgen, *Z. Electrochem.* **66**, 636 (1962).
10. R. Schachtner and H. G. Sockel, in *Reactivity of Solids*, J. Wood, O. Lindquist, C. Helgeson, and N. G. Vannerburg, eds., Plenum Press, New York, 1977, p. 605.
11. R. H. Doremus, *J. Electrochem. Soc.* **143**, 1992 (1996).
12. O. Dersch, A. Zouine, F. Rauch, and J. E. Ericson, *Fresenius J. Anal. Chem.* **358**, 217 (1997).
13. O. Dersch and F. Rauch, *Fresenius J. Anal. Chem.* **365**, 114 (1999).
14. Z. D. Sharp, B. J. Gilletti, and H. S. Yoder, *Earth Planet. Sci. Lett.* **107**, 339 (1991).

8

DIFFUSION AND REACTION OF WATER IN RHYOLITE GLASS (OBSIDIAN)

8.1 INTRODUCTION

In Chapters 6 and 7 it was shown that diffusion of water in silica glass and crystalline quartz can be explained with the diffusion–reaction model. The extensive experimental data on diffusion profiles of hydrogen and oxygen isotopes resulting from diffusion of dissolved molecular water and its reaction and exchange with the oxide network are consistent with the diffusion–reaction model. Another oxide system for which there are considerable data on water diffusion and reaction is rhyolite glass (obsidian). This aluminosilicate glass has been studied by two different research groups.

One group of research workers have studied the diffusion of water into obsidian (rhyolite glass) from water vapor, and in diffusion couples, measuring the resulting hydrogen concentration profiles with the ion microprobe (SIMS) [1–3]. See Table 8.1 for glass compositions. Measurements were made from 650 to 950°C; thus the materials were above the glass transition temperature of about 600°C during measurement, so they were highly viscous melts. A second group examined the diffusion of water out of samples of rhyolite glasses; they measured the concentration profiles of molecular water (H_2O) and OH groups bound in the silicate network with infrared spectra [4] in glasses held at temperatures between 400 and 550°C. Both of these groups analyzed their results in terms of concentration-dependent diffusion coefficients, without explicitly including in the diffusion description the reaction of water with the network:

$$H_2O + Si—O—Si = 2SiOH \qquad (8.1)$$

In both silica glass and quartz the concentration of dissolved molecular water is much smaller than the concentration of reacted OH groups. However, in the rhyolite

Table 8.1 Compositions of Rhyolite Glasses

	Los Posos [1]		Samples KS [4, 5]	
Oxide	wt %	mol %	atoms/cm^3 × 10^{21}	wt %
SiO_2	77.1	82.7	18.0	77.5
Al_2O_3	12.3	7.8	3.52	12.5
Fe_2O_3	1.1	1.0	0.2	1.0
TiO_2	0.1	0.1		
CaO	0.3	0.3	0.07	0.5
MgO	< 0.1			
Na_2O	4.2	4.4	1.66	3.6
K_2O	4.7	3.2	1.40	4.8
H_2O	0.2	0.7		

glass these two concentrations are not too different. This situation requires a somewhat different mathematical treatment, as given in the next section, than for the situation where the reacted concentration is much greater than the dissolved concentration. In subsequent sections the experimental data are compared to calculations from the model.

8.2 DIFFUSION–REACTION MODEL

The equation of continuity for diffusion plus chemical reaction is (Section 3.2)

$$\frac{\partial C}{\partial t} = D\frac{\partial^2 C}{\partial x^2} - \frac{\partial S}{\partial t} \tag{8.2}$$

in which D is the diffusion coefficient of mobile water in the glass, assumed to be independent of concentration, x is distance, and t is time.

The concentration of mobile molecular water is C, and the concentration of immobile reacted OH groups is S, *as* H_2O. In Refs. 1 to 3 the total water concentration

$$W = S + C \tag{8.3}$$

was measured from hydrogen profiles with the ion microprobe (SIMS). At 850°C it was found that W^2 was proportional to the surrounding pressure of water vapor (see Ref. 2 and references therein); since the concentration C of dissolved molecular water is proportional to this pressure:

$$W^2 = C\beta \tag{8.4}$$

in which β is a coefficient with units of concentration. From Eq. 8.3,

$$\frac{\partial W}{\partial t} = \frac{\partial S}{\partial t} + \frac{\partial C}{\partial t} \tag{8.5}$$

and from Eq. 8.4,

$$\frac{\partial C}{\partial x} = \frac{2W}{\beta} \frac{\partial W}{\partial x} \tag{8.6}$$

If Eqs. 8.5 and 8.6 are put in Eq. 8.2,

$$\frac{\partial W}{\partial t} = \frac{\partial}{\partial x} \left(\frac{2WD}{\beta} \frac{\partial W}{\partial x} \right) \tag{8.7}$$

Equation 8.7 has the form of the differential equation for diffusion with an effective concentration-dependent diffusion coefficient D_e:

$$D_e = \frac{2WD}{\beta} \tag{8.8}$$

If the glass initially contains a uniform concentration W_i of total water, this amount must be subtracted from the concentration W.

At temperature below about 600°C, reaction 1 in rhyolite glass does not reach local equilibrium during the initial diffusion times, as described below. From Eq. 8.1 the rate of reaction to form OH groups, at a particular distance X, is

$$\frac{\partial S}{\partial t} = \lambda C - \frac{\mu S^2}{S_i} \tag{8.9}$$

in which λ and μ are rate coefficients with units of reciprocal time. In the experiments in Refs. 4 and 5, the samples of rhyolite glass were annealed at long times to produce uniform initial concentrations of dissolved molecular water and reacted water and to reach local equilibrium. This initial uniform concentration of reacted water is S_i. A factor of S_i is included in the last term in Eq. 8.9 so that the reaction coefficient μ has units of reciprocal time. Then the glass was heated in dry nitrogen. Under these conditions the concentration of dissolved molecular water at the glass surface is zero, as shown by the experimental measurements. The concentration of reacted water at the glass surface S_0 is not necessarily equal to zero, although this condition was assumed in the analysis of Ref. 4. If Eq. 8.9 is integrated with respect to t at $x = 0$ (the glass surface) and the condition $C_0 = 0$, the result is (see Section 4.4)

$$S_0 = \frac{S_i}{1 + \mu t} \tag{8.10}$$

Thus the surface concentration S_0 of reacted groups starts at S_i and depends on the time of diffusion; it equals zero only at long times (local equilibrium). The value μ can be calculated from measured values of S_0 and Eq. 8.10, as shown below.

At early times of diffusion ($\mu t < 2$), the concentration of dissolved molecular water C has an exponential profile (Section 4.2):

$$\frac{C}{C_i} - 1 = \exp\left(-x\sqrt{\frac{\lambda}{D}}\right) \tag{8.11}$$

in which C_i is the initial concentration of molecular water and the profile of reacted groups S is also exponential:

$$\frac{S_i - S}{S_i - S_0} = \exp\left(-x\sqrt{\frac{\lambda}{D}}\right) \tag{8.12}$$

since

$$\frac{S_i - S}{S_i - S_0} = \frac{C_i - C}{C_i} \tag{8.13}$$

as is found experimentally. See Chapter 4 for more discussion of the relationship between S and C in Eq. 8.13.

8.3 ANALYSIS OF HIGH-TEMPERATURE DIFFUSION PROFILES

In Refs. 1 and 2, experiments on diffusion of water into rhyolite glass from surrounding water vapor were reported. The profiles of hydrogen in Refs. 1 to 3 are equivalent to profiles of total water W in the glasses measured. These profiles were measured with an ion microprobe (SIMS) after the glasses had been heated at high temperatures and held at high pressures. An example of a profile of hydrogen concentration is shown in Fig. 8.1; other profiles in Refs. 1 and 2 had the same shape as the one in Fig. 8.1. The statistical counting error for the measurements in Refs. 1 and 2 was about $\pm 3\%$ of the full ordinates in Figs. 8.1 and 8.2. The conditions of these experiments at temperatures from 650 to 950°C fit those for the validity of Eq. 8.2 (binary decomposition reaction) and local equilibrium; thus the treatment of Section 8.2 applies. The line in Fig. 8.1 was derived from the diffusion–reaction model with an effective diffusion coefficient that is directly proportional to the total water concentration in the glass (Eqs. 8.7 and 8.8). The diffusion coefficient of mobile molecular water is independent of concentration, although the profile of mobile molecular water is not an error function. The fit of the calculated line to the experimental data is excellent, confirming the applicability of the diffusion–reaction model to these experiments. An appropriate measure of the goodness of fit can be found with a sum-of-squares analysis, as described in the

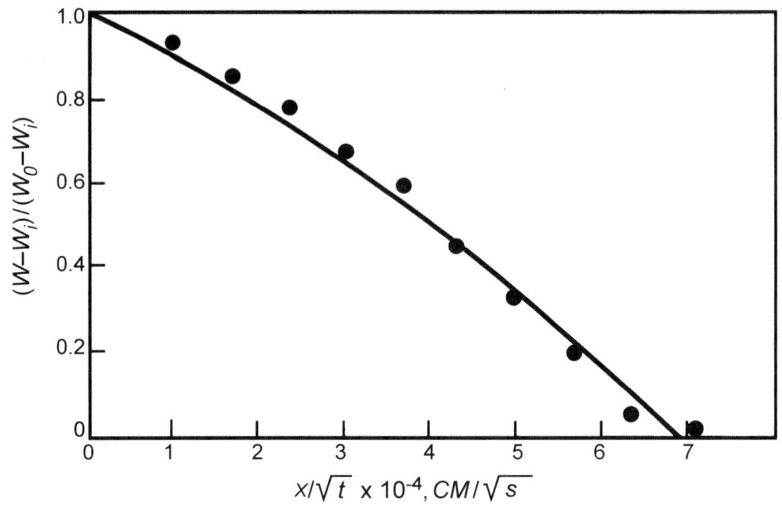

Figure 8.1 Profile of concentration W of total water in a rhyolite glass after diffusion for 1 h at 850°C and 700 bar pressure. Points from Fig. 4 of Ref. 1; line for a diffusion coefficient proportional to concentration (Eq. 8.8) (see Fig. 2.3). (From Ref. 9.)

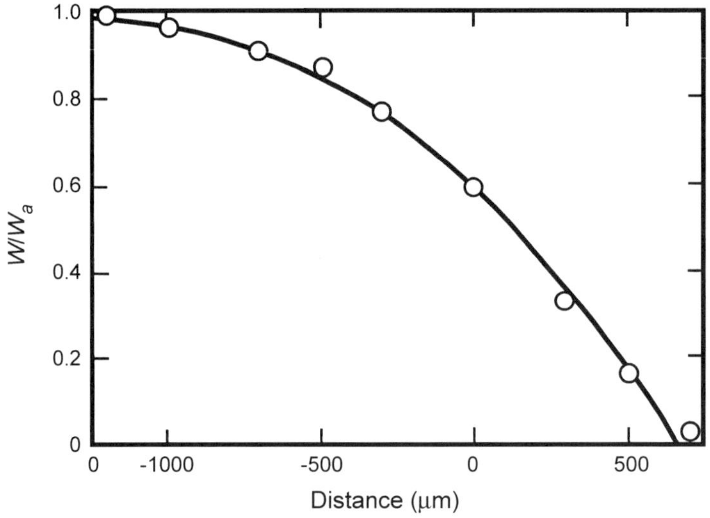

Figure 8.2 Profile of concentration W of total water in a couple of wet and dry rhyolite glass after diffusion at 70 MPa and 850°C for 6 h. Points from Fig. 1 of Ref. 3; line from Fig. 2.2. (From Ref. 9.)

appendix to this chapter. The result is a value of $R^2 = 0.985$, where $R^2 = 1.0$ would be a perfect fit. The R value is analogous to a correlation coefficient, but not identical to it, as described in the appendix.

In Ref. 3 diffusion couples of dry and wet sections of rhyolite glass were held together and heated at high temperatures. The resulting profiles of hydrogen were measured with an ion microprobe. These profiles had the shape expected for an effective diffusion coefficient proportional to concentration in infinite geometry (Fig. 2.2), with $(W - W_i)/(W_a - W_i) = 0.59$ at $x = 0$, as shown in Fig. 8.2, again showing the applicability of the diffusion–reaction model to these experiments with diffusion couples. The value of 0.59 is just that found in Fig. 2.2 for these conditions. The initial uniform concentration of total water in the wet region is W_a, and the initial residual uniform concentration of total water in the dry region is W_i.

Effective diffusion coefficients calculated from the profiles in Refs. 1 to 3 are listed in Table 8.2. The D_e values correspond to the surface concentrations W_0 given in the table. These W_0 values were normalized to a solubility of total water of 3.15 wt% at 850°C and 699 bar pressure from Ref. 6; this value is judged to be more reliable than that of 3.7 wt% given in Ref. 3, because in Ref. 6 a larger number of samples were studied, and quenching effects were allowed for. The W_0 values were inversely proportional to the absolute temperature. For Ref. 3 the D_e value was calculated from the measured profile (infinite geometry); the normalized dimensionless distance is $x/2\sqrt{D_e t}$, and the value of D_e corresponds to the concentration W_a initially in the wet region. The D_e value for the diffusion couple agrees within experimental error with those for semi-infinite geometry.

The diffusion coefficients of mobile molecular water can be calculated from the D_e values and Eqs. 8.4 and 8.8, with values of W_0 and C_0 at the sample surface and W_a and C_a in the water-containing half of the diffusion couple. Then

$$D = \frac{D_e W_0}{C_0} \tag{8.14}$$

Table 8.2 Diffusion Coefficients for Water in Rhyolite Glass from 650 to 950°C at 700 bar Pressure

T(°C)	t (min)	D_e (cm^2/s × 10^8)	W_0 (wt%)[a]	D (cm^2/s × 10^7)	Refs. for Experimental Data
950	362	20.2	2.89	6.54	2
950	165	21.1	—	6.84	2
950	60	21.5	—	6.97	2
850	163	12.2	3.15	3.61	2
850	60	18.1	—	5.36	1
850	360	11.2	—	3.31	3
750	360	6.87	3.62	1.85	2
750	169	8.32	—	2.25	2
750	61	8.75	—	2.36	2
650	360	2.31	3.75	0.55	2

[a]For the diffusion couple of Ref. 3 at 850°C, this concentration is the initial concentration W_a in the wet or hydrated part of the couple.

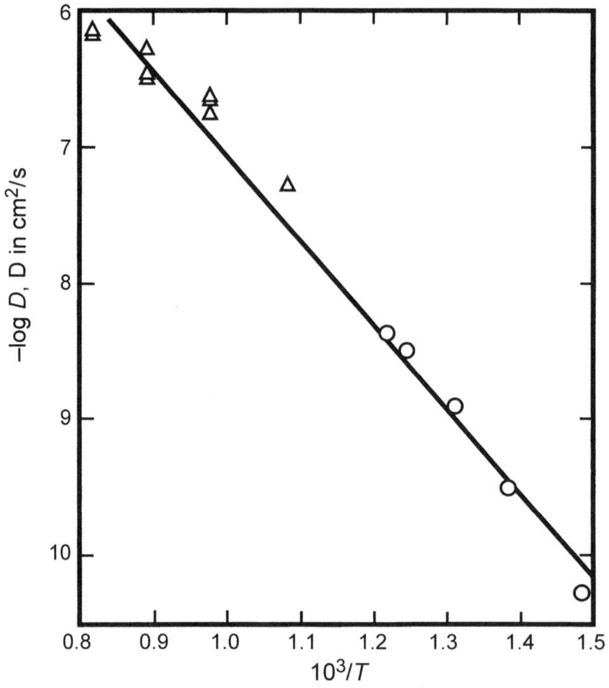

Figure 8.3 Log diffusion coefficients of water in obsidian in cm^2/s as a function of $1/T$. △ from Table 8.2 and Refs. 1 to 3; ○, from Table 8.3 and Refs. 4 and 9.

Values of W_0/C_0 are extrapolated from measurements in Refs. 5 and 6. Values of D are plotted in Fig. 8.3 as log D versus $1/T$. This plot is discussed further below.

The local rate of reaction of water is so fast above about 600°C in rhyolite glass that the distribution between dissolved and reacted water cannot be preserved by quenching from above 600°C [5–7]. Nevertheless, the profiles of total water concentration in Refs. 1 to 3 measured at room temperature after quenching should be characteristic of the high-temperature profiles. The maximum mean diffusion distance X during the quench can be estimated from the equation $X = \sqrt{2Dt}$, and for a time 10 s at 850°C, X is about 30 μm, which is small compared to the diffusion distances of about 0.5 mm in Refs. 1 to 3.

8.4 ANALYSIS OF LOW-TEMPERATURE PROFILES

In Ref. 4, experiments on diffusion of water out of rhyolite glasses held at temperatures from 400 to 550°C were reported. Concentrations of both OH groups and molecular water were measured with infrared spectroscopy. An example of measured profiles of OH groups and molecular water is shown in Fig. 8.4. According to the authors, the relative precision is better than about 2% at distances

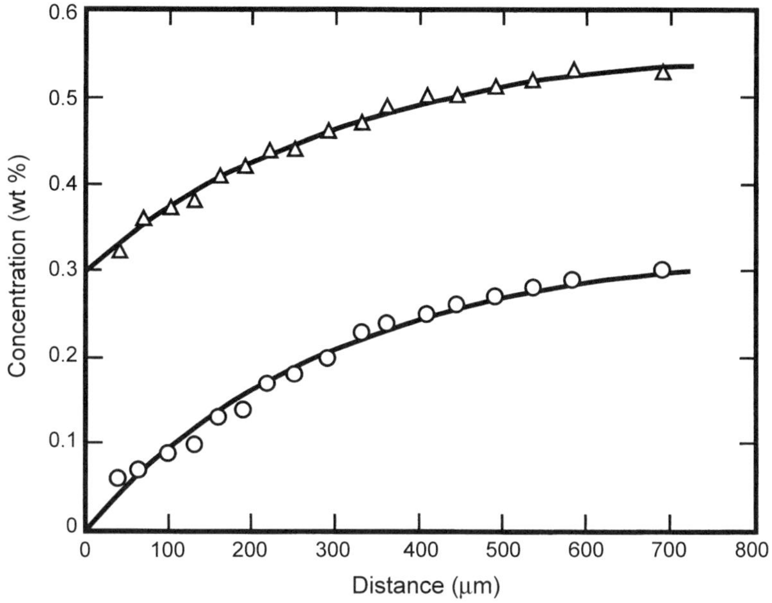

Figure 8.4 Profiles of molecular water (O) and reacted OH groups (△) in obsidian after 51.7 days at 450°C. Lines drawn from exponential profiles (Eqs. 8.11 and 8.12). (From Refs. 4 and 9.)

greater than 30 μm from the surface; this error is smaller than the sizes of the data points in Figs. 8.4, 8.5, and 8.6. The measurements of S, the triangles in the figure, extrapolate to a nonzero value at the sample surface. This result is evidence that local equilibrium between OH groups and dissolved molecular water has not been reached. Thus the analysis of Section 4.2 is appropriate to these measurements from Ref. 4.

The values of the concentrations of OH groups at the sample surface allow calculation of the reaction rate parameter μ (see Eq. 8.10). Values of S_0/S_i, the extrapolated surface concentrations of OH groups, divided by the concentrations of these groups far from the surface (the initial concentrations) are given in Table 8.3. Values of μ calculated from Eq. 8.10 are also shown in the table. These values when plotted as log μ versus $1/T$ fit on a straight line, with an activation energy from least squares of about 94 kJ/mol and a correlation coefficient of 0.93.

For values of μt less than about 2, exponential profiles of OH groups and of molecular water are expected (see Eqs. 8.11 and 8.12). Values of log $(S_i - S)/S_i - S_0)$ and $(C_i - C)/C_i$ at 403°C are plotted as a function of x in Fig. 8.5, and they show a linear relation, as expected from Eqs. 8.11 and 8.12. The line in Fig. 8.5 was drawn from a least-squares fit of the data with a correlation coefficient of 0.996. The lines in Figs. 8.4 and 8.6 are calculated from an exponential function, and they fit the experimental data very well. When plotted on a log scale as in Fig. 8.5, the data in Figs. 8.4 and 8.6 have a least-squares fit similar to that of Fig. 8.5.

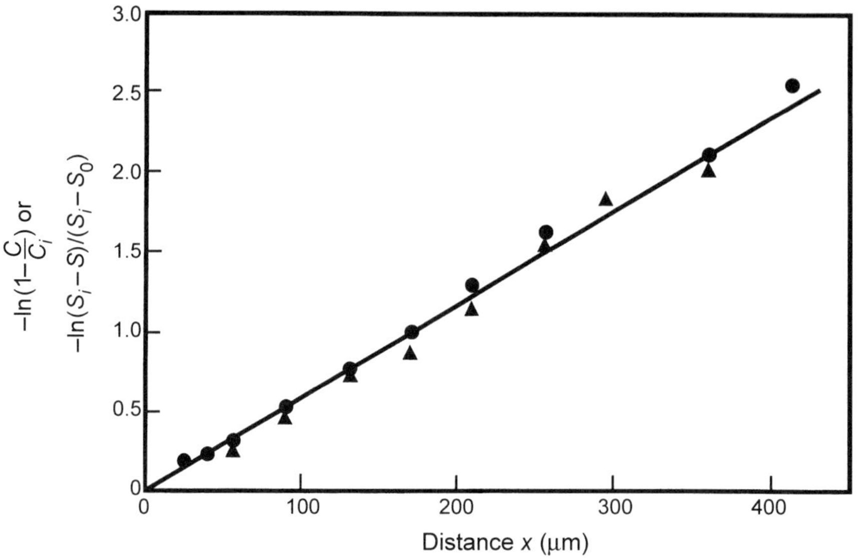

Figure 8.5 Log $(S_i - S)/(S_i - S_0)$ (●) and log $(C_i - C)/C_i$ (▲) as a function of distance x from the sample surface for obsidian held at 403°C for 39.9 days (see Eqs. 8.11 to 8.13). (From Refs. 4 and 9.)

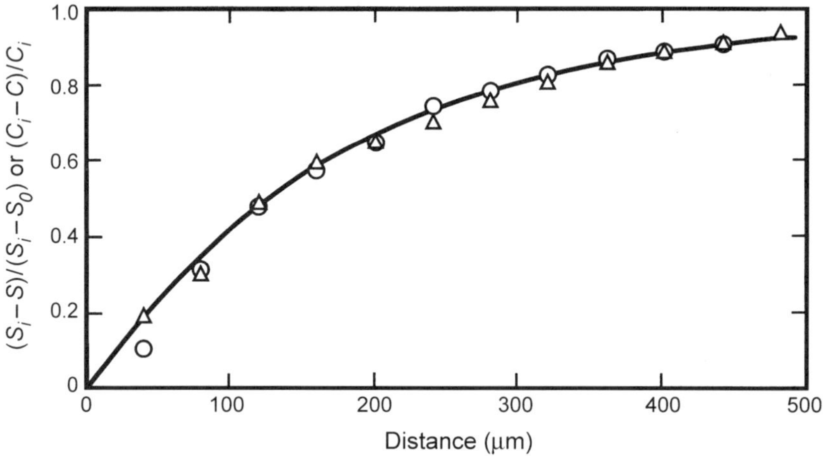

Figure 8.6 Profiles of OH groups (△) and molecular water (○) in obsidian after 18.6 days at 490°C; line drawn from Eq. 8.11 or 8.12. Data plotted as $(S_i - S)/(S_i - S_0)$ and $(C_i - C)/C_i$ as a function of the distance from the sample surface. (From Refs. 4 and 9.)

Diffusion coefficients of molecular water calculated from the profiles in Ref. 4 and exponential fits such as in Figs. 8.4 to 8.6 are shown in Table 8.3. The μ values in the table were calculated from Eq. 8.10 and surface concentrations of OH groups; and $\lambda = S_i \mu / C_i$ (Eq. 8.9 when $\partial S/\partial t = 0$). A value of D/λ was calculated

Table 8.3 Diffusion Coefficients and Reaction Rate Parameters μ for Diffusion of Water in Rhyolite Glass

T(°C)	S_0/S_i	t (days)	μ ($s^{-1} \times 10^7$)	S_i/C_i	D/λ ($cm^2 \times 10^4$)	D ($cm^2/s \times 10^{10}$)
403	0.560	39.9	2.28	0.901	2.71	0.56
450	0.536	51.7	1.94	1.697	9.92	3.26
490	0.333	18.6	12.5	2.35	4.41	13.0
530	0.323	11.4	21.3	2.70	5.66	32.0
550	0.200	13.8	33.5	4.87	2.02	44.4

Source: Data from Ref. 4.

from the fit to the profiles, and finally, D, the diffusion coefficient of molecular water, calculated. These D values are plotted in Fig. 8.3 and fit on a single line of log D versus $1/T$. A linear regression analysis gives the line in the figure and an activation energy of about 119 kJ/mol, with a correlation coefficient r of 0.992. The temperature dependence is $D = 0.13 \exp(-119{,}000/RT)\, cm^2/s$.

8.5 DISCUSSION

The comparisons in Sections 8.3 and 8.4 show that the diffusion–reaction model provides an accurate description of the diffusion profiles of water species in obsidian at both high and low temperatures. At temperatures above about 600°C, local equilibrium between mobile and immobile water is reached, and at lower temperatures there is not local equilibrium. The calculated diffusion coefficients of molecular water for both these temperature ranges and local equilibrium and nonequilibrium all fit on a single Arrhenius curve, both above and below the glass transition temperature, which is about 600°C for dry obsidian. The activation energy of 119 kJ/mol for diffusion of molecular water in rhyolite glass is greater than the value of 70 kJ/mol for diffusion of water in fused silica (Chapter 6). For crystalline quartz the activation energy for water diffusion is about 123 kJ/mol (Chapter 7). These activation energies are compared with those for many other crystalline, glassy, and molten oxides in Chapter 9. There is good correlation of these three activation energies with the ionic porosity or "open space" in their oxide structures. The results on obsidian show that this correlation is valid for both molten and glassy states (above and below the glass transition temperature) because the activation energy is the same for molten and glassy obsidian. The implication of these comparisons is that the ionic porosity or openness of obsidian does not change much between melt and glass.

It is of interest to compare the absolute values of the diffusion coefficients of molecular water in silica glass and in obsidian. At 950°C the diffusion coefficient of molecular water is about $1.5(10)^{-7}\, cm^2/s$, whereas in obsidian it is about $6.5(10)^{-7}\, cm^2/s$. Thus the rate of diffusion of water is considerably greater in obsidian than in silica glass, although the silica glass is more open. At 400°C, where both obsidian and silica are glassy, the diffusion coefficients are both about $5(10)^{-11}\,\mathrm{cm}^2/\mathrm{s}$. This change results from the higher activation energy for diffusion of water in obsidian than in silica glass.

The activation energy Q of diffusion for molecules in a glass has been described with the equation (Section 5.3)

$$Q = 8\pi G r_D (r - r_D)^2 \tag{8.15}$$

which is the elastic energy to dilate a spherical cavity from radius r_D to r. This equation is applied to gases in glass by assuming that the gas molecules dissolve in interstices in the glass and pass through "doorways" of radius r_D in jumping from one doorway to another; r is the radius of the diffusing molecule. In Eq. 8.15, G is the elastic shear modulus. From Section 5.3 and Figure 5.5 r_D is about 0.37 Å, and G is 20 Gpa, which is somewhat less than the measured value of 30.1 Gpa; a lower value of G was also calculated from Eq. 8.15 for silica glass in Section 5.3.

REFERENCES

1. J. R. Delaney and J. L. Karsten, *Earth Planet. Sci. Lett.* **52**, 191 (1981).
2. J. L. Karsten, J. R. Holloway, and J. R. Delaney, *Earth Planet. Sci. Lett.* **59**, 420 (1982).
3. K. E. Lapham, J. R. Holloway, and J. R. Delaney, *J. Non-Cryst. Solids* **67**, 179 (1984).
4. Y. Zhang, E. M. Stolper, and G. J. Wasserburg, *Geochim. Cosmochim. Acta* **55**, 441 (1991).
5. Y. Zhang, E. M. Stolper, and P. D. Ihinger, *Am. Mineral.* **80**, 593 (1995).
6. M. Novak and H. Behrens, *Geochim. Cosmochim. Acta* **59**, 3445 (1995).
7. A. Shen and H. Keppler, *Am. Mineral.* **80**, 1335 (1995).
8. N. R. Draper and H. Smith, *Applied Regression Analysis,* Wiley, New York, 1966, Chap. 1.
9. R. H. Doremus, *J. Non-Cryst. Solids* **261**, 101 (2000).

APPENDIX

An approximate measure of the goodness of fit of data to a nonlinear curve such as that in Fig. 8.1 results from a sum-of-squares treatment analogous to that for a linear regression. From Ref. 8, p. 14:

$$\begin{matrix}\text{sum of squares} \\ \text{about the mean}\end{matrix} = \begin{matrix}\text{sum of squares} \\ \text{about regression}\end{matrix} + \begin{matrix}\text{sum of squares} \\ \text{due to regression}\end{matrix} \tag{A8.1}$$

$$\sum(Y_i - \bar{Y})^2 = \sum(Y_i - Y_a)^2 + \sum(Y_a - \bar{Y})^2$$

In this equation Y_i is a measured dependent variable, and Y, the mean of all Y_a, is the value of the regression line at independent variable X_i. The quantity $\sum(Y_i - Y_a)^2$ is a measure of the deviations from the regression line; if it is zero, the fit is perfect. The quantity

$$R^2 = 1 - \frac{\sum(Y_i - Y_a)^2}{\sum(Y_i - \overline{Y})^2} \qquad \text{(A8.2)}$$

is therefore a measure of fit; if R^2 is 1.0, the fit is perfect. Equation A8.2 can also be used as an approximate measure of fit for a nonlinear curve, such as that shown in Fig. 8.1.

9

DIFFUSION AND REACTION OF WATER IN OXIDES

9.1 INTRODUCTION

In Chapters 6 to 8, experimental results on water diffusion and reaction in silica glass, crystalline quartz, and rhyolite glass have been compared with the diffusion–reaction model. For each of these systems there are experimental data from two or more research groups. The data fit the diffusion–reaction model very well, providing strong evidence for the validity of the model and its applicability to these systems. A wide variety of conditions in these three materials lead to good agreement with the diffusion–reaction model. Among these conditions are: local equilibrium between dissolved molecular water and OH groups in the silica network of the solid, glass, or melt (see Eq. 9.1); lack of such equilibrium; diffusion of water into an hydrated solid; exchange between deuterium isotope on the molecular water (D_2O) and the hydrogen in a hydrated solid; exchange between ^{18}O in dissolved molecule and ^{16}O in the silicon–oxygen network of the solid or melt; equilibrium and nonequilibrium isotopic exchange; time dependence of an apparent diffusion coefficient; proportionality of the apparent diffusion coefficient to water vapor pressure; and different geometries: semi-infinite; infinite couple, and plane layer. For all these different conditions the diffusion–reaction model provides a satisfactory explanation of diffusion profiles and quantities of water in silica glass, crystalline quartz, and rhyolite glass.

In many other crystals, glasses, and melts there have been less extensive experimental measurements of the diffusion of water. In this chapter the experimental activation energies for these systems are compared and shown to depend on the ionic porosity Z (openness) of the silica structure. Since the activation energies for water diffusion in silica glass fit the same correlation, the implication is that water diffusion in all these systems follows the diffusion–reaction mechanism, even though not as extensive data are available for comparison with the model as for the three other systems.

In Section 9.2 some equations from the diffusion–reaction model are repeated for emphasis and use in this chapter. Then experiments are discussed on diffusion of water in two systems: pyrope garnet and calcite. Pyrope garnet is a dense crystal, yet it fits the diffusion–reaction model; the pressure dependence of water diffusion in calcite also fits the model. Then experimental activation energies for diffusion of water in a wide variety of crystalline minerals and compounds, and in glasses, are compared to the openness of their structures.

9.2 EQUATIONS

In the diffusion–reaction model, dissolved molecular water reacts with oxygen in the oxide [1–4]:

$$H_2O + \text{—O—} = 2\,\text{—OH} \tag{9.1}$$

The oxygen atom —O— and the —OH groups are bonded to other groups in the oxide; for example, in silicates they are usually bonded to silicon or aluminum. In other oxides these groups can be bonded to other cations. The oxygen atoms and hydroxyl groups bonded in the oxide network are much less mobile than the dissolved molecular water, because the oxygen–cation bonds are strong (have a high bond enthalpy). From Eq. 9.1 an equilibrium constant K can be written

$$K_2 = \frac{S^2}{C} \tag{9.2}$$

in which S is the concentration of hydroxyl groups and C is the concentration of dissolved water. In this book S is in units of water concentration, so it is equal to one-half the actual concentration of OH groups. The bimolecular form of the reverse reaction of Eq. 9.1 implies that the hydroxyl groups diffuse at least a few times the intergroup separation distance, so neighboring groups can interact. The equilibrium concentration of hydroxyl groups in silica glass is proportional to the square root of the surrounding water vapor pressure above about 600°C [5], consistent with Eq. 9.2.

If molecular water diffuses into an oxide that contains a low concentration of hydroxyl groups, the dissolved water forms additional hydroxyl groups by Eq. 9.1. The combination of diffusion and reaction lead to a profile of hydroxyl groups resulting from an effective diffusion coefficient D_e that depends on concentration S [2; 3; 6, p. 326ff.] (see Sections 2.6 and 3.2):

$$D_e = \frac{2SD}{K_2} \tag{9.3}$$

in which D is the diffusion coefficient of molecular water, assumed to be independent of the concentration of molecular water.

In another kind of experiment deuterated or tritiated water is diffused into an oxide that has previously been equilibrated with untagged water at the same water vapor pressure, or contains a comparable concentration of OH groups. In this experiment the deuterium or tritium exchanges with hydrogen in the oxide. The effective diffusion coefficient associated with the results profile of deuterium or tritium in the oxide is [3]; Section 3.3

$$D_e = \frac{S_T D}{K_2} \tag{9.4}$$

in which S_T is the total concentration of hydrogen isotopes in the oxides, which is uniform in space and time, and K_2 is the equilibrium constant from Eq. 9.2. Thus in this experiment the effective diffusion coefficient is not a function of isotopic concentration.

It is also possible to diffuse molecular $H_2{}^{18}O$ into an oxide from water vapor with molecules tagged with ^{18}O. In this experiment the ^{18}O exchanges with ^{16}O in the oxide structure. The effective diffusion coefficient is [3]; Section 3.3

$$D_e = \frac{C_T D}{S_x} \tag{9.5}$$

in which S_x is now the total oxygen concentration in the oxide, again uniform in time and space, and C_T is the total (uniform) concentration of molecular water in the oxide. This concentration is proportional to the water vapor pressure. As for Eq. 9.4, the effective diffusion coefficient is independent of tracer concentration.

A measure of the openness of a structure is the ionic porosity Z (see Section 5.4):

$$Z = 1 - \frac{V}{V_c} \tag{9.6}$$

in which V is the volume of a molecule and V_c is the solid volume per molecule. V is calculated by summing the volumes of the individual atoms in the molecule (see Section 5.4) and V_c is calculated from the bulk density ρ and the molecular weight M of the molecule:

$$V_c = \frac{M}{\rho(0.602)} \tag{9.7}$$

with units of Å^3.

9.3 DIFFUSION OF WATER IN PYROPE AND CALCITE

Studies of diffusion of water in these two crystals show important features consistent with the diffusion–reaction model. Pyrope is a member of the garnet family of minerals with an ideal composition of $3MgO{\cdot}Al_2O_3{\cdot}3SiO_2$. However, natural crystals always contain some iron oxide; the Fe^{2+} content gives the garnet crystals their red color. The structure of pyrope consists of cubic close-packed oxygen ions arranged as a continuous network of corner-sharing SiO_4 tetrahedra and AlO_6 octahedra [32, p. 1037ff.]. Substitution of other divalent and trivalent ions for Mg and Al ions is common. The value of ionic porosity Z calculated from the ideal density of $3.56\ g/cm^3$ and molecular weight of 403.15 g/mol is 0.268, which is close to the Z value for ideal close packing of spheres (0.260). Thus the ideal pyrope structure is highly dense.

Wang *et al.* [33] measured profiles of OH groups in pyrope crystals from infrared absorption after they were heated in air or nitrogen. An example is shown in Fig. 9.1. This dehydration profile does not fit the erf profile for a constant diffusion coefficient but does fit the profile (Fig. 2.5) for a diffusion coefficient directly proportional to concentration. This dependence is expected from the diffusion–reaction model (Eq. 3.3 and Section 3.2). Thus even in the dense pyrope crystal water diffuses molecularly and reacts with the silicon and aluminum oxygen network by Eq. 9.1.

Calcite ($CaCO_3$) crystals have trigonal symmetry [32, p. 426ff.]. The carbonate groups are planar, with a central carbon atom surrounded by three equally spaced oxygen atoms. These carbonate groups form layers alternating with layers of calcium ions in octahedral coordination with six oxygen atoms from the carbonate groups. All layers are perpendicular to the axis, with carbonate triangles of successive layers pointing in opposite directions. The ionic porosity Z of this structure is 0.398, implying an intermediate degree of openness between pyrope and silica glass (Z=0.509). Even with this structure, which is quite different from silicate structure, the diffusion of water in calcite is consistent with the diffusion reaction model. For example, the effective diffusion coefficient is proportional to the water vapor fugacity [19], as expected for dissolution and diffusion of water molecules.

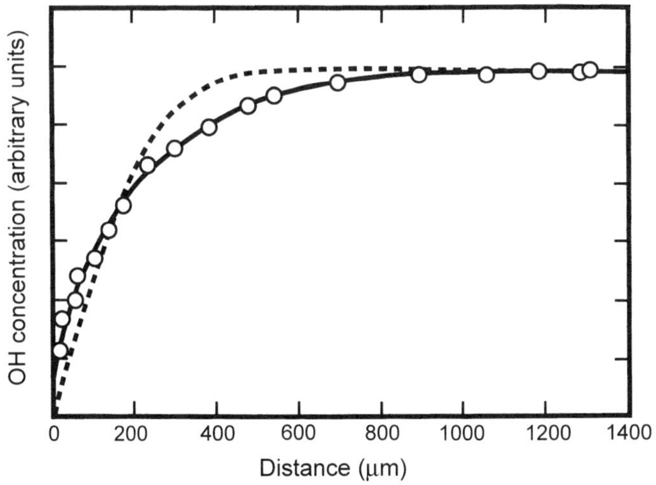

Figure 9.1 Profile of OH groups in a pyrope crystal heated 10830 s at 850°C in air. Points, experimental measurements with infrared absorption; solid line from Eq. 9.3 and Fig. 7.6 of Ref. 6 (see Section 2.6); dashed line from erf solution (Eq. 2.17).

9.4 COMPARISON OF EXPERIMENTAL ACTIVATION ENERGIES FOR DIFFUSION OF WATER IN OXIDES WITH IONIC POROSITY

The results of experimental measurements of diffusion, reaction, and exchange of water in the five oxides described in Chapters 6 to 8 and Section 9.3 provide strong evidence for the applicability of the diffusion–reaction model to these systems. Experimental measurements of water diffusion in a wide variety of other oxides are consistent with various aspects of the diffusion–reaction model, although more detailed measurements are needed to demonstrate conclusively the validity of the model for these oxides. In this section these data are listed and graphed and show a good correlation with the ionic porosity or openness of these oxides, both crystalline and glassy.

An important aspect of the diffusion–reaction model is the molecular dissolution of water in the oxide. Infrared spectra show that molecular water dissolves in many different silicate glasses, with the following compositions: soda-lime, rhyolite, basalt, albite, anorthite, jadeite, orthoclase, and other calcium and aluminum silicates (see Ref. 18 and references therein).

Often, effective diffusion coefficients are calculated by assuming that the profiles fit solutions of the diffusion equation for a constant diffusion coefficient, such as erf and erfc solutions. If the process is actually following the diffusion–reaction mechanism, these diffusion coefficients can show a concentration dependence (Eq. 9.3). Furthermore, if the system is not in local equilibrium, diffusion coefficients calculated in this way will show a dependence on diffusion time (see Chapter 4). At short times the calculated effective diffusion coefficient should be greater than that at long times. This time dependence is clearly shown in silica glass [3] and quartz [4]. In most other systems the time dependence of diffusion coefficients of water has not been examined thoroughly, although in a few oxides this time dependence is clearly present. In magnetite at 600°C, D_e was $3.9(10)^{-17}$ cm^2/s after 6.8 days and $0.27(10)^{-17}$ cm^2/s after 32 days [23], and in phlogopite mica at 600°C, D_e was $3.4(10)^{-17}$ cm^2/s after 29 days and $0.56(10)^{-17}$ cm^2/s after 73 days. These results are consistent with the time dependence expected from the diffusion–reaction model before local equilibrium is established.

Fortier and Giletti compared the effective diffusion coefficient of water in silicate minerals to their ionic porosity Z [24]. They found a fairly good correlation between D_e at 500 and 700°C and water pressure of 100 MPa for eight crystalline minerals and Z. Diffusion coefficients for three micas did not fit this correlation. I suggest that the activation energy for water diffusion is a better basis for comparison than the diffusion coefficient. The diffusion–reaction model shows that the absolute values of apparent diffusion coefficients can be influenced by a variety of factors that should not influence the activation energy, such as solubility of water in the oxide, the time of diffusion, the dependence of D_e on water concentration, and the type of diffusion experiment. If these factors are accounted for in terms of the model, the activation energy is the same for different conditions, as shown in Refs. 3 and 4.

In Table 9.1 the activation energies for diffusion of water in 18 glassy and 16 crystalline oxides are listed. The Z values in the table are those given in Ref. 24 or

Table 9.1 Activation Energy for Diffusion of Water in Crystalline and Glassy Oxides

Material	Symbol for Fig. 9.4	Nominal Chemical Composition	Z	Q (kJ/mole)	Refs.
Fused silica	Fs	SiO_2	0.509	70	3, 5, 9, 14–17
Obsidian	Ob	[a]	0.488	119	7, 8, 17, 34
16 Silicate glasses	SG	Multicomponent (see Ref. 25)	0.481	104 (mean)	25
K-feldspar	KF	$K_2O \cdot Al_2O_3 \cdot 6SiO_2$	0.452	107	20, 26
Anorthite	An	$CaO \cdot Al_2O_3 \cdot 2SiO_2$	0.445	110	26
Albite	Al	$Na_2O \cdot Al_2O_3 \cdot 6SiO_2$	0.441	89	26
β-Quartz	BQ	SiO_2	0.443	123	4, 10–13
Phlogopite	Ph	$K_2O \cdot 6MgO \cdot Al_2O_3 \cdot 6SiO_2 \cdot 2H_2O$	0.403	175	27
Calcite	Ca	$Ca\ CO_3$	0.398	173	19
Biotite	Bi	$K_2O \cdot 3MgO \cdot 3FeO \cdot Al_2O_3 \cdot 6SiO_2 \cdot 2H_2O$	0.396	142	27
Hornblende	Ho	[b]	0.384	172	21
Tremolite	Tr	$2CaO \cdot 5MgO \cdot 8SiO_2 \cdot H_2O$	0.380	163	21
Muscovite	Mu	$K_2O \cdot 3Al_2O_3 \cdot 6SiO_2 \cdot 2H_2O$	0.382	159	27
Richerite	Ri	$CaO \cdot 5MgO \cdot Na_2O \cdot 8SiO_2 \cdot H_2O$	0.375	157	21
Fluorapatite	Fl	$Ca_5(PO_4)_3F$	0.376	205	22
Magnetite	Mg	Fe_3O_4	0.350	188	23
Diopside	Di	$CaO \cdot MgO \cdot 2SiO_2$	0.343	226	28
Zircon	Zi	$ZrO_2 \cdot SiO_2$	0.302	210	29
Rutile	Ru	TiO_2	0.273	300	30
Pyrope	Py	$3MgO \cdot Al_2O_3 \cdot 3SiO_2$	0.268	254	33

[a] Wt. % $77SiO_2$, $12Al_2O_3$, $2FeO$, $4Na_2O$, $5K_2O$.
[b] Wt. % $46SiO_2$, $14Al_2O_3$, $14FeO$, $14MgO$, $9CaO$, $3Na_2O$.

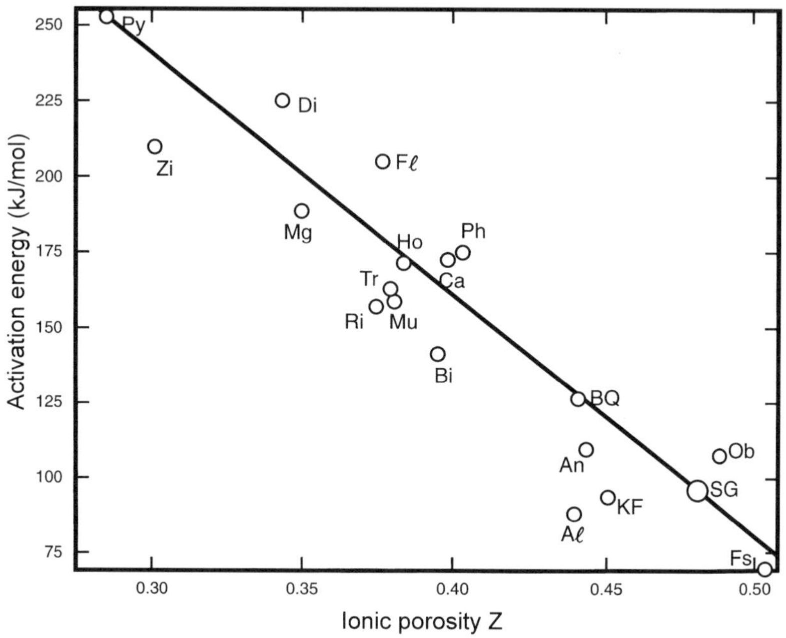

Figure 9.2 Activation energy for diffusion of water in oxides as a function of the ionic porosity (openness) Z, from Table 9.1. Line from linear regression (Eq. 9.8).

were calculated from the density ρ and composition of the oxide, with molecular weight M (see Eq. 9.7). This method is exactly the same as working with the unit cell and has the advantages that it does not require knowledge of the crystal structure and can be used for glasses. The ionic radius of oxygen was taken as 0.138 nm, to be consistent with Ref. 24. Other ionic radii were the effective ionic radii (IR) [The designation IR is for a particular radius in the tables of Shannon] from Table 1 of Shannon [31]. The data in Table 9.1 are plotted in Fig. 9.2. The means for 16 silicate glasses are plotted as a large circle, to reduce confusion in the plot. A linear regression between activation energy Q and Z gives

$$Q = 471 - 795Z\,\text{kJ/mol} \tag{9.8}$$

with a correlation coefficient of 0.943. This value and Fig. 9.2 demonstrate a good correlation between Q and Z. The results for the 16 silicate glasses were arbitrarily given a weighting of 1/2, so that they did not overbalance the results on different crystals. It is especially significant that the micas, fluorapatite, and calcite all fit with the other glasses and crystals; the diffusion coefficients of water in the micas [24] and in fluorapatite, calcite, and magnetite do not fit on the same Z line as other minerals [19]. Activation energies for ziron and rutile fall somewhat off the correlation line, but not too far off. In these dense minerals the effective diffusion coefficient was not a function of vapor pressure above about 1 atm. It is possible that the number of sites for dissolution of water is so small that saturation of sites occurs at 1 atm or below; more experiments are needed to test this possibility. The activation energy for pyrope, which is highly dense, fits close to the regression line.

Equation 9.8 can predict the activation energy for diffusion of water in an oxide for which measurements have not been made. The diffusion coefficient of molecular water can be estimated from the equilibrium coefficient in a similar material, and the total oxide content at higher temperatures where local equilibrium holds (see Eq. 9.5). At lower temperatures nonequilibrium must be included.

REFERENCES

1. R. S. MacDonald, *J. Phys.* **62**, 1168 (1958).
2. R. H. Doremus, *Reactivity of Solids*, in J. W. Mitchell and R. C. DeVries, eds., Wiley, New York, 1969, p. 667.
3. R. H. Doremus, *J. Mater. Res.* **10**, 2379 (1995).
4. R. H. Doremus, *Earth Planet. Sci. Lett.* **163**, 43 (1998).
5. A. J. Moulson and J. P. Roberts, *Trans. Faraday Soc.* **57**, 1208 (1961).
6. J. Crank, *Mathematics of Diffusion*, 2nd ed., Oxford University Press, Oxford, 1975.
7. J. R. Delaney and J. L. Karsten, *Earth Planet. Sci. Lett.* **52**, 191 (1981).
8. K. E. Lapham, J. R. Holloway, and J. R. Delaney, *J. Non-cryst. Solids* **67**, 179 (1984).
9. I. Burn and J. P. Roberts, *Phys. Chem. Glasses* **11**, 106 (1970).
10. A. Kats, *Philips Res. Rep.* **17**, 133–195, 201–279 (1962).
11. B. J. Giletti and R. A. Yund, *J. Geophys. Res.* **89B**, 4039 (1984).
12. P. F. Dennis, *J. Geophys. Res.* **89B**, 4047 (1984).
13. J. R. Farver and R. A. Yund, *Chem. Geol.* **90**, 55 (1991).
14. G. J. Roberts and J. P. Roberts, *Phys. Chem. Glasses* **7**, 82 (1966).
15. H. Wakabayashi and M. Tomozawa, *J. Am. Ceram. Soc.* **12**, 1850 (1989).
16. M. Helmich and F. Rauch, *Glastech. Ber.* **66**, 195 (1993).
17. Y. Zhang, E. M. Stolper, and G. J. Wasserburg, *Geochim. Cosmochim. Acta* **55**, 441 (1991).
18. L. A. Silver, P. D. Ilhinger, and E. M. Stolper, *Contrib. Mineral Petrol.* **104**, 142 (1990).
19. J. R. Farver, *Earth Planet. Sci. Lett.* **121**, 575 (1994).
20. R. A. Yund and T. F. Anderson, *Geochim. Cosmochim. Acta* **42**, 235 (1978).
21. J. R. Farver and B. J. Giletti, *Geochim. Cosmochim. Acta* **49**, 1403 (1985).
22. J. R. Farver and B. J. Giletti, *Geochim. Cosmochim. Acta* **53**, 1621 (1989).
23. B. J. Giletti and K. C. Hess, *Earth Planet. Sci. Lett.* **89**, 115 (1988).
24. S. M. Fortier and B. J. Giletti, *Science* **245**, 1481 (1989).
25. H. Scholze and H. O. Mulfinger, *Glastech. Ber.* **32**, 381 (1959).
26. B. J. Giletti, M. P. Semet, and R. A. Yund, *Geochim. Cosmochim. Acta* **42**, 45 (1978).
27. S. M. Fortier and B. J. Gilett, *Geochim. Cosmochim Acta* **55**, 1319 (1991).
28. J. R. Farver, *Earth Planet. Sci. Lett.* **92**, 386 (1989).

29. E. B. Watson and D. J. Cherniak, *Earth Planet. Sci. Lett.* **148**, 527 (1997).
30. D. K. Moore, D. J. Cherniak, and E. B. Watson, *Am. Mineral.* **83**, 700 (1998).
31. R. D. Shannon, *Acta Crystallogr.* **A32**, 751 (1976).
32. R. V. Gaines *et al.*, Dana's *New Mineralogy*, Wiley, New York, 1997.
33. L. Wang, Y. Zhang, and E. J. Essene, *Am. Mineral.* **81**, 706 (1996).
34. R. H. Doremus, *J. Non-cryst. Solids* **261**, 101 (2000).

10

SPECIATION OF WATER IN SILICATE GLASSES AND MELTS

10.1 INTRODUCTION

In Chapters 6 to 9, much evidence is presented that shows that water dissolves in silicate glasses and melts as molecules. Many references also demonstrate molecular dissolution of water in silicate glasses and melts [1–6]. These dissolved water molecules react with the silicon–aluminum–oxygen network of the glass to form —OH groups:

$$—\mathrm{O}—(n) + \mathrm{H_2O}(d) = 2\mathrm{OH}(n) \tag{10.1}$$

In this equation (n) means "part of the solid glass network" (but see the discussion below), and (d) means "dissolved molecularly in the glass". The symbol —O— designates an oxygen ion in the glass network; it can be bonded either to silicon or aluminum ions. The exact configuration of the OH groups is controversial. From a nuclear magnetic resonance (NMR) study, Kohn *et al.* [7] concluded that in nepheline ($NaAlSiO_4$) and $CaAl_2Si_{6.6}O_{17.3}$ melts, one hydrogen atom is associated with a bridging oxygen atom between aluminum and silicon atoms, and the remaining OH forms molecular sodium hydroxide (NaOH) with a sodium ion that compensated for the excess negative charges around the aluminum atom substituted for a silicon atom. The more traditional view of the formation of two hydroxyl groups bonded to aluminum or silicon atoms was advocated in a model of Sykes and Kubicki [8] based on infrared and Raman spectroscopy and molecular orbital calculations. These models were discussed further by Kohn *et al.* [9] and Sykes and Kubicki [10]. For the present purpose, the important result from Eq. 10.1 is the equilibrium constant (Eq. 10.2), which should be the same for both of these models. Richet and Polian [11] have additional ideas on the structure of water in silicate glasses. The distribution of water between dissolved molecules and reacted OH groups, called *speciation*, helps to determine the outgassing tendency of glasses and melts, and so is important in understanding the influence of water on glass properties such as viscosity, strength, and chemical durability.

Sowerby and Keppler [12] measured the concentrations of molecular and reacted water in a rhyolite glass at a range of temperatures from 124 to 900°C and pressures from ambient to 10 kbar. The important feature of this study is the measurement of infrared absorption at temperature and pressure, so that changes in concentrations during temperature and pressure quenching were avoided. These authors measured the dependence of speciation of water as a function of total water concentration at temperatures between 124 and 434°C.

The speciation of water in a variety of quenched silicate glasses has been measured from infrared absorption [13–17]. There has been some question about the validity of measurements for samples quenched from above the glass transition temperature and at high water concentrations [18, 19], thus emphasizing the value of measurements at temperature and pressure.

Equation 10.1 can be described with an equilibrium constant K:

$$K = \frac{[\mathrm{OH}]^2}{[\mathrm{H_2O}][\!-\!\mathrm{O}\!-\!]} \tag{10.2}$$

in which the brackets represent thermodynamic activities. If ideal mixing of components can be assumed, concentrations can be substituted for activities, but this assumption of ideality leads to K values that are a function of total water concentration for speciation of water in silicate glasses. To correct for this variation, a regular solution model with four arbitrary parameters has been fit to speciation data [15, 20].

In this chapter a Langmuir model is invoked to explain the dependence of speciation on total water concentration. In this model there are a fixed number of sites at which the reaction in Eq. 10.1 can take place, so at higher total water concentrations the concentration of OH groups saturates and becomes constant. There are two arbitrary parameters with this model, and it fits the experimental data as well as a regular solution model with four arbitrary parameters. In Section 10.2 the Langmuir model is described, and then it is compared to the experimental speciation data of Sowerby and Keppler [12]. Then the Langmuir model is used to describe experimental data [21] on solubility of water in aluminosilicate melts. There are also some comments on the partial molar volume of water in silicate glasses and melts, and finally, the model and comparisons are discussed.

10.2 LANGMUIR MODEL FOR SPECIATION

The Langmuir model [22] was devised for adsorption from a gas onto a solid surface with a limited and fixed number of adsorption sites (monolayer). In this model the fraction of adsorption sites θ is related to the fugacity f of the gas by

$$\theta = \frac{kf}{1 + kf} \tag{10.3}$$

in which k is a constant. Solving Eq. 10.3 for k gives

$$k = \frac{\theta}{f(1 - \theta)} \tag{10.4}$$

The fugacity of the gas is used instead of its pressure to allow for nonideal gas behavior at high gas pressures. Equations 10.3 and 10.4 can be derived from statistical mechanics [23, p. 427ff.].

To apply these equations to the reaction of Eq. 10.1 and the equilibrium expression of Eq. 10.2, it is assumed that concentrations can be used for the bracketed quantities in Eq. 10.2 instead of activities, which means that the constituents behave as if they are in ideal solutions. The OH groups in the glass and melts are at low concentrations (dilute solutions), so they should not interact significantly, justifying the assumption of ideality for them. The water molecules dissolve in individual interstices in the glass or melt [5, Chap. 8], so their interactions are shielded by network constituents. If the available interstices for water solubility are exhausted, an additional Langmuir parameter can be introduced [24–26]. The reacting network oxygens are highly immobile, so they do not interact. These assumptions of ideality are contrary to the treatment in terms of a regular solution model.

The usual units of concentration in an equilibrium expression such as Eq. 10.2 are moles or molecules per unit volume. These units derive from the reactions and collisions of molecules in the gas phase. One can define the thermodynamic activity a_i of a component i in the equation (see also Section 2.2)

$$\mu_i = \mu_0 + RT \ln a_i \tag{10.5}$$

The limiting condition is also required, in terms of a reference function $g(z)$ of a reference variable z [27, p. 400ff.]:

$$\lim_{z \to z^*} \frac{a_i}{g(z)} = 1 \tag{10.6}$$

For the present purposes the most convenient limiting function is the concentration C of dissolved water molecules in moles or molecules/volume and the concentration S of OH groups, with the ideal limiting condition being infinite dilution:

$$\lim_{c_i \to 0} \frac{a_i}{c_i} = 1 \tag{10.7}$$

Then the ideal solutions of water molecules and OH groups in glasses and melts are expected, as described above. One can always define the activity in terms of different reference functions, but such definitions can lead to peculiar nonidealities even at low concentrations. Thus the concentrations in this paper are usually molecules per unit volume. In the geological literature the units of OH concentration are given in terms of the number of water molecules required to form the OH groups in Eq. 10.1; this convention is followed in this chapter, so that the actual concentration of OH groups per unit volume is $2S$.

The Langmuir model suggests that the number of reactive sites in the glass is limited and saturates at high water concentration. Thus the factor [—O—] in Eq. 10.2 is replaced by $R-S$. R is the concentration of reactive sites and S is the concentration of reacted OH groups. Then Eq. 10.2 becomes

$$K = \frac{S^2}{C(R-S)} \tag{10.8}$$

in which $C = [H_2O]$ is the concentration of dissolved water and $S = [OH]$ is the concentration of reacted OH groups. A similar equation has successfully described the solubilities of inert gases in fused silica at high pressures [24–26].

10.3 COMPARISON WITH EXPERIMENT

With the assumption of ideality, Eq. 10.2 becomes

$$K = \frac{S^2}{C[4.28(10)^{22} - S]} \tag{10.9}$$

in which $2S$ is the concentration of OH groups, C is the concentration of dissolved molecular water in the glass, and $4.28(10)^{22}$ atoms/cm^3 is the concentration of network oxygen atoms in the glass studied by Sowerby and Keppler [12]. There are two network oxygen atoms for each aluminum and each silicon atom. The concentrations of atoms in their glass are given in Table 10.1. The aluminum ions are four-coordinated in the glass network, leaving an unsatisfied negative charge. This charge is compensated by neighboring alkali or alkaline earth ions, not bound in the glass network. The total oxygen concentration, which is not measured directly but calculated from stoichiometry, is close to double the concentration of silicon plus aluminum ions, showing that all or nearly all oxygen atoms are bonded to two other atoms, either silicon or aluminum. Thus there are few, if any, nonbridging oxygen atoms in this glass.

Values of the equilibrium constant K are shown in Table 10.2. These values of K from ref. 12 decrease by a factor of about 2 as the total water concentration increases from 1 to 4 wt%. Thus the assumption of Eq. 10.9 with simple ideal behavior gives a poor fit to the data, even at low water concentrations. The concentration of reacted OH groups from ref. 12 is plotted as a function of the

Table 10.1 Concentrations of Cations in Silicate Melts (ions/cm^3 × 10^{-21})

Cation	Rhyolite [12]	Haplogranite [21]
Si	17.51	17.55
Al	3.88	3.68
Na	1.33	2.08
K	1.58	1.67
Ca	0.29	
Mg	0.36	

Table 10.2 Calculated Values of the Equilibrium Constant *K* in Eq. 10.9 (Ideal Solution) for 124 to 434°C

Sowerby and Keppler [12, Table 2]		Holtz *et al.* [21], 800°C	
Total wt % Water	*K*, Mean for Six Temperatures	Pressure (bar)	*K* (from Table 10.5)
3.93	0.0148	300	0.26
3.08	0.0163	500	0.27
2.05	0.0247	700	0.36
1.00	0.0308	1000	0.28
		1300	0.34
		1400	0.33
		2000	0.28
		3000	0.27
		4000	0.24
		4800	0.16
		5000	0.18

concentration of dissolved molecular water in rhyolite glass in Fig. 10.1, at temperatures of 124 to 434°C and total water concentration up to 4 wt % at 1 atm external pressure; the composition of their glass is given in Table 10.3. The concentrations in the figure were calculated from the infrared absorption data in Table 2 of Sowerby and Keppler and their measured extinction coefficients of $2.638(10)^5$ and $2.139(10)^5$ cm^2/mol for the OH and H_2O bands, respectively. The line in the figure was calculated from Eq. 10.8 with $R = 1.4(10)^{21}$ molecules/cm^3 OH as H_2O and $K = 1.60$ (dimensionless). The fit is good to within the experimental error of the data, which show little or no temperature dependence over this range of temperatures, implying that the parameters K and R are also independent of temperature in this range. These data are often plotted as S and C separately as a function of total water content, as shown in Fig. 10.2; again the fit is good. In careful experiments, Sowerby and Keppler [12] showed that the extinction

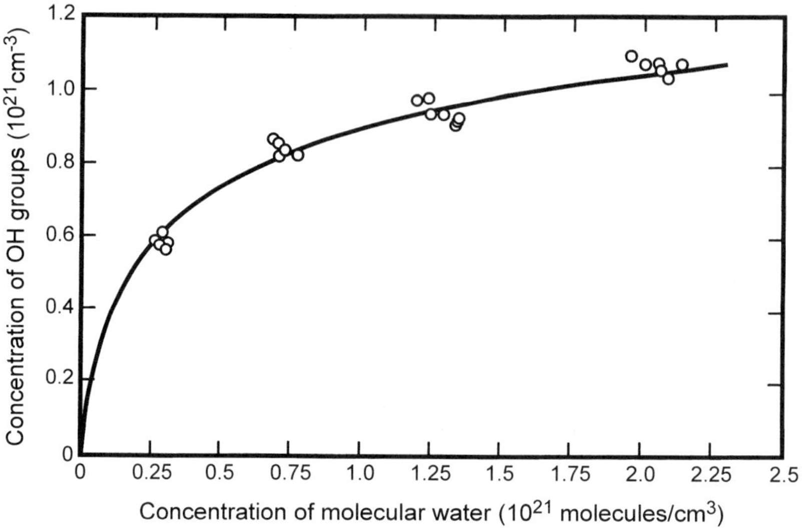

Figure 10.1 Concentration of OH groups as H_2O as a function of concentration of dissolved molecular water at six different temperatures from 124 to 434°C in rhyolite glass, from Table 2 of Ref. 12; line from Eq. 10.8 with $R = 1.4(10)^{21}\,\text{cm}^{-3}$ and $K = 1.60$. (From Ref. 6.)

Table 10.3 Compositions of Glasses That Show Saturation of Reacted OH Groups (mol %)

	Rhyolitic [12]	AOQ [16,21]	Haplogranitic [35–37]	Soda-Potash [13]	Orthoclase [15]	Calcium [15]
SiO_2	81.24	82.54	75.0	76.9	75.3	66.59
Al_2O_3	9.02	8.64	12.5	1.3	12.1	8.79
Na_2O	3.08	4.89	12.5	10.8	—	—
K_2O	3.66	3.93	—	3.0	12.5	—
CaO	1.36	—	—	—	—	24.62
MgO	1.65	—	—	—	—	—
ZnO	—	—	—	1.3	—	—

coefficients for infrared absorption of OH groups (2.22-μm band) and molecular water (1.9-μm band) were independent of temperature up to high temperatures (perhaps 800°C). The good agreement between concentrations at different temperatures, as shown in Figs. 10.1 and 10.2, supports this independence of extinction coefficients with temperatures up to 500°C. At temperatures above the glass transition temperature of about 400°C, for a total water concentration of about 3.8 wt %, Sowerby and Keppler found that the K value of Eq. 10.9, using concentrations for [OH] and [$2H_2O$] and the total of unreacted network oxygen for [—O—], increased as a function of temperature. These authors apparently

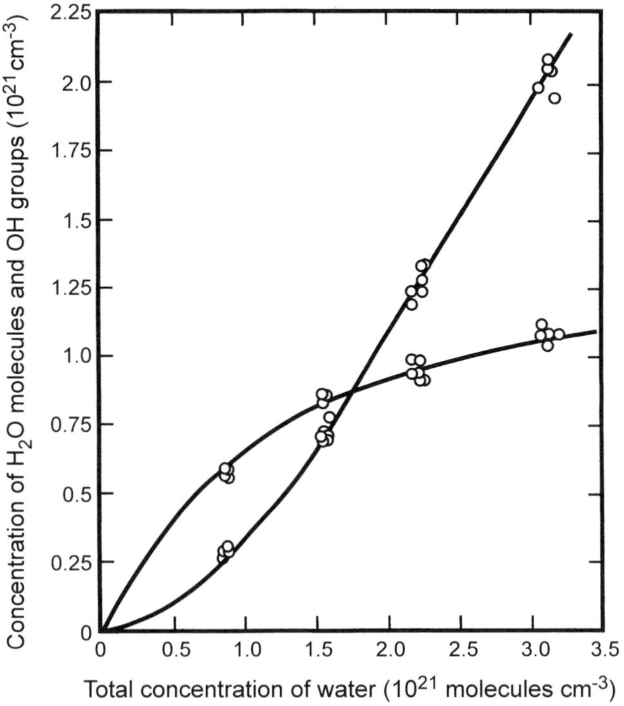

Figure 10.2 Concentrations of OH groups as H_2O and dissolved molecular water as a function of total water concentration in rhyolite glass; lines from Eq. 10.8 with $R = 1.4(10)^{21}\ \mathrm{cm}^{-3}$ and $K = 1.60$. (From Refs. 6 and 12.)

measured the concentrations of OH and H_2O at only one total water concentration for temperatures above 434°C, so it is not possible to calculate K and R values from Eq. 10.8 for these higher temperature data. At 900°C and 10 bar the concentration of reacted OH groups was $4.4(10)^{21}\ \mathrm{cm}^{-3}$, showing that the saturation concentration of reactive sites is much greater at this temperature than at temperatures below the glass transition temperature. Thus both K and R increase as the temperature increases above the glass transition temperature. The value of $R = 2.8(10)^{21}$ OH molecules/cm^3 (2×1.4) is about double the aluminum ion concentration of $1.6(10)^{21}$ atoms/cm^3 in the glass of Sowerby and Keppler. Thus there is apparently no direct relation of the R value to the composition of the glass.

Speciation in a variety of other compositions of silicate glasses has been measured [13–17, 28]. See Table 10.3 for some compositions. In all of these studies it was found that the concentrations of reacted OH groups saturated (became constant) at high total water concentrations. Some values of saturation concentrations from studies of quenched glasses are given in Table 10.4. These values may not be quantitatively exact, because of possible changes during quenching. They can be compared with the value of $R = 2.8(10)^{21}$ OH groups as deduced from the in situ study of Sowerby and Keppler [12]. In the samples quenched from above the

Table 10.4 Saturation Concentrations of OH Groups for Quenched Silicate Glasses (Langmuir Model)

Melt	Saturation Concentration[a] (OH groups/cm^3 × 10^{21})	Ref.
Calcium aluminosilicate	4.4	15
Orthoclase	3.4	15
Jadeite	3.2	15
Albite	3.4	28
Soda-potash	3.5	13

[a]OH concentrations are actual values (two times concentrations as H_2O).

glass transition temperature, an exact quantitative comparison with the Langmuir model is not possible, because rapid reaction during quenching can change the speciation, as mentioned in Section 10.1. It seems likely that the ideal model (Eq. 10.9) is inadequate for describing these experiments. For example, for the measurements of OH and H_2O concentrations of Ihinger *et al.* [17] for rhyolite melts at water concentrations below 4.5 wt % at 500°C, the values of K in Eq. 10.9 decrease by about 35% as the water concentration increases from 1.4 wt % to 4.5 wt %. A constant K value over this range results with the Langmuir model for the data of Ihinger *et al.* [17]. In other studies [14, 28] much larger changes of K were found, especially at high water concentrations.

10.4 SOLUBILITY OF WATER

In all the experiments described in Section 10.3, the speciation of water was studied in samples in which the total water concentration was fixed at different values by treatment in water vapor at high temperature and pressure. In a different kind of experiment [21], the total amount of water incorporated in the glass as a function of temperature and pressure of water vapor is studied. These measurements of total water solubility should not be influenced by quenching problems, because the diffusion of water out of the sample during rapid quench is negligible. These measurements can be examined in terms of the dissolution of molecular water in the glass and the Langmuir model for speciation.

In a variety of silicate glasses it is found that the ratio C_d/C_g, in which C_d is the concentration of molecularly dissolved gas (unreacted) in the glass and C_g is the concentration of this gas in the gas phase, is constant over wide ranges of temperature and gas pressure. In the chemical literature this ratio is sometimes called the *Ostwald solubility* α. Many measurements of helium, neon, hydrogen, and argon in silica glass give a constant ratio over wide ranges of temperature and pressure (see Section 5.3). If the gas behaves nonideally, one must use the ratio

$$\alpha = \frac{C_d}{C_g \gamma} \tag{10.10}$$

in which γ is the activity or fugacity coefficient for the gas:

$$\gamma = \frac{f}{p} \tag{10.11}$$

in which f is the fugacity for the gas pressure p. For molecular water in fused silica, calculations of diffusion coefficients at different temperatures and pressures also imply that the ratio of Eq. 10.10 is constant [6]. The validity of Eq. 10.10 as a function of pressure is supported by models of gas solubility derived from statistical mechanics [5]. The near constancy of α with temperature implies that the dissolved molecules interact very little with the surrounding glass network. Data on the solubilities of inert gases in a variety of glass compositions are summarized by Shelby [29].

The value of α in Eq. 10.10 can be estimated from the infrared absorption of molecular water at temperature and pressure as measured by Nowak and Behrens [16] and Sowerby and Keppler [12], and fugacity of water from Pitzer and Sterner [30] as calculated by Zhang [31]. At 823°C and a pressure of 5 and 4.5 kbar and 800°C and 1.5 kbar, a mean value of α is found to be about 0.10 ± 0.02. The concentrations of dissolved molecular water at different pressures can be calculated from Eq. 10.10 with $\alpha = 0.10$; these concentrations are given in Table 10.5. The total water concentrations are then calculated from the experimental water solubilities measured by Holtz *et al.* [21], and are also listed in the table. These authors did not report melt densities, so a value of 2.3 g/cm^3 was assumed for 800°C. This density was estimated as follows. For an aluminosilicate glass containing 82.5 mol% SiO_2 , balance Al_2O_3, the density is about 2.37 g/cm^3

Table 10.5 Calculated Concentrations of Dissolved Molecular Water and OH Groups at 800°C[a]

Pressure (bar)	Fugacity[b] (bar)	Concentrations (molecules/cm^3 × 10^{-21})		
		Total Water	Molecular Water	OH Groups
300	286	1.638	0.19	1.44
500	461	2.153	0.31	1.84
700	626	2.888	0.42	2.46
1000	858	3.145	0.58	2.57
1300	1078	3.860	0.73	3.1
1400	1150	3.968	0.77	3.19
2000	1592	4.476	1.08	3.40
3000	2407	5.729	1.62	4.10
4000	3387	6.875	2.29	4.59
4800	4330	7.213	2.92	4.29
5000	4541	7.667	3.10	4.57

[a]Calculated from solubility data of Holtz *et al.* [21].
[b]Calculated by Zhang [31] from Pitzer and Sterner [30].

[32, p. 57]. With a thermal expansion coefficient of about $7(10)^{-6}$ [32, p. 132], this value is reduced to $2.34\,\mathrm{g/cm^3}$ at 800°C from Eq. 10.14. At higher pressure above 1 kbar the density decreases [12]. Thus an estimate of the density at 800°C is $2.3 \pm 0.4\,\mathrm{g/cm^3}$. A small change in this value does not influence the conclusions from the following calculations but only shifts all the concentrations by a small constant factor. The density should vary somewhat with pressure, but the exact values are uncertain. The concentration of reacted OH groups can be calculated from the difference between the total water concentration $[H_2O]_t$ and the dissolved molecular water concentration $[H_2O]_m$:

$$[OH] = [H_2O]_t - [H_2O]_m \qquad (10.12)$$

Calculated values of reacted OH concentrations are shown in Table 10.5 and are plotted in Figs. 10.3 and 10.4. They show the same saturation of OH concentration as found in direct speciation measurements. The curves in Figs. 10.3 and 10.4 are calculated from Eq. 10.8 with $K = 3.0$ and $R = 7(10)^{21}$ molecules/cm^3. The fit is within experimental error. The values of K calculated for an ideal solution with $[\text{—O—}] = 4.23(10)^{22}$ in Eq. 10.9 vary by more than a factor of 2, as shown in Table 10.3. These comparisons show that the solubility of water in silicate melts results from the dissolution of molecular water in the melt (Eq. 10.10) and its reaction to form OH groups (Eq. 10.8), and that the reaction saturates (Langmuir model).

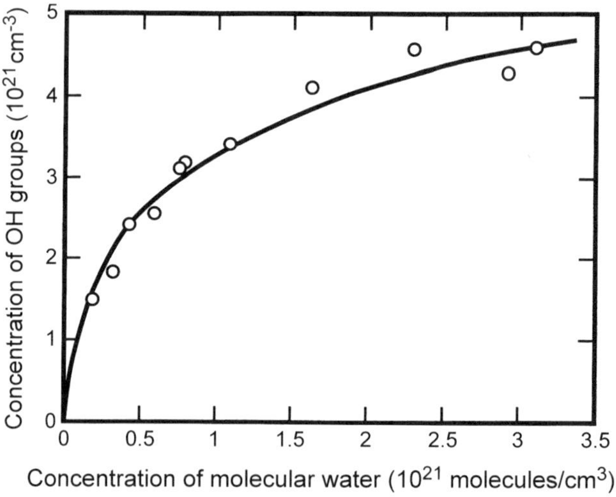

Figure 10.3 Concentration of OH groups as H_2O as a function of concentration of dissolved molecular water at 800°C and different pressures of water vapor, calculated as described in the text; see Table 10.5 for data, from solubility measurements of Holtz *et al.* [21] at 800°C in a haplogranitic melt. Line from Eq. 10.8 with $K = 3.0$ and $R = 7(10)^{21}$ molecules/cm^3. (From Ref. 6.)

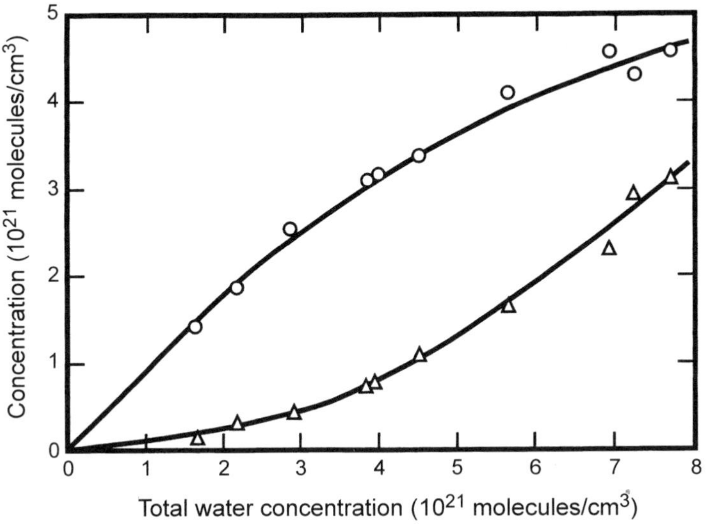

Figure 10.4 Concentrations of OH groups as H_2O and dissolved molecular water as a function of total water concentration; see Table 10.5 for data, from solubility measurements of Holtz *et al.* [21] at 800°C in a haplogranitic melt. Lines from Eq. 10.8 with $K = 3.0$ and $R = 7(10)^{21}$ molecules/cm^3. (From Ref. 6.)

Table 10.6 Solubility of Water in Silica Glass

Pressure (bar)	Fugacity (bar)	Temperature (bar)	Concentration (molecules $H_2O/cm^3 \times 10^{21}$) Total	Molecular	OH Groups
1010	867	1350	1.82	0.06	1.77
1500	1232	1275	2.22	0.09	2.13
1550	1273	1200	2.32	0.09	2.24
2020	1608	1250	2.82	0.11	2.70
2520	2000	1200	3.15	0.15	3.00
4030	3412	1200	3.91	0.25	3.66
4100	3472	1350	4.25	0.23	4.02
6000	5740	1200	5.92	0.42	5.50

Source: Ref. 34.

Moulson and Roberts [33] measured the concentration S of OH groups in silica glass as a function of water vapor pressure from $1.33(10)^4$ Pa (0.132 atm) to $9.30(10)^4$ Pa (0.92 atm). They found that S^2 was proportional to the water vapor pressure, as expected from Eqs. 10.9 and 10.11. At these low water vapor pressures, the fugacity coefficient in Eq. 10.11 is 1. Holtz *et al.* [34] measured the solubility of water in silica glass from a pressure of 100 to 600 MPa (1010 to 6060 bar). Their experimental data are given in Table 10.6 as total water concentration. The

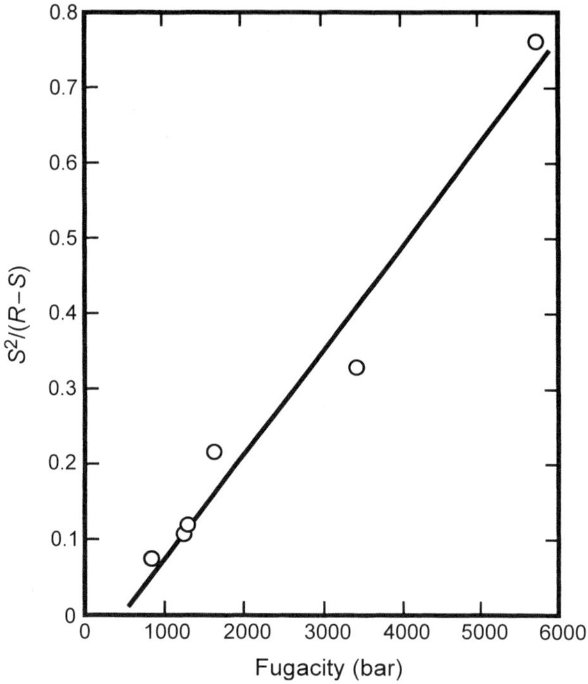

Figure 10.5 Solubility of water in silica glass; data of Holtz *et al.* [34]. $S^2/(R-S)$ from Eq. 10.9 plotted as a function of fugacity of water vapor (see Eqs. 10.10 and 10.11).

concentration of molecular water dissolved in the glass was calculated from Eqs. 10.10 and 10.11 using a value of $\alpha = 0.015$. This value is deduced from measurements of the solubilities of other gases in silica glass, as described in Chapters 5 and 6. Then from Eq. 10.12 the concentration of reacted SiOH groups (S) is just the difference between total and molecular water. In silica glass the concentration of reactive sites R is the concentration of oxygen atoms in the silica network, or $4.41(10)^{22}$ atoms/cm^3. In Fig. 10.5 the term $S^2/[4.41(10)^{22}-S]$ is plotted as a function of fugacity of water vapor. A regression analysis gives the line in the figure with a correlation coefficient of 0.989. Thus Eq. 10.9 describes the solubility of water in silica glass up to a pressure of 600 MPa without any Langmiur correction for limiting reaction sites. The intercept at zero pressure is not zero concentration in Fig. 10.5; the reason for this discrepancy is unclear.

10.5 PARTIAL MOLAR VOLUME

In a thermodynamic system containing several components, partial molar quantities are a convenient way to describe extensive properties of the system. Thus the partial molar volume of a component i in a mixture is defined as the change in total volume of the system with an incremental addition of n_i moles of i to the mixture; the amounts of all other components, as well as temperature and pressure, must remain the same. In mathematical terms,

$$\text{partial molar volume of } i = \left(\frac{\partial V}{\partial n_i}\right)_{T,p,n_j \neq n_i} \tag{10.13}$$

The relationship of partial molar volumes to the total volume V of the system can be found from Euler's theorem on homogeneous functions [27, Chapter 9]. A function $f(x,y,z)$ is homogeneous of degree n in the variables x, y, and z if

$$f(\lambda x, \lambda y, \lambda z) = \lambda^n f(x, y, z) \tag{10.14}$$

for all values of λ. Therefore, an extensive property of a system is a homogeneous function of the amounts of the system, in the first degree.

If the homogeneous function of Eq. 10.14 is differentiated partially with respect to λ:

$$x\frac{\partial f}{\partial(\lambda x)} + y\frac{\partial f}{\partial(\lambda y)} + z\frac{\partial f}{\partial(\lambda z)} = n\lambda^{n-1}f \tag{10.15}$$

Since this equation is valid for all values of λ, it is valid for $\lambda = 1$. With $\lambda = 1$, Eq. 10.15 becomes

$$x\frac{\partial f}{\partial x} + y\frac{\partial f}{\partial y} + z\frac{\partial f}{\partial z} = nf \tag{10.16}$$

which is Euler's theorem on homogeneous functions and is valid for any number of independent variables. If this theorem is applied to the total volume V of a system,

$$V = \sum_i n_i \left(\frac{\partial V}{\partial n_i}\right)_{T,p,n_j \neq n_i} \tag{10.17}$$

Thus the total volume of a multicomponent mixture is the sum of terms of n_i times the partial molar volumes. In principle the partial molar volume can be calculated from measurements of the densities of mixtures of different composition, but such measurements for silicate melts at high temperatures and pressures are difficult. Burnham and Davis [35] measured the densities of melts of $NaAlSi_3O_8$ plus water at temperatures from 700 to 950°C and pressures from 1 to 10 kbar. They were able to measure melts containing 0, 8.25, and 10.9 wt % water. From plots of specific volume (reciprocal of density) as a function of wt % water at constant pressure [35, Figs. 12 and 13] they concluded that the plots were linear from 0 to 10.9 wt % water, so that the partial molar volume of water was constant over this composition range. However, these plots represent a substantial extrapolation from the two high water concentrations to lower concentrations, and some of their plots, especially at high pressures, are actually not linear. Thus it is not certain that the partial molar volumes of water in these melts are constant.

Ochs and Lange [36] measured the densities of cooled melts of $NaAlSi_3O_8$ with different amounts of water at ambient temperature and pressure. They measured the linear coefficients of thermal expansion α_L of the resulting glasses and calculated the densities at the glass transition temperature, where they assumed that glass and melt densities were the same. The result is a series of measurements of density at different temperatures for each different water concentration. Their densities at different temperatures are shown in Table 10.7, with the thermal expansion volume calculated from the relation

$$\frac{1}{\rho} = \frac{1}{\rho_r}[1 + 3\alpha_L(T - 298)] \tag{10.18}$$

in which ρ is the density at temperature T and ρ_r is the density at 298 K, the reference temperature. Terms including higher powers of the temperature difference are negligible.

To calculate partial molar volumes, the melts are considered to be binary mixtures of dry $NaAlSi_3O_8$ and water. Then from Eq. 10.17, dividing both sides by V, we obtain

$$C_A\left(\frac{\partial V}{\partial n_A}\right)_{n_B} + C_B\left(\frac{\partial V}{\partial n_B}\right)_{n_A} = 1 \tag{10.19}$$

at a particular temperature and pressure. Subscripts A and B refer to $NaAlSi_3O_8$ and water, respectively. The partial molar volume of $NaAlSi_3O_8$ is assumed to be constant with water composition. Then $\partial V/\partial n_A = M_A/\rho_0$, where M_A is the molecular weight of $NaAlSi_3O_8$ and ρ_0 is the density of the dry melt. If W is the

Table 10.7 Partial Molar Volumes of Water in $NaAlSi_3O_8$ at 300°C[a]

Weight Fraction Water	Glass Transition Temperature (°C)	Density at T_g (g/cm^3)	Partial Molar Volume (cm^3/mol)
0.0187	397	2.343	5.00
0.0212	334	2.361	3.31
0.0228	419	2.348	2.57
0.0244	403	2.316	8.94
0.0333	331	2.307	10.22
0.0355	309	2.317	9.27
0.0480	341	2.273	11.50
0.0538	232	2.262	12.27
0.0607	257	2.262	11.45
0.0607	254	2.258	11.87

[a] Calculated from densities of Ochs and Lange [36, Table 2] and Eq. 10.21.

weight fraction of water in the melt, then

$$C_A = \frac{(1 - W)\rho}{M_A} \qquad C_B = \frac{W\rho}{M_B} \tag{10.20}$$

in which ρ is the density of the melt containing W fraction of water and $M_B = 18.015$ is the molecular weight of water. Then from Eqs. 10.19 and 10.20,

$$\frac{\partial V}{\partial n_B} = \left(\frac{1}{\rho} - \frac{1 - W}{\rho_0}\right)\frac{M_B}{W} \tag{10.21}$$

Equation 10.21 is valid at a particular temperature and pressure. The pressure is 1 bar and the temperature is the glass transition temperature. Thus values of ρ_0 must be corrected by Eq. 10.18 to T_g. Partial molar volumes of water as calculated from the measurements of Ochs and Lange on a $NaAlSi_3O_8$ melt are given in Table 10.7 and plotted in Fig. 10.6. To compare these values at the same temperature of 300°C, a small temperature correction was made based on the measurements of Burnham and Davis [35]. The resulting partial molar volumes of water in $NaAlSi_3O_8$ melts are constant with water content above about 0.04 wt fraction water [about $3(10)^{21}$ molecules/cm^3] at an average value of 12.9 cm^3/mol, but at lower water contents the partial molar volumes decrease.

Ochs and Lange [37] also measured densities of a potassium calcium silicate glass (59.5 mol % SiO_2, 30.5% CaO, and 10.0% K_2O) and a rhyolite glass

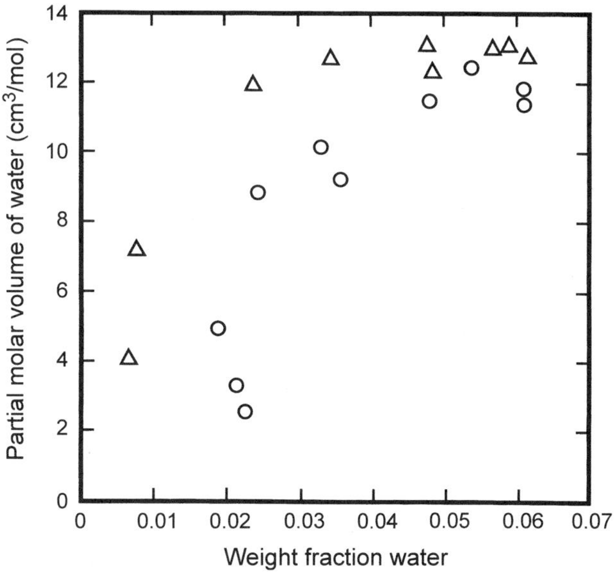

Figure 10.6 Partial molar volume of water in silicate melts at 300°C as a function of total water weight fraction from the, data of Ochs and Lange [36,37] (see Tables 10.7 and 10.8). ○, $NaAlSi_3O_8$; △, 0.1 K_2O, 0.305 CaO, 0.595 SiO_2.

Table 10.8 Partial Molar Volumes of Water in Potassium Calcium Silicate and Rhyolite Glasses at 300°C[a]

Weight % Total Water	Glass Transition Temp. T_g (°C)	Density at T_g (g/cm^3)	Partial Molar Volume (cm^3/mol H_2O)
	Potassium Calcium Silicate 59.5 mol% SiO_2, 30.5% CaO, 10.0% K_2O		
0	697	2.497	
0.65	485	2.575	4.1
0.75	429	2.580	7.2
2.38	367	2.554	11.9
3.41	324	2.533	12.7
4.75	282	2.505	13.1
4.85	277	2.516	12.3
5.64	257	2.490	13.0
5.89	235	2.485	13.1
6.14	246	2.463	12.8
	Rhyolite Glass 83.6 mol% SiO_2, 8.0% Al_2O_3, 0.7% CaO, 4.3% Na_2O, 3.4% K_2O		
0.05	817	2.324	
1.29	422	2.259	15.78
1.91	317	2.232	19.8
2.06	355	2.232	18.4
2.57	332	2.239	15.5
4.71	258	2.203	15.4
5.30	222	2.191	15.6
5.92	175	2.133	15.8

[a]Calculated from densities of Ochs and Lange [37] and Eq. 10.21.

(83.6% SiO_2, 8.0% Al_2O_3, 0.7% CaO, 4.3% Na_2O, and 3.4% K_2O). Partial molar volumes calculated for these densities at 300°C and 1 atm pressure are given in Table 10.8. The values for the potassium calcium silicate glass show the same trend as those for albitic glass in Table 10.7 and Fig. 10.6; at water concentrations above about 3 wt % the partial molar volume is constant at an average value of 12.8 cm^3/mol and decreases at lower water contents. However, for the rhyolitic glass there is no evidence for a decrease in partial molar volume down to a water content of 1.29 wt %.

The decrease in partial molar volume of water at low water contents can be understood from the measurements of speciation of water in silicate melts. At low water contents most of the water reacts with the oxygen network to form OH groups. This reaction breaks up the network and makes room for the OH groups, so that the specific volume is not increased much. At higher water contents, more water is dissolved as molecular water, which leads to an increase in the total volume and the constant partial molar volume of about 12.8 cm^3/mol. In the rhyolite glass less water reacts to form OH groups, which is more like silica glass.

10.6 DISCUSSION

The comparisons above show that the solubility and speciation of water in silicate glasses and melts can be understood as resulting from the dissolution of molecular water from the gas phase into the melt of glass (Eq. 10.10), and the reaction of this molecular water with the oxide network by Eq. 10.1. The number of reactive sites is limited, so the amount of reacted OH groups saturates at high water concentrations (Langmuir model, Eq. 10.8). The measurements of Sowerby and Keppler [12] show that the equilibrium constant and the number of reactive sites are constant at up to the glass transition temperature, but both increase with increasing temperature above this temperature.

Mott [38] has suggested that the viscosity of network liquids such as silicates results from the formation of network defects (broken oxygen bonds) in the network. As the number of defects increases, flow is easier. The glass transition can be considered to occur at the temperature at which the melt has a particular viscosity of 10^{12} Pa-s. Below this temperature there are few defects, and the glass structure relaxes slowly compared to experimental times; the structure is "frozen in." Above the glass transition temperature there are an increasing number of defects as the temperature increases, and flow becomes easier as the temperature increases. Water leads to network defects (broken bonds) that cause lower melt viscosity and a lower glass transition temperature. As the temperature increases above the glass transition temperature, the ease of breaking bonds by reaction 1 increases, as reflected in the increase in the equilibrium constant and the number of reactive sites.

Structural features of the reaction of water with $Na_2O{\cdot}4SiO_2$ glasses have been revealed with Raman, infrared, and nuclear magnetic resonance spectroscopies by Zotov and Keppler [39]. Their results show that reaction 1 occurs preferentially at bridging oxygen atoms bonded to two silicon atoms bonded to four other silicon–oxygen tetrahedra (Q_4 bonding). Only at high water contents (above about 6 wt % water) does the reaction start to occur at bridging oxygens bonded to a Q_3 group (which is bonded to three silicon–oxygen tetrahedra). Thus a nonbridging oxygen on a silicon atom seems to make it less reactive. These results are consistent with the speciation measurements; the reaction of Eq. 10.1 occurs preferentially at particular bridging oxygens, giving constant K and R values as a function of water content. Measurements of speciation, solubility, and partial molar volume of water in silicate melts and glasses provide a consistent set of parameters that can be understood in terms of molecular solubility and the Langmuir limited-site model.

REFERENCES

1. R. H. Doremus, in *Reactivity of Solids*, J. W. Mitchell *et al.*, eds., Wiley, New York, 1969, p. 667.
2. F. M. Ernsberger, *J. Am. Ceram. Soc.* **60**, 91 (1992).

3. R. F. Bartholomew, in *Treatise on Materials Science and Technology*, Vol. 22, M. Tomozawa and R. H. Doremus, eds., Academic Press, San Diego, CA, 1982, p. 75.
4. E. M. Stolper, *Geochim. Consmochim. Acta* **46**, 2609 (1982).
5. R. H. Doremus, *Glass Science*, 2nd ed., Wiley, New York, 1994.
6. R. H. Doremus, *Am. Min.* **85**, 1674 (2000).
7. S. C. Kohn, R. Dupree, and M. Golan-Mortuza, *Chem. Geol.* **96**, 399 (1992).
8. D. Sykes and J. D. Kubicki, *Geochim. Cosmochim. Acta* **57**, 1039 (1993).
9. S. C. Kohn, M. E. Smith, and R. Dupree, *Geochim. Cosmochim. Acta* **58**, 1377 (1994).
10. D. Sykes and J. D. Kubicki, comment by S. C. Kohn, M. E. Smith, and R. Dupree on *Geochim. Cosmochim. Acta* **58**, 1381 (1994).
11. P. Richet and A. Polian, *Science* **281**, 396 (1998).
12. J. R. Sowerby and H. Keppler, *Am. Mineral.* **84**, 1843 (1999).
13. R. F. Bartholomew, B. L. Butler, H. L. Hoover, and C. K. Wu, *J. Am. Ceram. Soc.* **63**, 481 (1980).
14. E. M. Stolper, *Am. Mineral.* **74**, 1247 (1989).
15. L. A. Silver, P. D. Ihinger, and E. Stolper, *Contrib. Mineral. Petrol.* **104**, 142 (1990).
16. M. Nowak and H. Behrens, *Geochim. Cosmochim. Acta* **59**, 3445 (1995).
17. P. D. Ihinger, Y. Zhang, and E. M. Stolper, *Geochim. Cosmochim. Acta* **63**, 3567 (1999).
18. D. B. Dingwell and S. L. Webb, *Eur. J. Mineral.* **2**, 427 (1990).
19. Y. Zhang, E. M. Stolper, and P. D. Ihinger, *Am. Mineral.* **80**, 593 (1995).
20. Y. Zhang, E. M. Stolper, and G. J. Wasserburg, *Geochim. Cosmochim. Acta* **55**, 441 (1991).
21. F. Holtz, H. Behrens, D. B. Dingwell, and W. Johannes, *Am. Mineral.* **80**, 94 (1995).
22. I. Langmuir, *J. Am. Chem. Soc.* **40**, 1361 (1918).
23. R. Fowler and E. A. Guggenheim, *Statistical Thermodynamics*, Cambridge University Press, Cambridge, 1939, p. 427.
24. R. M. Barrer and D. E. W. Vaughn, *Trans. Faraday Soc.* **63**, 2275 (1967).
25. J. E. Shelby, *J. Appl. Phys.* **47**, 135 (1976).
26. J. F. Shackelford and J. S. Masaryk, *J. Non-Cryst. Solids* **30**, 127 (1978).
27. F. J. Wall, *Chemical Thermodynamics*, 3rd ed., W.H. Freeman, San Francisco, 1974, Chap. 19.
28. L. Silver and E. M. Stolper, *J. Petrol.* **30**, 667 (1989).
29. J. E. Shelby, *Handbook of Gas Diffusion in Solids and Melts*, ASM International, Materials Park, OH, 1996.
30. K. S. Pitzer and S. M. Sterner, *J. Chem. Phys.* **101**, 3111 (1994).
31. Y. Zhang, *Rev. Geophys.* **37**, 493 (1999).
32. N. P. Bansal and R. H. Doremus, *Handbook of Glass Properties*, Academic Press, San Diego, CA, 1986.
33. A. J. Moulson and J. P. Roberts, *Trans. Faraday. Soc.* **57**, 1208 (1961).
34. F. Holtz, J. Roux, H. Behrens, and M. Pichavant, *Am. Mineral.* **85**, 682 (2000).

35. C. W. Burnham and N. F. Davis, *Am. J. Sci.* **267A**, 70 (1971).
36. F. A. Ochs and R. A. Lange, *Contrib. Mineral. Petrol.* **129**, 155 (1997).
37. F. A. Ochs and R. A. Lange, *Science* **283**, 1314 (1999).
38. N. F. Mott, *Philos. Mag.* **B56**, 257 (1987).
39. N. Zotov and H. Keppler, *Am. Mineral.* **83**, 823 (1998).

11

HYDROGEN IN OXIDES

11.1 INTRODUCTION

Hydrogen reacts with oxides to form hydroxyl and hydride groups. As an example, hydrogen reacts with the silicon–oxygen–silicon bridge in a silicate glass:

$$\text{Si—O—Si} + H_2 = \text{SiOH} + \text{HSi} \tag{11.1}$$

The hydroxyl group has been identified from optical absorption in the near infrared; the main band is at about 3650 cm^{-1} (2.74 μm wavelength). The SiH group absorbs at about 2250 cm^{-3} (4.44 μm wavelength). Hydrogen can also reduce an ion in a solid or melt that has multiple oxidation states. The reduction of trivalent iron to divalent in a silicate can be written as

$$2Fe^{+3}O + H_2 = 2Fe^{+2}OH \tag{11.2}$$

A possible reaction sequence for iron in a silicate is that in the trivalent state iron substitutes for silicon in the silicate network or structure. In this four-coordinated position the iron has an extra negative charge, which must be compensated by a positive ion such as sodium or hydrogen. Reduction of the iron to the divalent state leads to the equivalent of two ionic bonds between oxygen and iron, with the two other oxygen ions bonded to hydrogen ions. The iron does not change its position in the structure; in its divalent state it links the two oxygen ions and can coordinate to two or four additional oxygens. In a glass the bonding oxygen ions are called nonbridging. Energetically, the scheme of Eq. 11.2 is favorable because the atoms do not change positions. The details of this reaction depend on the sites at which particular ions occur in crystalline and amorphous structures; Eq. 11.2 is given as one possible example.

Traditionally, the diffusion of hydrogen in an oxide has been described by motion of hydrogen or hydroxyl ions through the structure. In this book the mechanism emphasized is the solution of molecular hydrogen in the oxide and its motion by jumping between interstitial positions in the oxide structure. Experimental support for this view comes from the pressure dependence of diffusion coefficients, from profiles of hydrogen as compared to those derived from the diffusion–reaction model, and from absolute values of the diffusion coefficients and their activation energies for different diffusing species. In the diffusion of molecular species one expects a dependence on the molecular size and on the openness of the solid or melt structure;these dependencies are found experimentally for diffusion of hydrogen in oxides.

Titles of articles on diffusion of hydrogen, water, and oxygen in oxides can be misleading. Often, the kind of diffusion is presumed to follow from the isotope or atom analyzed. Thus diffusion from water into an oxide is called hydrogen diffusion if the hydrogen in the water is measured, and oxygen diffusion if the oxygen (isotope) is measured. In this book the focus is on the diffusing species. *Hydrogen diffusion* means diffusion of molecular hydrogen (H_2); water diffusion means solution and diffusion of water molecules. *Water diffusion* can be followed by measuring hydrogen, deuterium, and oxygen isotope concentrations. *Oxygen diffusion* means diffusion of molecular oxygen or some ion containing oxygen in the solid or melt.

In Section 11.2, diffusion of hydrogen in silica glass is discussed. This subject has become of great practical importance, because hydrogen diffuses rapidly enough in optical fibers of silica glass at room temperature to cause reactions that lead to optical absorption that can degrade the performance of the fibers.

In Section 11.3 the diffusion of hydrogen into various silicate glasses with a variety of reactions, including reduction, colloid formation, and reaction with defects is discussed. In Section 11.4 the diffusion and reaction of hydrogen in some crystalline minerals such as olivine and diopside is treated.

11.2 HYDROGEN IN SILICA GLASS

Molecular hydrogen and deuterium dissolve and diffuse in silica glass with a diffusion coefficient D in cm^2/s by [1–4]

$$D = 5.6(10)^4 \exp\left(-\frac{5240}{T}\right) \tag{11.3}$$

which corresponds to an activation energy of 43.5 kJ/mol. This activation energy fits closely with those for diffusion of other gases in silica glass as a function of molecular size, as shown in Table 5.2 and Fig. 5.4. Deuterium has close to the same dependence on temperature as given by Eq. 11.3 with a value about 0.79 times smaller than that for hydrogen [1–4]. This ratio is not far from the theoretical one of 0.73. Diffusion coefficients for the reactive gases hydrogen, water, and oxygen in silica glass are compared in Table 11.1. The values at 25 and 200°C for oxygen represent a long extrapolation from experimental data at higher temperatures. The equation for oxygen is based on only two experimental measurements of the diffusion coefficient by Norton [7], although these measurements are consistent with more measurements of permeation rates.

The data for diffusion of hydrogen in silica are accurate enough to show that a preexponential temperature factor is needed in Eq. 11.3 [4, 5, Section 5.3]. Thus a more accurate equation is

$$D = 1.9(10)^{-7} T \exp(-4093T) \tag{11.4}$$

in which the units of $1.9(10)^{-7}$ are cm^2/s-deg. However, for most purposes Eq. 11.3 is adequate. A preexponential temperature factor is also needed for diffusion of helium and neon in silica [5, p. 132ff.], and probably also for water (Section 6.3).

In their measurements of diffusion coefficients of hydrogen from the input and decay transients during permeation measurements (Chapter 2), Lee and his collaborators found that these transients did not fit those expected for simple diffusion [1–3]. An example is shown in Fig. 11.1 for evolution of hydrogen from the outlet side of a silica tube after the gas on the inlet side is pumped out. At long times the rate of evolution is higher than expected from simple diffusion, and the deviation becomes more marked as the temperature increases. The explanation for this result can be understood in terms of the reaction of the dissolved hydrogen with the silica by Eq. 11.1. When hydrogen is introduced into the input side of the permeation tube, there is a delay before steady-state permeation is reached that is much longer than expected for simple diffusion. During this time hydrogen is reacting by Eq. 11.1 with the silicon–oxygen network. When this reaction is complete, the hydrogen permeates through the silica at the expected rate, without further removal by reaction. When the hydrogen is pumped out of the inlet side of the tube, the dissolved molecular hydrogen diffuses out of the tube rapidly, as expected. At longer times the reacted hydrogen begins to recombine to molecular hydrogen, so the rate of permeation of hydrogen out of the tube is greater than expected from simple diffusion only.

A complete mathematical treatment of this situation is difficult. The initial profile of reacted hydrogen is linear from the input side to zero at the output surface. After pumping out the hydrogen on the inlet side, recombined hydrogen escapes from both sides of the tube. Lee *et al.* found a linear dependence of the logarithm of the escape rate as a function of time. If the reaction is first order, the concentration at any point decreases logarithmically with time; thus it is possible that the reaction rate is first order. If the slopes of the log rate versus time curves are a measure of the rate of reaction, the results at temperatures of 950 and 700°C give an activation energy of roughly 70 kJ/mol. Lee [2] reported measurements of the exchange rate of deuterium for the reacted hydrogen;the details of these measurements were not provided. He reported an activation energy of about 52 kJ/mol for this exchange.

Table 11.1 Diffusion Coefficients of Molecules in Silica Glass

		D (cm^2/s) at:				Mean Diffusion Distance $\sqrt{2Dt}$ (cm) in a Day [t=8.64(10)^4 s]		
Molecule	D (cm^2/s)	25°C	200°C	1000°C	Refs.	25°C	200°C	1000°C
H_2	$5.6(10)^{-4}\exp(-5240/T)$	$1.2(10)^{-11}$	$8.6(10)^{-9}$	$9.2(10)^{-6}$	1–4	$1.4(10)^{-3}$	0.039	1.3
H_2O	$1.7(10)^{-4}\exp(-8.770/T)$	$2.8(10)^{-17}$	$1.5(10)^{-12}$	$1.7(10)^{-7}$	6	$2.2(10)^{-6}$	$5.1(10)^{-4}$	0.29
O_2	$2.9(10)^{-4}\exp(-13600/T)$	$5.2(10)^{-20}$	$1.0(10)^{-16}$	$6.9(10)^{-9}$	7	$9.5(10)^{-8}$	$4.2(10)^{-6}$	0.035

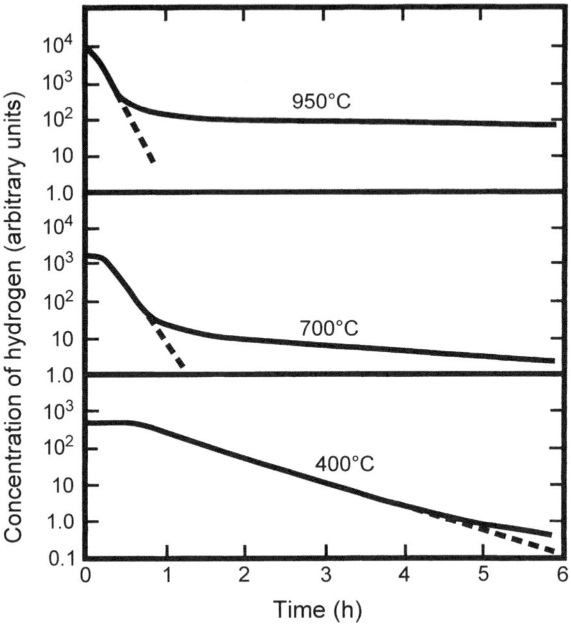

Figure 11.1 Log of evolution rate of hydrogen from silica glass after a permeation experiment as a function of time at different temperatures. Solid lines, experimental; dashed lines, simple diffusion. (Data from Ref. 1.)

The experiments of Lee *et al.* show that the rate of reaction of hydrogen is slow compared to diffusion for their silica samples. Below about 400°C the rate of exchange is barely measurable, and even at 950°C the reacted hydrogen is recombining at a measurable rate only after about 30 min. This situation is very different from the reaction rates of water with silica glass described in Chapter 6. The reaction is

$$\text{Si—O—Si} + H_2O = \text{SiOH} + \text{HOSi} \tag{11.5}$$

This reaction rapidly goes to equilibrium above about 600°C and precedes at an appreciable rate even at 90°C. Thus water is much more reactive with the silicon–oxygen bonds than hydrogen is.

The reaction of water with silica glass does not depend much on the impurities or defects in the glass, apparently because Eq. 11.5 occurs at the uniform and high concentration of bridging oxygen ions. In contrast, the reaction of hydrogen with silica depends strongly on the concentrations of impurities and defects in the glass. In Table 6.1 four types of commercially available silica glass are listed. Lee *et al.* first studied type I silica (GE 204). Presumably, the reaction of hydrogen was associated with an impurity in the silica, probably aluminum.

Thus the total number of reactive sites is limited in this material. The results of Lee *et al.* [1–3] show that the reaction was slow compared to diffusion, so that the system was far from the condition of local equilibrium. Lee *et al.* also measured the permeation and diffusion of hydrogen in Spectrosil silica (type III). In this material the extent of reaction was much smaller than in type I silica, probably because of the lower aluminum concentration in this silica. There was still exchange of deuterium and hydrogen in the silica in OH groups, which are present at high concentration in type III silica. Shelby [4] found no reaction of hydrogen with type III silica.

In type IV silica, hydrogen reacts to form only SiOH groups with no SiH groups. Shelby explained this result [8] from the reaction of hydrogen with residual oxygen in this silica. The manufacturing process eliminates hydroxyl groups but leaves excess dissolved molecular oxygen. This oxygen is immobile compared to the hydrogen (see Table 11.1) and reacts with the hydrogen to form water, which then forms SiOH groups by Eq. 11.5. This reaction occurs rapidly even at low temperatures, in marked contrast to the slow reaction (Eq. 11.1) of hydrogen with the silicon–oxygen network.

The small concentration of residual immobile molecular oxygen acts as a fixed concentration of reactive sites. Thus the appropriate mathematical description of this situation gives a sharp boundary between silica with OH groups silica without these groups. This boundary moves into the silica glass proportional to the square root of time (see Fig. 11.2). Shelby [8] suggested the equation describing this

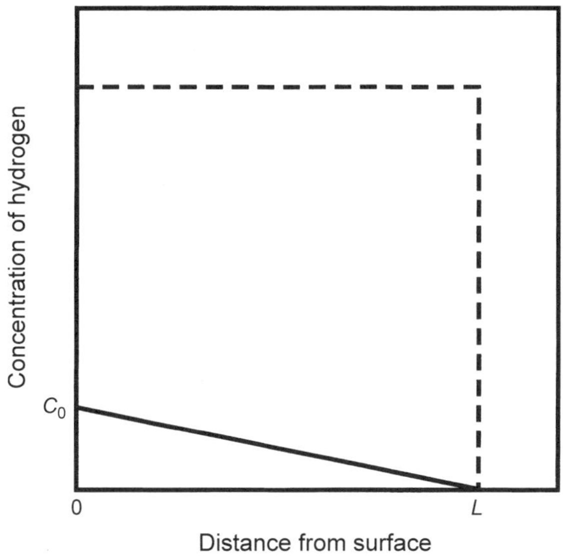

Figure 11.2 Schematic concentration profiles of reacted OH groups and dissolved molecular hydrogen reacting at a fixed number of sites. Solid line, concentration of dissolved molecular hydrogen; dashed line, concentration of reacted OH groups (see Eq. 11.6).

boundary motion (see Eq. 3.31):

$$L^2 = \frac{2C_0Dt}{P} \tag{11.6}$$

in which L is the thickness of the reacted layer, C_0 the concentration of dissolved hydrogen at the glass surface, D the diffusion coefficient of hydrogen in the glass, P the concentration of oxygen molecules dissolved in the glass, and t the time of diffusion. The *tarnishing reaction* [9; 10, p. 305 ff.] of Eq. 11.6 is valid when $C_0/P \ll 1$, as is the case here. Shelby estimated the concentration P of oxygen to be about $2.7(10)^{18}$ molecules/cm^3, and the solubility of hydrogen is about $1.7(10)^{17}$ molecules/cm^3 at 800°C [8]. Thus the condition $C_0/P \ll 1$ is met, and Eq. 11.6 is valid. Shelby found a value of C_0D close to that derived from permeation measurements for hydrogen. Thus when there are oxygen molecules dissolved in the glass, they react rapidly with hydrogen by the reaction

$$2H_2 + O_2 = 2H_2O \tag{11.7}$$

and the water formed reacts with the silicon–oxygen network by Eq. 11.5 to form SiOH groups, which is also a fast reaction. See also Section 3.5.

Hydrogen also reacts with certain point defects in silica glass. As an example, consider the aluminum–oxygen hole center [AOHC]. This center can be formed during processing of vitreous silica or by irradiation by ionizing radiation such as γ-rays or x-rays. This center consists of a hole (missing electron) on the oxygen atom bridging silicon and aluminum atoms [11]. When four-coordinated aluminum substitutes for silicon in the silicon–oxygen network of a silicate, it has a net negative charge, because it supplied only three bonding electrons. Normally, this net negative charge is compensated by a positive ion such as an alkali or hydrogen, but the AOHC supplies this missing positive charge, so the positive ion is not needed. Hydrogen reacts with this center to form an OH group on the bridging oxygen:

$$\text{Si—O—Al—O} + \tfrac{1}{2}\,H_2 = \text{SiO—Al—OH} \tag{11.8}$$

The AOHC has an optical absorption band in the visible at about 0.55 μm, which gives a purple color to the silica glass. Equation 11.8 removes the color. Since there are only a limited number of defect centers, there is a sharp edge between the regions containing defects and that without them, so the conditions for Eq. 11.6 are fulfilled and the color boundary moves into the silica proportional to the square root of time [12]. The process is controlled by the permeation of hydrogen; the value of C_0D from Eq. 11.6 is just that expected from permeation measurements of hydrogen.

11.3 HYDROGEN IN FIBER OPTIC CABLES OF SILICA

Hydrogen in silica cables can cause optical absorption at the wavelengths of use in the cables, which degrades their performance. This absorption results from two different factors. The first is absorption of dissolved molecular hydrogen in the silica cable. In the gas phase the symmetrical hydrogen molecule does not absorb in the infrared, but when it is dissolved in the silica glass hydrogen interacts with the surroundings and has a fundamental absorption band at 2.42 μm ($4130\,cm^{-1}$) in the infrared [13,14]. The first overtone at 1.24 μm is in the wavelength range of fiber signals. Thus dissolved hydrogen must be minimized to reduce this absorption. See Refs. 15 and 16 for reviews and additional references.

Hydrogen can also react to form hydroxyl groups in the cables. This reaction occurs at impurities or defects in the silica, as described in Section 11.2. Fiber optic cables contain added germanium oxide to control the refractive index of the fiber, and hydrogen can react at germanium sites by a reaction similar to Eq. 11.1:

$$\text{Ge—O—Ge} + H_2 = \text{GeOH} + \text{HGe} \tag{11.9}$$

The GeOH group has an overtone at 1.41 μm wavelength, as compared to the SiOH group at 1.39 μm, and the GeH group has a fundamental absorption band at 4.59 μm ($2180\,cm^{-1}$) wavelength, whereas the SiH band is at 4.37 μm ($2290\,cm^{-1}$), so in principle it is possible to distinguish between hydroxyl and hydride groups bonded to silicon or germanium, although sometimes these bands overlap.

Experiments with the purest silica glass (see Section 11.4 and Ref. 15) show that the reaction of hydrogen to form SiOH and SiH (Eq. 11.1) is slow or negligible, except possibly at melting temperatures above 1800°C (see below). Thus formation of hydroxyl groups usually occurs only at impurity sites. In highly pure silica fibers containing germanium oxide, the extent of reaction depends strongly on the amount of phosphorus (P_2O_5) in the fiber. With no phosphorus, the reaction (Eq. 11.9) is negligible at 200°C; as phosphorus is added to the fiber the reaction becomes quite rapid at 200°C [15–17]. The exact mechanism of the influence of phosphorus is uncertain. Presumably, a phosphorus group in the silicon–oxygen network increases the reaction rate at neighboring sites:

$$\text{Ge—O—Ge—O—P} + H_2 = \text{GeOH} \cdot \text{HGe—O—P} \tag{11.10}$$

Germanium hydride (GeH) groups form in irradiated silica glass [18,19]; it is not clear if they form in the presence of neighboring phosphorus bonds. There are also studies of the GeOH and GeH bands in unirradiated silica [20–23].

There have been few measurements of OH profiles in hydrogen-treated silica. The profile of OH in a high-purity silica rod containing 1.6 wt% P_2O_5 was measured with microbeam infrared absorption [24, 25] after treatment in hydrogen at temperatures from 200 to 900°C. The profiles at 900 and 700°C are shown in Fig. 11.3 and show the shape expected for an effective diffusion coefficient D_e proportional to concentration, as modified by the cylindrical geometry (see Section 2.6 and Eq. 3.12):

$$D_e = \frac{2SD}{K_2} \tag{11.11}$$

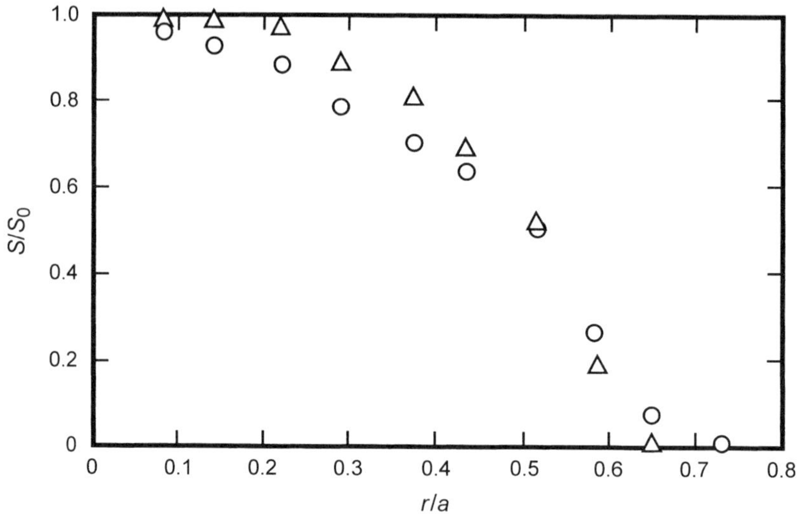

Figure 11.3 Profiles of concentration S of OH groups in high-purity silica glass containing 1.6 wt % P_2O_5 after treatment in 1 atm hydrogen at 700°C for 48 h (○), and 900°C for 19 h (△). *r/a*, distance *r* into the fiber divided by fiber radius *a*. S_0 is the hydrogen concentration at the fiber surface. (From Ref. 25.)

in which S is the concentration of OH groups (as H_2O), D the diffusion coefficient of molecular hydrogen (which is not a function of concentration), and K_2 the equilibrium constant for reaction:

$$K_2 = \frac{S^2}{C} \tag{11.12}$$

in which C is the concentration of molecular hydrogen. These profiles show zero concentration of OH groups at a finite depth into the fiber, just as expected for a diffusion coefficient proportional to concentration (Eq. 11.11). This edge at zero concentration moves into the fiber proportional to the square root of time except at long times. Agreement with the effective diffusion coefficient of Eq. 11.11 is evidence for local equilibrium between dissolved and reacted hydrogen (Eq. 11.12). At lower temperature the OH concentrations at the fiber surface were lower than the equilibrium solubility, and the profiles did not show a finite intercept of the depth axis, but approach this axis asymtotically, as shown in Fig. 11.4. This situation is expected when the reaction is not at local equilibrium (see Chapter 4). These profiles are therefore in close agreement with trends expected from the diffusion–reaction model described in Chapters 3 and 4.

The change to no local equilibrium below about 650°C is strikingly similar to the diffusion–reaction behavior of water in silica glass (Chapter 6). Two possibilities are:

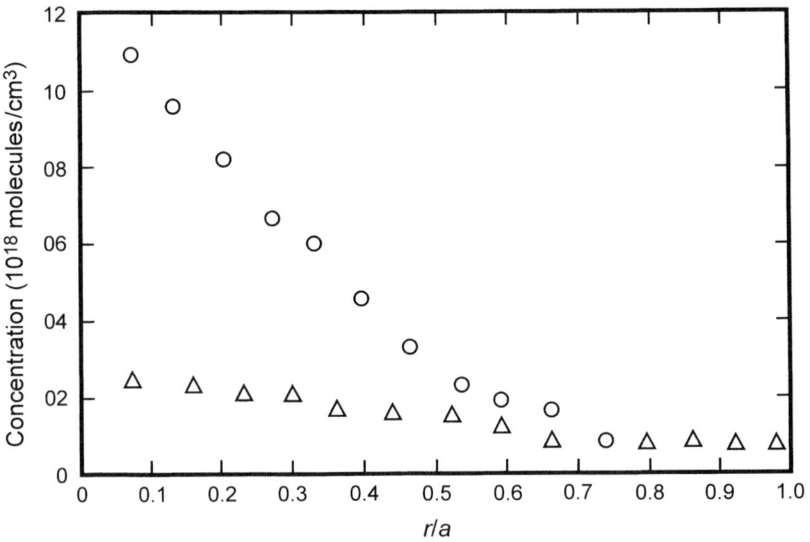

Figure 11.4 Profiles of concentration of OH groups in high-purity silica glass containing 1.6 wt % P_2O_5 after treatment in 1 atm hydrogen at 588°C (○) and 500°C (△) for 48 h. *r/a*, distance or into the fiber divided by fiber radius *a*. (From Ref. 25.)

1. The reaction of hydrogen at phosphorus groups has the same temperature dependence as the reaction of water at Si–O–Si bonds, by coincidence. Four-coordinated phosphorus can substitute in the silicon–oxygen lattice, but one oxygen bond is double. One possibility is

$$\text{Si—O—P—O—Si} + H_2 = \text{SiH} + \text{HOPOSi} \tag{11.13}$$

The reaction with phosphorus might go as depicted in Eq. 11.13, but there is no direct spectral evidence for it that I have found.

2. A second possibility is that the processing of silica to incorporate phosphorus leads to some residual oxygen in the silica, and this oxygen reacts with the hydrogen to form water by Eq. 11.6, and the water then reacts with the silica to form OH groups by the reaction

$$\text{Si—O—Si} + H_2O = \text{SiOH} + \text{HOSi} \tag{11.14}$$

The change to local equilibrium above 650°C is strong evidence for this mechanism.

To summarize the reactions discussed in this section and Section 11.2, rates of different reactions of hydrogen with groups in silica glass are compared in Table 11.2. Of course, temperature speeds all reactions; the comparisons of Table 11.2 are effectively at a particular temperature.

Table 11.2 Qualitative Comparison of Reaction Rates of Hydrogen with Different Groups in Silica Glass

Rapid reaction:
Aluminum oxygen hole center
Phosphorus above 650°C (possibly water is reacting)
Defects from irradiation
Germanium with phosphorus
Slower reaction:
Aluminum in silicon site
Phosphorus below 650°C (possibly water is reacting)
Very slow reaction:
Si–O–Si bonds in pure silica

11.4 HYDROGEN IN MULTICOMPONENT SILICATE GLASSES

Hydrogen molecules dissolve and diffuse in multicomponent silicate glasses just as in silica glass, as summarized by Shelby [26]. The activation energy for hydrogen diffusion depends mainly on the amount of glass former (SiO_2, B_2O_3) in the glass composition, but there are also specific effects; for example, diffusion seems to be slowed by alkaline earth ions. It seems likely that there is a correlation between the ionic porosity (open space) in the glass structure and the activation energy for diffusion, just as found for water in a wide variety of oxides (see Chapter 9), but there are no data over a wide range of porosity to test this possibility.

Hydrogen reacts with ions in these glasses, to reduce them to a lower valence state or to neutral atoms. These atoms then diffuse together to form colloidal particles. Some examples of these reduction reactions are shown in Table 11.3. In a detailed study of the reduction of silver ions in soda-alkaline earth silicate glasses, Barton and Morain [27] showed from their yellow color that the reduced silver formed colloids. After the reduction reaction to silver atoms, these atoms diffused

Table 11.3 Reaction of Hydrogen with Ions in Multicomponent Silicate Glass by Oxidation–Reduction

Ion	Reaction	Glass Type	Refs.
Silver	$Ag^{+} \rightarrow Ag^{0}$	Soda-alkaline earth	27
Iron	$Fe^{3+} \rightarrow Fe^{2+}$	Soda-lime	28
Arsenic	$As^{5+} \rightarrow As^{3+}$	Potassium	29, 30
	$As^{3+} \rightarrow As^{0}$	borosilicate,	
Antimony	$Sb^{5+} \rightarrow Sb^{3+}$	potassium	30
	$Sb^{3+} \rightarrow Sb^{0}$	borate	
Bismuth	$Bi^{3+} \rightarrow Bi^{0}$		30

to nucleation sites to form silver particles. The amount of reduced silver in the glass increased proportional to the square root of time, as expected from Eq. 11.6, but also from a variety of other diffusion-controlled processes, such as described in Chapter 4. The authors were able to calculate the diffusion coefficients of silver and the permeation coefficient of hydrogen from their measurements. The analytical treatment for this calculation was promised for a future publication, but apparently was not published. The activation energy for hydrogen permeation in a soda-lime silicate glass (72 wt % SiO_2, 15% CaO, 13% Na_2O) was about 82 kJ/mol, which is more than twice the activation energy for permeation in silica glass. The activation energy for diffusion of silver was about 109 kJ/mol, which is not far from the curve in Fig. 5.4 for a silver atom of 1.45 Å radius. It is also possible that the silver ions diffuse to the surfaces of growing particles and are reduced there, as happens for gold ions [31]. In this case the comparison should be with interionic diffusion of silver and sodium ions, as made by Barton and Morain, which has the same activation energy of 109 kJ/mol in a similar glass [32].

The reduction of iron in a silicate glass from Fe^{3+} to Fe^{2+} is shown in Eq. 11.2. This reaction was followed by changes of optical absorption spectra in the visible range and by electron spin resonance [28]. The amount of reaction was proportional to the square root of time and the square root of hydrogen pressure, as expected from Eq. 11.2. The results were interpreted with Eq. 11.6; the product C_0D (permeation rate) of hydrogen had an activation energy of about 113 kJ/mol in a soda-lime silica glass (72 wt % SiO_2, 14% Na_2O, 8% CaO, 4% MgO, 2% other) which is somewhat higher than found by Barton and Morain in a similar glass.

The formation of colloids of arsenic, antimony, and bismuth in a potash borosilicate glass (65 mol % SiO_2, 20 mol % B_2O_3, 13 mol % K_2, 1.7 mol % Al_2O_3) was followed from the darkening of the glass as the colloids formed. The optical absorption increased from a wavelength of 1 μm to shorter wavelengths after treatment in about 1 atm of hydrogen for 64 h at 455°C. The absorption increased in the sequence As $<$ Bi $<$ Sb. From the optical properties of antimony [33, 34] and the equation for the optical absorption of small spherical particles [35], one can calculate that antimony particles should show a strong absorption with a maximum at about 0.31 μm, decreasing at higher wavelengths. A dark region moved into the glass proportional to the square root of heat treatment time at 455°C for the glass containing 0.5 wt % arsenic oxide. At the same time, absorption from hydroxyl groups increased proportional to the square root of time. Thus as the arsenic ions (+3 or +5) are reduced to elemental arsenic, the hydrogen forms hydroxyl groups with the oxygen ions that were bonded to the arsenic before reduction.

11.5 HYDROGEN IN CRYSTALLINE OXIDES

The diffusion of hydrogen in certain crystalline oxides has been studied, as summarized by Shelby [26, Chap. 4]. Hydrogen transport can result from diffusion of molecular hydrogen or water, or hydrogen ions or atoms. In many of these studies there is not enough information to tell what is the diffusing entity. For example, often, diffusion profiles or dependence on gas pressure were not measured, and both water and hydrogen gas can be present in the gas phase.

In rutile (TiO_2) the diffusion coefficient of hydrogen (tritium) as measured appears to be the same at different hydrogen and water pressure and mixtures of these gases [36, 37]. This conclusion depends critically on whether or not equilibrium was reached in "compacts" of metals and oxides (Cu/Cu_2O, Ni/NiO, Fe/FeO) that controlled the amount of hydrogen and water in the gas phase. If equilibrium was reached, the results suggest that the diffusion coefficient of hydrogen in the rutile was controlled by diffusion of hydrogen ions or atoms from one site to another in the rutile. The reaction to form hydroxyl groups in the rutile, identified from infrared spectra [38], can be written

$$2O{-}Ti^4{-}O{-} + H_2 = 2O{-}Ti^3{-}OH^+{-} \qquad (11.15)$$

The hydrogen reduces the titanium from +4 to +3 valence and forms a neighboring hydrogen ion (H^+) to compensate for the resulting net negative charge. It is not clear exactly where this hydrogen ion is located. This reaction involves minimum rearrangement of the constituent ions, with only hydrogen moving. The diffusion coefficient of hydrogen is about eight orders of magnitude greater than the diffusion coefficient of oxygen (^{18}O) in rutile (see Chapter 12).

In diopside ($CaMgSi_2O_6$ with 7% Fe), hydrogen is also involved with an oxidation–reduction reaction of the iron (Equation 11.2). Diffusion of hydrogen in dopside is about 10 orders of magnitude faster than diffusion of ^{18}O [39, 40]. These large differences cannot be explained by a difference in hydrogen and oxygen concentrations in the minerals, as was valid for hydrogen and oxygen in silica glass, quartz, and silicon (see Chapters 6, 7, and 14). Thus tentatively, the mechanism of hydrogen diffusion can be identified as the motion of hydrogen molecules or ions from one site to another. Both these minerals are electronic conductors because of the titanium and iron ions of different valence states, so the transport of hydrogen ions could be compensated by transport of electrons. More experiments are needed to confirm this mechanism.

REFERENCES

1. R. W. Lee, R. C. Frank, and D. E. Swets, *J. Chem. Phys.* **36**, 1062 (1965).
2. R. W. Lee, *J. Chem. Phys.* **38**, 448 (1963).
3. R. W. Lee and D. L. Fry, *Phys. Chem. Glasses* **7**, 19 (1966).
4. J. E. Shelby, *J. Appl. Phys.* **48**, 3387 (1977).
5. R. H. Doremus, *Glass Science*, Wiley, New York, 1994.
6. R. H. Doremus, *J. Mater. Res.* **10**, 2379 (1995).

7. F. J. Norton, *Nature* **191**, 701 (1961).
8. J. E. Shelby, *J. Appl. Phys.* **51**, 2589 (1980).
9. F. Booth, *Trans. Faraday Soc.* **44**, 796 (1948).
10. J. Crank, *Mathematics of Diffusion*, 2nd ed., Oxford University Press, Oxford, 1975.
11. E. J. Friebele and D. L. Griscom, in *Treatise on Materials Science and Technology*, Vol. **17**, M. Tomozawa and R. H. Doremus, eds., Academic Press, San Diego, CA, 1979, p. 257ff.
12. J. E. Shelby, *J. Appl. Phys.* **60**, 4325 (1986).
13. M. Fox and S. J. Stannard-Powell, *Electron. Lett.* **19**, 916 (1983).
14. K. J. Beales, D. M. Cooper, and J. D. Rush, *Electron. Lett.* **19**, 917 (1983).
15. N. Uchida and N. Yesugi, *J. Lightwave Technol.* **LT-4**, 1132 (1986).
16. J. Stone, *J. Lightwave Technol.* **LT-5**, 712 (1987).
17. K. W. Plessner and S. J. Stannard-Powell, *Electron. Lett.* **20**, 250 (1984).
18. Q. Zeng, J. F. Stebbins, A. D. Heaney, and T. Erdogan, *J. Non-Cryst. Solids* **258**, 78 (1999).
19. C. Dalle, P. Cordier, C. Depecker, P. Niay, P. Bernage, and M. Doway, *J. Non-Cryst. Solids* **260**, 83 (1999).
20. H. Itoh, Y. Ohmori, and M. Nakahara, *J. Lightwave Technol.* **LT-4**, 116 (1986).
21. M. K. Schurman and M. Tomozawa, *J. Non-Cryst. Solids* **202**, 93 (1996).
22. B. I. Greene, D. M. Krol, S. G. Kosinski, P. J. Lemaire, and P. N. Saeta, *J. Non-Cryst. Solids* **168**, 195 (1994).
23. F. M. Araujo, E. Joanni, M. B. Margues, and O. G. Okhatnikov, *Appl. Phys. Lett.* **72**, 3109 (1994).
24. J. Stone, J. M. Wiesenfeld, D. Marcuse, C. A. Burrus, and S. Yang, *Appl. Phys. Lett.* **47**, 328 (1985).
25. J. M. Wiesenfeld, J. Stone, D. Marcuse, C. A. Burrus, and S. Yang, *J. Appl. Phys.* **61**, 5447 (1987).
26. J. E. Shelby, *Handbook of Gas Diffusion in Solids and Melts*, ASM International, *Materials Park*, OH, 1996, Chap. 2.
27. J. L. Barton and M. Morain, *J. Non-Cryst. Solids* **3**, 115 (1970).
28. J. E. Shelby and J. Vitko, *J. Non-Cryst. Solids* **45**, 83 (1981).
29. J. T. Kohlic, B. M. Wright, J. M. Jewell, and J. E. Shelby, *Phys. Chem. Glasses* **31**, 133 (1990).
30. M. R. Tuzzolo and J. E. Shelby, *J. Non-Cryst. Solids* **143**, 181 (1992).
31. R. H. Doremus, in *Symposium on Nucleation and Crystallization in Glasses and Melts*, M. K. Reser, ed., *American Ceramics Society*, Columbus, OH, 1962, p. 119.
32. R. H. Doremus, *J. Phys. Chem.* **68**, 2212 (1964).
33. M. Cardona and D. L. Greenway, *Phys. Rev.* **133A**, 1685 (1964).
34. E. D. Palik, *Handbook of Optical Constants of Solids*, Vol. **3**, Academic Press, San Diego, CA, 1984, p. 174ff.
35. H. C. Van de Hulst, *Light Scattering by Small Particles*, Wiley, New York, 1957, p. 267.

36. O. W. Johnson, S. H. Park, and J. W. DeFord, *J. Appl. Phys.* **46**, 1026 (1975).
37. J. V. Cathcast, R. A. Perkins, J. B. Bates, and L. C. Manley, *J. Appl. Phys.* **50**, 4110 (1979).
38. J. B. Bates and R. A. Perkins, *Phys. Rev.* **B16**, 3713 (1977).
39. S. Hercule and J. Ingrin, *Am. Mineral.* **84**, 1577 (1999).
40. S. C. Wood, S. Mackwell, and D. Dyor, *Am. Mineral.* **85**, 480 (2000).

12

DIFFUSION OF OXYGEN IN NOMINALLY DRY OXIDES

12.1 INTRODUCTION

Diffusion of oxygen in oxides is important in many practical processes. Oxidation and corrosion of metals and semiconductors often involve oxygen diffusion in oxides. Mechanical properties of oxides, such as creep, sintering rates, and annealing of defects and radiation damage, and oxidation–reduction reactions can be influenced by oxygen. The oxidation state of a solid or melt depends on the chemical potential of oxygen in the solid or melt, as discussed in Chapter 13; to change this potential, oxidizing and reducing agents must diffuse.

In many research papers no distinction is made between transport of oxygen by water or by other means. In some experiments water is used as the source of isotopes for diffusion studies. These "hydrothermal" conditions can lead to dissolution and diffusion of molecular water in a solid or melt. This diffusion type was discussed in Chapters 6 to 10, where evidence was presented to show that the dissolution and diffusion of molecular water in a solid or molten oxide is often the fastest process to transport oxygen. Thus a careful distinction is made in this book between this process of water transport and experiments in which water is not added, which I am calling "nominally dry" conditions. Nevertheless, water is almost always present in small quantities in the gas-handling systems of nominally dry experiments, as shown directly by the discussion in Section 12.2. Furthermore, all natural solids and melts contain water in the range 0.01 to 1%, often at the higher end of this range. Thus truly "dry" conditions are not present unless experimenters take extraordinary precautions to remove water and to keep its concentration low.

In this chapter experiments on diffusion of oxygen in silica glass are discussed first, because they are extensive and informative. Then possible mechanisms for molecular and ionic diffusion of oxygen in solid and molten oxides are outlined, and finally, diffusion data on a large variety of crystalline oxides and in melts are examined.

12.2 DIFFUSION OF OXYGEN IN SILICA GLASS

In the 1950s and early 1960s, Frank Norton measured permeation and diffusion of different gases in a variety of glasses at the General Electric Research Laboratory, where I was privileged to be his colleague. His measurements of the permeation and diffusion of oxygen in silica glass [1] are the standard for these important parameters, for example, for controlling oxidation of silicon; it is remarkable that no one has thought it necessary to repeat these experiments. They show that oxygen dissolves and diffuses molecularly (O_2) in the silica glass. The activation energy for diffusion of oxygen in silica glass is related directly to its molecular size, as shown in Chapter 5, especially in Table 5.2 and Fig. 5.4. The equation for diffusion of molecular oxygen in silica glass from Norton's measurements is

$$D = 2.7(10)^{-4} \exp\left(\frac{113}{RT}\right) \qquad (12.1)$$

in cm^2/s and an activation energy in kJ/mol. These experiments show that oxygen dissolves and diffuses in silica glass with little interaction with the silicon–oxygen network of the glass, just as for inert gases.

Another kind of experiment involves diffusion and exchange of isotopically labeled oxygen into glass. The ^{18}O concentration in the gas phase can be followed with a mass spectrometer [2–4] and the profile of the isotopically tagged ^{18}O [5–16] in the network oxygen can be measured with a nonresonant nuclear reaction or from secondary ion mass spectroscopy (SIMS). Much of the profiling work was done on thin silica films growing on silicon [5–9,12–14]. This work was valuable for demonstrating that molecular oxygen was the diffusing species during oxidation of silicon, because ^{18}O accumulated at the silicon–oxide interface (see Fig. 12.1).

In contrast to the straightforward interpretation of Norton's experiments, the mechanisms involved in the exchange experiments have been uncertain. The effective diffusion coefficients found in these experiments are much lower than the diffusion coefficients measured by Norton, and the results of different researchers differ widely with respect to absolute values of the effective diffusion coefficients and their activation energies. A variety of mechanisms have been proposed to explain these isotopic exchange experiments, including molecular diffusion, network diffusion, two diffusing species, traces of water, and a slow surface exchange reaction.

The ingenious interdiffusion tracer experiments of Mikkelson [11] demonstrate diffusion of network oxygen at temperatures from 1200 to 1400°C with an activation energy of 453 kJ/mol. The extrapolated network diffusion coefficient at

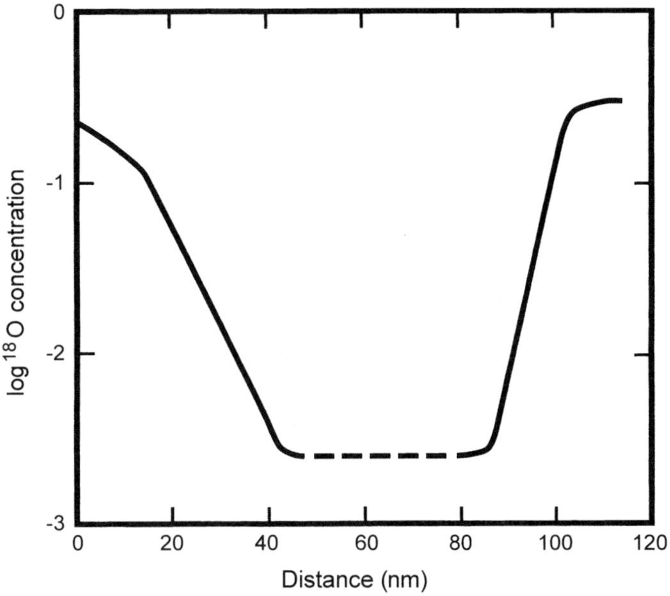

Figure 12.1 Profile of ^{18}O in a silicon dioxide layer grown on silicon, first in ^{16}O and then in ^{18}O. (From Ref. 12.)

1000°C is $6.7(10)^{-19}\,cm^2/s$. The mean diffusion distance, $y = \sqrt{2Dt}$, for this diffusion coefficient D and a time of 10 min is about 0.3 nm, and for 15 h is 3 nm. These distances are much smaller than the diffusion distances at half surface concentration in profiles measured [8, 9] at these times, showing that it is highly unlikely that lattice diffusion contributes to these profiles.

The most useful experiments for determining the mechanisms in the exchange are those in which the profile or distribution of ^{18}O in the silica was measured [5–11], so these experiments are considered first. The experiments in which the total amount of exchanged ^{18}O was measured [2–4] are then discussed, although they were the first in time. The experiments of the French research groups [6, 7, 12, 13] with nuclear reaction analysis showed that at 1000°C and lower temperature the tagged oxygen diffuses through the silica without appreciable exchange (middle distance range of about 40 to 80 nm in Fig. 12.1). Nevertheless, some ^{18}O is found in the silicon–oxygen network near the silica–gas surface. These results have been confirmed in measurements[8–10, 14] of ^{18}O profiles in the silica films by SIMS. The ^{18}O has low concentration away from the silica–gas and silica–silicon interfaces, but increases near these interfaces.

A profile of ^{18}O is given in Fig. 12.1, and shows these features. At the silicon–oxide interface, ^{18}O is incorporated into the growing oxide; in the center of the oxide there is a region of low ^{18}O concentration. This profile shows that tagged molecular oxygen diffuses through the oxide without exchanging with network oxygen. Near the oxide–gas interface there is a profile of tagged oxygen. I propose that this profile results from diffusion and exchange of a small concentration of residual water into the oxide, as described in the following paragraphs.

The treatment of exchange experiments of tagged water with the diffusion–reaction model, as described in Chapter 3 for local equilibrium, found that the profile of exchanged ^{18}O isotope in the solid followed the profile expected (Eq. 12.3) for an effective diffusion coefficient D_e independent of isotope concentration. In these experiments the important exchange is between ^{18}O in the residual dissolved molecular water and the ^{16}O in the silicon–oxygen network. There is also formation of a low concentration of SiOH groups in the silica by reaction of the dissolved water with the silicon–oxygen network. This reaction contributes a negligible amount of ^{18}O to the network as compared with the exchange reaction. Because of the exchange reaction between dissolved molecules and oxygen atoms in the oxide, the effective diffusion coefficient is lower than the diffusion coefficient D of the dissolved molecules (water or oxygen) and is given by (see Section 3.3)

$$D_e = \frac{DC_T}{S_T} \tag{12.2}$$

in which S_T is the total concentration of oxygen atoms in the oxide and C_T is the total concentration of dissolved molecular oxygen, assumed to be uniform and constant with time. The profile of ^{18}O at the oxide–gas surface results from the diffusion and exchange of dissolved residual water from the gas phase. With local equilibrium the profile of exchanged isotope is just that for diffusion into a semi-infinite solid with constant surface concentration, so that

$$S = S_0 \operatorname{erfc}\left(\frac{x}{2\sqrt{D_e t}}\right) \tag{12.3}$$

in which S is the concentration of exchanged isotope in the solid, S_0 this concentration at the solid surface ($x = 0$), x the distance, and t the time of diffusion. See Section 3.3 for more details and for derivations.

To examine the origin of these surface profiles, effective diffusion coefficients calculated from them with Eq. 12.3 (and other means) are compared in Table 12.1. In Refs. 8, 9, and 14 to 16, the samples were silica films of differing thicknesses on silicon; in Ref. 10 the samples were bulk silica glass with low water content. In all the studies except for Ref. 18, the $^{18}O/^{16}O$ ratio in the silicon–oxygen network at the oxide–gas surface equaled that in the gas, showing that local equilibrium had been reached at the silica surface, so that Eqs. 12.2 and 12.3 were applicable. There is a striking difference of a factor of more than 3000 in these D_e values for nominally identical experiments by different investigators.

The SIMS profiles in these studies do not fit the erfc profile of Eq. 12.3 very well. In the profiles of Ref. 14, the calculated D_E values varied by a factor of 2 over the measured depth. The profiles in Ref. 9 fit Eq. 12.3 better, but the calculated D_E values still are about $\pm 20\%$ in range. The diffusion coefficients reported in Ref. 10

Table 12.1 Effective Diffusion Coefficients and Concentrations of Water in Oxygen Diffusion in Silica Glass at 1000°C

Effective Diffusion Coefficient D_e (cm^2/s)	Water Concentration in Oxygen Gas [ppm by weight (calculated)]	Ref.
$2(10)^{-15}$	3400	14
$8(10)^{-16}$	1400	9
$3(10)^{-17}$	51	10
(See the text)	69	8
$6(10)^{-19}$	10	15

for different times at a particular temperature vary by an order of magnitude. These differences possibly result from the nonlinear proportionality of SIMS intensities to concentration in these early SIMS data. Thus these effective diffusion coefficients in Table 12.1 are averages and have considerable uncertainty.

The profiles in Ref. 8 do not fit Eq. 12.3. The low surface concentrations of exchanged isotope in these experiments show that local equilibrium was not reached in them because of the short diffusion times of less than 30 min. The surface concentrations of isotopes were about 0.03 to 0.07 of those expected for local equilibrium, so from the analysis of Chapter 4 one would expect an exponential profile of exchanged isotope. The profiles in Ref. 8 fit an exponential dependence better than Eq. 12.3, but still show higher-than-expected tracer concentrations at the silica–gas surface and at very low concentrations.

Calculation of the residual water concentration in the gas phase required to give the effective diffusion coefficients in Table 12.1 was carried out as follows. The concentration S_T of oxygen in the silica network is $S_T = 2\ (2.20)(6.02)(10)^{23}/60.1 = 4.41(10)^{22}$ O atoms/cm^3. The diffusion coefficient D of water molecules in silica is $1.7(10)^{-7}$ cm^2/s at 1000°C (see Chapter 6 and Ref. 17). With these numbers and the D_e values in Table 12.1 the concentration C_T of dissolved molecular water can be calculated from Eq. 12.2. The concentration of water in the gas phase to give this surface concentration dissolved in the silica is $C_T/0.015$ (see Chapter 6). In all the studies in Table 12.1 except the last, the nominal gas pressure of oxygen was apparently atmospheric, so the gas concentration was $5.7(10)^{18}$ molecules O_2/cm^3. In Ref. 15 the pressure was 0.1 atm. Then the values of water concentration by weight were calculated from $[C_T/(0.015)(5.7)(10)^{18}\ P] \times 18\ (10)^6/32$, where P is the pressure in atmospheres.

The water concentrations in the gas phase required to give the effective diffusion coefficients in Table 12.1 are also given in the table. There is a direct relation between the care taken in the studies listed in the table to exclude residual water from the gas phase and the water concentrations calculated. From the experimental description in the papers, one can conclude from Ref. 14 that no special precautions were taken to remove water, from Ref. 9 that some drying was attempted, from Ref. 10 that the oxygen gas was quite dry, and from Refs. 8 and 15, a residual water concentration of about 10 ppm was claimed. These correlations are strong evidence that trace water is the main source of the ^{18}O profiles measured at the silica–gas interface.

For the profiles in Ref. 8, the low surface concentration of exchanged ^{18}O means that $S_0/S_T \ll 1$, where S_0 is the surface concentration of ^{18}O and S_T the total oxygen concentration. These are conditions for Eqs. 4.14 to 4.16 to be valid, giving an exponential profile of exchanged ^{18}O:

$$S^* = \lambda C_T t \exp\left(-x\sqrt{\frac{\lambda}{D}}\right) \tag{12.4}$$

in which $S_0{}^* = \lambda C_T t$ is the surface concentration of exchanged ^{18}O, λ a forward reaction coefficient, and t the diffusion time. The measured profiles are linear on a log S^* versus distance plot [8, Fig. 6] over the middle range of concentrations. Furthermore, the slopes of these lines are about the same for times of 10, 20, and 30 min, as expected from Eq. 12.4 but not from Eq. 12.3. These profiles therefore support the present diffusion–exchange model. These slopes give a value of $\sqrt{\lambda S_T/D}$ of about $1.0(10)^6$ per centimeter. Thus $\lambda = 1.0(10)^{12}\ D/S_T$. The surface concentrations, S^*, of exchanged ^{18}O in the network at different times are given in Table 12.2, as derived from extrapolation of the linear parts of the log S^*–distance profiles. The linearity of S^* on time is also expected from the diffusion–exchange model and Eq. 12.4. From this equation and the data, a value of $\lambda C_0{}^*$ of $3.0(10)^{-6}$ mol/cm is found. Thus with the value of λ above,

$$C_T{}^* = \frac{3.0(10)^6 S_T}{1.0(1.0)^{12} D} \tag{12.5}$$

The total lattice concentration of oxygen $S_T = 0.0732\ \text{mol/cm}^3$, and the diffusion coefficient of molecular water at 1000°C is $1.7(10)^{-7}\ \text{cm}^2/\text{s}$,[15] so that $C_T{}^* = 1.8\ (10)^{-11}\ \text{mol/cm}^3$. This concentration in the glass requires about 69 ppm by weight of water in the gas phase. A similar calculation was made by Rochet *et al.* [7] for their analysis of the total amount of ^{18}O near the silica–gas surface.

Kalen *et al.* [10] found a long "tail" of ^{18}O in the silica lattice in their profile after heating at 1100°C. I propose that this tail results from exchange of ^{18}O from the oxygen gas with the oxygen in the silica lattice. There is good evidence from

Table 12.2 Surface Concentrations of Exchanged ^{18}O in Silica Films

Diffusion Time (min)	$S_0{}^*/S_T$	$S_0{}^*/S_T t$ (min^{-1})
10	0.027	$2.7(10)^{-3}$
20	0.046	$2.3(10)^{-3}$
30	0.071	$2.4(10)^{-3}$

Source: Ref. 8.

the other studies that this exchange rate is low at 1000°C and below, but Costello and Tressler [14] found complete exchange equilibrium at 1300°C. Thus it seems reasonable that exchange from the oxygen gas should be appreciable at 1100°C. An exact quantitative comparison is not possible because of the interference of the water exchange near the silica gas interface. Nevertheless, the tail in these profiles is consistent with lattice exchange of the tagged oxygen.

Three sets of studies of oxygen diffusion in silica glass in the early 1960s by isotopic exchange gave discordant results [2–4] and effective diffusion coefficients much lower than those measured by Norton [1]. Researchers have puzzled over these results for many years. The conclusion that residual water makes an important contribution to exchange with lattice oxygen makes these results more understandable. The measurements were made by following the changes in the isotopic ratio in the gas phase, so the total amount of exchange into the lattice was measured. Without profile measurements and water concentrations, a quantitative comparison with the present model is difficult, but different amounts of water and oxygen exchange can probably explain these results. In some of these experiments the amount of exchange was proportional to the square root of time, corrected for the decrease in isotope in the gas phase. This is the result expected from Eq. 12.3, so the effective diffusion coefficient for oxygen exchange is given by Eq. 12.2. At 1000°C and 1 atm pressure, $C_T \approx 10^{-8}\,\mathrm{mol/cm^3}$ and $S_0 = 0.073\,\mathrm{mol{\cdot}cm^3}$, so $D_E \approx 10^{-6}D$. There is about six orders of magnitude difference between the diffusion coefficient of molecular oxygen at 1000°C and those in Refs. 2 to 4, consistent with this rough calculation.

In Refs. 2 and 4, a surface exchange reaction was invoked to explain deviations from the amount of total exchange expected from an integration of Eq. 12.3. However, the measured effective diffusion coefficients were proportional to gas pressure; this dependence is not expected for a slow surface exchange reaction and network diffusion, but it is the dependence predicted by the diffusion–reaction mechanism. The total amount of exchange as a function of time is quite similar in both the diffusion–reaction equation and equations with surface exchange, as shown in Chapter 4, so this particular dependence is not evidence for surface exchange.

Some workers have assumed that molecular diffusion does not contribute to isotope exchange profiles, because it is so fast. The present treatment shows that this conclusion is incorrect. Even if D is large, the effective diffusion coefficient DC_T/S_T can be small, because the concentration of dissolved molecular oxygen or water, C_T, is so much smaller than the total concentration of oxygen, S_T, in the silicon–oxygen network.

The experiments of Mikkelson [11] show that network diffusion is too slow to contribute appreciably to the isotope exchange experiments with oxygen, and results on water also exclude network diffusion (Chapter 6). Surface exchange of dissolved and gaseous oxygen and water is rapid and is not slow enough to reduce the concentrations of these gases dissolved in the glass. The proportionality of the effective diffusion coefficients on pressure strongly supports these two conclusions.

The experiments of Pfeffer and Ohring [18, 19] on diffusion of implanted oxygen ^{18}O in silica agree with the diffusion–reaction model and show that dissolved molecular water controls the diffusion of the ^{18}O, even in nominally dry nitrogen. Gusev *et al.* [20] have measured ^{18}O profiles in silica layers with medium-energy ion-scattering spectroscopy and have found results similar to those in earlier studies [13], again supporting the role of residual water in these experiments.

In two studies of oxygen exchange in silica films on silicon with very low residual water in the oxygen, it was found that the amount of exchange increased as the film thickness decreased [8, 13]. In a study with higher water content, no such dependence was found [9]. In the former two papers the hypothesis that water was responsible for exchange near the silica–gas interface was rejected because of this dependence. However, there are some possible explanations for this dependence that do not require rejecting the water hypothesis. In a silica film on silicon there is a gradient of oxidant (water or oxygen) from the silica–gas interface to the silica–silicon interface. In the linear-parabolic model of oxidation of silicon, this gradient is much smaller in thin films than in thick, so that in thin films the average concentration of dissolved oxygen or water in the film is greater than in a thick film. In these experiments with low water content, both oxygen and water exchange can contribute to the ^{18}O profile near the silica–gas interface, even at 1000°C or lower temperatures. Thus it is likely that water is the main contributor to these profiles, especially for high water contents.

I conclude that the diffusion–exchange mechanism provides a satisfactory explanation for isotope exchange experiments on oxygen diffusion in silica glass. It is unnecessary to invoke network diffusion or a slow surface exchange reaction to understand these results. Furthermore, residual water in the gas phase can lead to dissolution of water and exchange of ^{18}O with the silicon–oxygen network near the silica–gas interface. This possibility exists in other experiments on diffusion of oxygen, and should be checked with experiments with different controlled amounts of water in the gas phase.

12.3 MOLECULAR TRANSPORT OF OXYGEN

To help judge which mechanisms of diffusion are most likely for diffusion of oxygen in oxides, some different possible mechanisms are discussed in this section and the next. All mechanisms can be considered to result from motion over an activation barrier of Gibbs free energy Δg:

$$\Delta g = \Delta h - T\Delta s \tag{12.6}$$

in which Δh is an enthalpy of activation, equal to the experimental activation energy Q, T the absolute temperature, and Δs an entropy of activation. For the diffusion of a dissolved molecule from one to site to another a distance λ away, the diffusion coefficient D is given by

$$D = \gamma\lambda^2\nu \exp\left(\frac{\Delta s}{R} - \frac{\Delta h}{RT}\right) \tag{12.7}$$

in which γ is a geometrical constant of order unity, ν a frequency factor (often taken as about 10^{13} s^{-1}), and R the gas constant.

An equation for the activation energy $Q = \Delta h$ for molecular diffusion in a solid or melt was given in Eq. 5.1 as

$$Q = 8\pi G r_D (r - r_D)^2 \tag{12.8}$$

in which r is the effective molecular radius, r_D the radius of a doorway between sites for dissolution, and G the shear modulus. In Chapter 5 activation energies for a number of molecules diffusing in silica glass fit the functional dependence of $\sqrt{Q}$ proportional to r from Eq. 12.8 with a reasonable doorway radius.

Possible molecular species to transport oxygen are water, oxygen, carbon monoxide, and carbon dioxide. A large number of experimental data which show that diffusion of molecular water in oxides can control the transport of oxygen in them were discussed in Chapters 6 to 10. When water (hydrothermal conditions) is the source of diffusion, reaction and exchange of oxygen in a solid or melt involves the dissolved molecular water. In Section 12.2 it was shown that trace amounts of water dissolve, diffuse, and exchange at the gas–silica surface even when the system is nominally dry, and that this residual water can dominate the exchange of isotopic oxygen near this surface. This discussion showed that amounts of water as small as 10 ppm in the gas phase can cause appreciable diffusion and exchange of ^{18}O in the solid. Although the experimenters took extensive efforts to limit water in the gas phase, enough was present to dominate exchange near the silica surface at 1000°C. Since molecular water diffuses much more rapidly than molecular oxygen or network ions in many oxides, residual water is always a possible source of oxygen transport in diffusion experiments. Experimenters should test for this possibility by changing the water concentration in the gas phase over wide limits.

Another possible source of oxygen transport by water is the water content of natural and synthetic oxides. Natural oxides deposit from aqueous solution or crystallize from magmas; in either case, incorporation of significant amounts of water in the solid is possible. This water is often held in the solid as reacted hydroxyl groups: for example, in silicates as SiOH groups or SiOH pairs. At the temperature of a diffusion experiment, this bound water forms a small concentration of molecular water in the solid that diffuses rapidly and can transport oxygen. Experimentally, one should synthesize oxides with very low water contents to see if they have different diffusion and exchange properties from materials with higher water contents.

Another possible diffusing molecule is oxygen, O_2. In silica glass oxygen has an activation energy for diffusion of about 113 kJ/mol compared to that of 70 kJ/mol for water and a diffusion coefficient about 30 times smaller at 1000°C. The effective radius of the water molecule is about 1.38 Å, and oxygen about 1.6 Å at 600°C. These differences are accentuated in more dense crystalline oxides, so the diffusion of molecular oxygen should be much slower than water in these materials. I do not know of any direct measurements of diffusion of oxygen molecules in solids other than silica glass.

The equilibrium ratio between oxygen molecules and atoms at 1000°C is about 10^8. Thus the concentration of oxygen atoms in oxides should be very small, and probably does not contribute to diffusion of oxygen, although their rate of diffusion should be comparable to that of water, because the oxygen atom has about the same effective size in a solid as a water molecule.

Carbon monoxide has a gas kinetic size smaller than that of water. The carbon–carbon distance in carbon monoxide is about 1.13 Å, so the CO molecule should have a size similar to that of a water molecule. There are some data on diffusion of carbon in calcite compared to that of oxygen from an atmosphere of carbon dioxide [21]. At temperatures from 600 to 800°C, the diffusion coefficient of oxygen was about 100 times greater than that of carbon. The mechanisms in these experiments are uncertain. I know of no direct measurement of diffusion of carbon monoxide in any material.

12.4 OXYGEN TRANSPORT BY IONS

The mechanism of oxygen transport in oxides favored by most workers has been diffusion of network ions of hydroxyl or oxygen, the latter assisted by oxygen ion interstitials or lattice vacancies. The experiments described in previous chapters show strong evidence that oxygen can diffuse as water molecules in many oxides, both crystalline or amorphous, and that oxygen molecules can dissolve and diffuse in silica glass. In these situations the diffusion of network hydroxyl and oxygen is much slower than that of molecular species.

Nevertheless, there is good evidence that in certain oxides network oxygen ions are the fastest way to transport oxygen. Zirconium oxide containing foreign ions such as calcium or yttrium in solid solution is an ionic oxygen conductor, as shown by experiments on electrical conductivity as compared to diffusion [22, 23]. In recent years other oxygen ion conductors such as $SrTiO_3$, have been confirmed [69, 70]. Thus the question is how extensive is oxygen transport by ions as compared to other means, such as molecules.

In ionic conduction in solids one must make a distinction between intrinsic and extrinsic material. *Intrinsic* material is stoichiometric and pure, so that the concentrations of defects are not influenced by impurities or lack of stoichiometry. In *extrinsic* material these influences can be present. In intrinsic oxides two kinds of defects are possible. In the first kind, called a *Schottky defect*, lattice vacancies of positive and negative ions are formed. For example, in MgO a pair of vacancies are formed, one a missing magnesium ion and the other a missing oxygen ion, to maintain electrical neutrality. In ZrO_2 two oxygen ion vacancies are formed for each zirconium vacancy. *Frenkel defects* are another type, in which a lattice ion is displaced into an interstitial position, leaving behind a vacancy in a site of this ion. The concentration n of either Schottky or Frenkel defects is [24, 25]

$$\frac{n}{N} = \exp\left(-\frac{\Delta g}{2RT}\right) = \exp\left(\frac{\Delta s}{2R}\right)\exp\left(-\frac{\Delta h}{2RT}\right) \tag{12.9}$$

See Eqs. 12.6 and 12.7 for definitions of the quantities in this equation; N is the total concentration of lattice sites. The enthalpy of formation Δh in this equation is the important factor that determines the concentration of defects. In Ref. 25, Table 2.2, p. 109, there are values of energies of formation for Schottky defects in a variety of ionic compounds. These Δh values range from about 240 kJ/mol (2.5 eV) for alkali halides to 742 kJ/mol (7.7 eV) for an Mg–O vacancy pair in MgO to 839 kJ/mol (8.7 eV) for an oxygen vacancy–interstitial pair in rutile (TiO_2). In a detailed calculation of the concentration of Mg–O vacancy pairs in intrinsic MgO in Ref. 25, it is shown that the concentration of these pairs is negligibly small. Thus the high energy of formation of lattice defects in most oxides means that these defects have very low concentrations. In Table 12.3 estimates are given of bond enthalpies for different oxygen–cation bonds. They show that these strong bonds are the reason for the high energy of formation of defects in oxides.

In extrinsic oxides, substitutional impurities that replace lattice ions can cause the formation of lattice vacancies. Thus in zirconia many cations have relatively high solid solubility. Yttrium or calcium ions substituted for zirconium ions lead to compensating oxygen vacancies to maintain electrical neutrality. These oxygen vacancies then lead to a high diffusion coefficient of oxygen ions. The diffusion coefficient in a defect mechanism can be written

$$D = n\gamma\lambda^2\nu \exp\left(\frac{\Delta s_m}{R}\right) \exp\left(\frac{-\Delta h_m}{RT}\right) \tag{12.10}$$

In this equation the enthalpy and entropy of motion have m subscripts to distinguish them from the enthalpy and entropy of defect formation. The concentration n is fixed by the concentration of impurities and is not a function of temperature, so the experimental activation energy for diffusion should be equal to the enthalpy of motion Δh_m of the vacancy. In zirconia the diffusion coefficient of oxygen is

Table 12.3 Estimates of Bond Enthalpies for Some Oxygen–Cation Bonds[a]

	Bond Enthalpy	
Bond	kJ/mol	eV
OH	460	5.0
OSi	470	5.1
OC	350	3.6
O=C	610	6.3
OAl	560	6.1
OMg	600	6.2
OCa	650	6.6
OZr	550	6.0

[a]Values given to two significant figures.

directly proportional to the concentration of dissolved impurity (e.g., Ca [22]), as required by this mechanism.

A number of oxygen ion conductors have been established by experiment. They appear to fall in five different structural categories, as listed in Table 12.4 [76, 82]. The mechanism of oxide transport presumably involves defect diffusion of oxygen ion vacancies. Either the energy of formation of the vacancies is low, or ions of different valence substitute in the structure and must be compensated by additional oxygen ion vacancies. In many simple oxides the solid solubility of cations in the pure oxide is limited. Thus the formation of high defect concentrations by extrinsic impurities is restricted in these materials.

A neglected defect for possible oxygen transport in oxides is the hydroxyl ion OH^-. In natural and synthetic oxides there is always some residual water unless extensive measures are taken to dry them. This water is present in natural and laboratory raw materials for making oxides and is tenaciously held in the oxide even with long heating at temperatures up to 1000°C. In oxides, oxygen ions are bound to cations by two bonds because of the two negative charges on the oxygen ions. Hydroxyl ions are singly bonded to cations, so the activation energies for motion and defect formation of hydroxyl groups should be roughly half of lattice oxygen ions. If there is 0.1 wt % of water in quartz as hydroxyl groups, their concentration is about $2(10)^{20}$ cm^{-3}. Thus in oxides without large amounts of dissolved substitutional cations, hydroxyl groups are likely to be the most prevalent ionic defects.

In silica glass [26] and obsidian [27], the concentration of hydroxyl groups is proportional to the square root of the oxygen pressure. This result implies that the hydroxyl groups can diffuse far enough to sample neighboring hydroxyl groups. The reaction to form hydroxyls can be written:

$$H_2O + SiOSi = 2\,SiOH \tag{12.11}$$

Table 12.4 Structural Categories of Oxygen Ion Conductors

Structural Type	Examples	Refs.
Fluorite	$ZrO_2+Y_2O_3+CaO$	22, 23
Perovskite	CeO_2	76
	δ-Bi_2O_3	77, 78
	$SrTiO_2$	69, 70
	$Ba_2In_2O_5$	75, 76
	$LaGaO_3$	79
Perovskite–Bi_2O_3 intergrowth layers	$Bi_4V_2O_{11}$	80
Pyrochlores	$Gd_2Zr_2O_7$	81
	$Gd_2Ti_2O_7$	
Monozite or β-$SnWO_4$	$La_2Mo_2O_9$	82

If the hydroxyl groups were completely immobile, the reverse reaction to form water should be first order rather than second order, and the number of hydroxyl groups should be directly proportional to the water vapor pressure. If a mean diffusion distance x of 50 Å in 10 h is assumed, the diffusion coefficient of hydroxyl groups is $D \approx x^2/2t = 3.5(10)^{-18}\,\text{cm}^2/\text{s}$. This diffusion coefficient is within the range of some measured values for oxygen diffusion in oxides.

12.5 EXPERIMENTAL ACTIVATION ENERGIES

There have been many experimental measurements of oxygen diffusion in nominally dry crystalline oxides. From five reviews [28–32] I chose results that seemed to be the most reliable and whose articles I could consult. The activation energies from these results are given in Table 12.5. Also in the table are calculated values of the ionic porosities Z of the oxides studied. See Chapter 5 for the method of their calculation and values of the ionic radius used in the calculations. The most important ionic radius in the calculations is oxygen; I used a radius of 1.38 Å for oxygen for all the compounds in the table, as was done for porosities given in Chapters 5 and 9. This value is larger than that of 1.35 Å used in Ref. 30, so the present values of Z are smaller than those in Ref. 30. The relative values are significant, but the absolute values are of less importance. The first group of activation energies in the table (Leucite through sapphive) were taken from Ref. 30, and the second group (Litharge through zircon) were taken from the references in the table. These two groups are plotted in Fig. 12.2. The activation energies in the last group in Table 12.5 are not included in the figure for the following reasons. The ionic porosity of Cu_2O is unreasonably high. Thus for this compound the hard-sphere model for atoms breaks down, just as it does for the diamond structure (see Chapter 5). The structure of Cu_2O is two interpenetrating diamond lattices; to calculate a useful value of porosity, the bonds must be taken into account; this calculation would be arbitrary and so was not attempted. Zirconia and $SrTiO_2$ are definitely oxygen ion conductors, and their activation energies fall well below the others in the figure. Measured activation energies for diffusion of oxygen in zinc oxide vary over a wide range from 124 to 684 kJ/mol, and there seems to be no rational way of choosing a suitable activation energy. In one study [74] the activation energies for samples of zinc oxide of close to the same composition from two different sources differed by nearly a factor of 2, although they were measured in the same way. Sublimation in zinc oxide is often a problem, but was apparently taken into account in Ref. 74. Diffusion coefficients in olivines depend on the oxidation state and show a wide spread in activation energies.

The 27 points in Fig. 12.2 fit the linear regression equation

$$Q = 765 - 1303Z \tag{12.12}$$

in which Q is the activation energy in kJ/mol and Z is the ionic porosity. The fit is fairly good, with a correlation coefficient of 0.869. The fit in Fig. 12.2 suggests that

Table 12.5 Activation Energies for Diffusion of Oxygen in Nominally Dry Crystalline Oxides, as Compared to Ionic Porosities

Mineral Name	Chemical Formula	Ionic Porosity Z	Activation Energy (kJ/mole)	Refs.
Leucite	$K_2O{\cdot}Al_2O_3{\cdot}4SiO_2$	0.474	59	30
Nepheline	$K_2O{\cdot}Na_2O{\cdot}4Al_2O_3{\cdot}2SiO_2$	0.443	105	33
Melilite	$2CaO{\cdot}Na_2O{\cdot}Al_2O_3{\cdot}4SiO_2$	0.415	140	34
α-Quartz	SiO_2	0.414	221	35
Anorthite	$CaO{\cdot}Al_2O_3{\cdot}2SiO_2$	0.445	199	36, 51
Perovskite	$CaTiO_3$	0.308	313	30
Periclase	MgO	0.328	370	37
Hematite	α-Fe_2O_3	0.302	405	37
Diopside	$CaO{\cdot}MgO{\cdot}2SiO_2$	0.343	431	33, 51
Spinel	$MgO{\cdot}Al_2O_3$	0.302	419	38, 39, 51
Sapphire	Al_2O_3	0.195	615	40
Litharge	PbO	0.535	95	41
	Y_2O_3	0.525	139	42
β-Quartz	SiO_2	0.443	159	43
	PuO_2	0.378	176	44
	ThO	0.427	209	45, 47
Bursenite	NiO	0.334	226	46
	HfO_2	0.349	233	47
Uraninite	UO_2	0.389	237	48–50
Calcite	$CaCO_3$	0.398	242	21
Sanidine	$K_2O{\cdot}Al_2O_3{\cdot}6SiO_2$	0.449	245	63
	CdO	0.442	248	64
Akermanite	$2CaO{\cdot}MgO{\cdot}2SiO_2$	0.433	278	51
	CoO	0.338	397	65
Eskolaite	Cr_2O_3	0.277	423	66
Zircon	$ZrO_2{\cdot}SiO_2$	0.210	448	55
Not in Fig. 12.2				
Cuprite	Cu_2O	0.626 (?)	167	68
Zincite	ZnO	0.463	124–684	71–74
	ZrO	0.314	114	22, 23
	$SrTiO_2$	0.355	140	69, 70
Rutile	TiO_2	0.264	251	2, 57, 67
Olivine	$2(Mg_1Fe)O{\cdot}SiO2$	0.348	266–416	60–62

the mechanism of oxygen diffusion is related to the openness of the oxide structure. A similar fit to activation energies of diffusion of water in oxides, with a better correlation coefficient of 0.943, is given in Chapter 9. In the case of water diffusion there is good evidence for silica glass, crystalline quartz, and rhyolite glass that the diffusion–exchange model with diffusion of molecularly dissolved water is the operative mechanism. The implication of the correlation of these results for water diffusion with the others is that they all diffuse by a similar mechanism, although more evidence is needed to confirm this conclusion. The implication of the

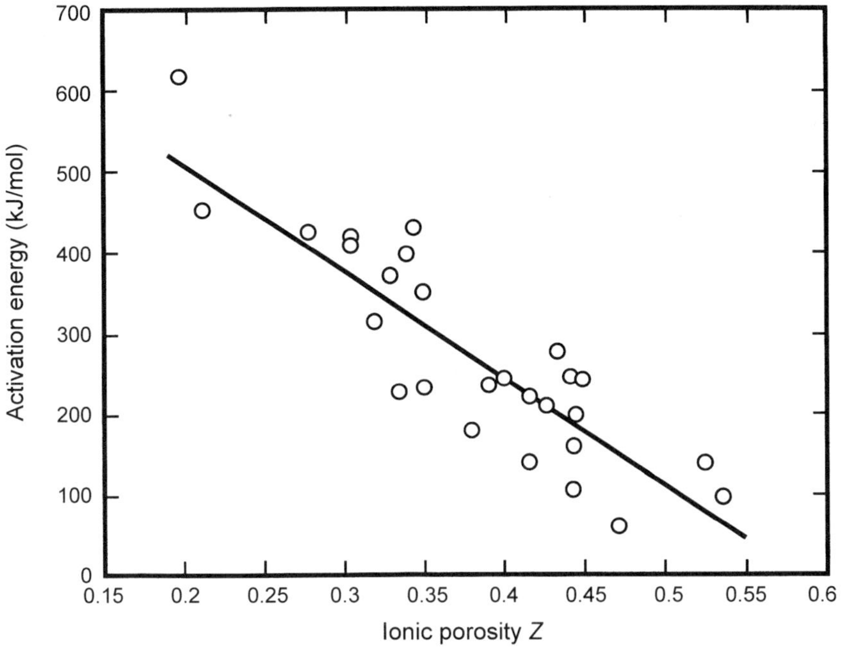

Figure 12.2 Activation energies for diffusion of oxygen in nominally dry oxides as a function of ionic porosity *Z*, from Table 12.4. Line from linear regression.

correlation of Fig. 12.2 is also that the mechanism of oxygen diffusion is the same for all the oxides in the figure.

12.6 FURTHER CONSIDERATIONS OF MECHANISMS

The mechanism of oxygen diffusion in crystalline oxides is almost always assumed to involve the diffusion of oxygen ions via either oxygen ion lattice vacancies or oxygen ion interstitials. In Section 12.4 several other possible mechanisms were suggested. In my opinion no mechanism has been established for diffusion of oxygen in most dry oxides. In this section I give some factors that must be explained by a mechanism of diffusion and some evidence for different mechanisms. I also suggest some experiments to test different mechanisms.

For the oxides in Table 12.4 there is evidence for the diffusion of oxygen ions aided by oxygen ion vacancies. In ZrO_2 and $SrTiO_3$ impurities of lower valence than zirconium or titanium ions generate vacancies. One question is whether or not similar materials show oxygen diffusion by this mechanism. The most likely candidates would be materials of similar chemical behavior, especially containing a heavy ion of valence four. Possibilities in Table 12.5 are HfO_2, PuO, UO_2, and TiO_2, all of which (except UO_2) have activation energies of oxygen diffusion well below the correlation line in Fig. 12.2. To examine the mechanism of ionic transport of oxygen, one can use the Einstein equation (e.g., Ref. 58) to show that electrical transport and ionic diffusion follow the same mechanism. Furthermore, as more cation impurities are added the number of vacancies increases, so the diffusion coefficient increases, which is additional evidence for ionic diffusion.

The correlation in Fig. 12.2 suggests that diffusion is related to the openness of the oxide structure. Thus diffusion of molecules is a possible mechanism of oxygen diffusion in oxides. Hydrogen, probably molecular, diffuses many orders of magnitude faster than oxygen in rutile and diopside, as described in Section 11.5. Again this result suggests molecular diffusion because it should depend on molecular size. The most likely candidate for transporting oxygen is residual water, even in nominally dry oxides. All natural and synthetic oxides contain small quantities of water in the form of OH groups, which at equilibrium will react to form a small concentration of dissolved molecular water, as described in Chapters 6 to 10. This water can diffuse rapidly in the oxide, as shown in Chapter 9, and could be the mechanism for transporting oxygen in oxides. This possibility can be tested experimentally by making oxides that are very dry. To do this one must start with raw materials that are extremely dry, such as dry silica sand and anhydrous compounds that have been thoroughly dried. The sintering of these oxides must be done in a dry atmosphere. Such dry oxides can have properties much different from oxides with a few tenths percent residual water. For example, in my laboratory we made a dry soda-lime silicate glass that normally can be melted at 1100°C or lower temperature. We could not melt this glass at 1500°C for long times, because of the much higher viscosity of water-free melts.

In Table 12.6 diffusion coefficients of oxygen under wet and nominally dry conditions in selected oxides are compared. In all materials except rutile, the diffusion of oxygen as water is much more rapid than in oxygen diffusion in dry

Table 12.6 Comparison of Diffusion of Oxygen in Crystalline Oxides under Wet and Dry Conditions

Mineral	Ionic Porosity Z	Activation Energy (kJ/mol)		Ratio of D values at 1100°C	Refs.
		Dry	Wet		
Anorthitre	0.445	199	110	1500	51, 52
Calcite	0.398	242	173	174 (700°C)	21, 53
Diopside	0.343	431	226	224	33, 51, 54
Zircon	0.210	448	210	20	55
β-Quartz	0.443	230	129	$88(10)^5$ (800°C)	2, 56
Rutile	0.264	251	258	≈1 (900°C)	2, 57

crystals. This result possibly suggests a role of residual water for transporting oxygen in many oxides. In all the oxides except rutile, the activation energy for oxygen diffusion is much greater in dry than in wet oxygen diffusion. This result would not be expected for residual water diffusion in dry oxides because in this mechanism the activation energies should not be too different in wet and dry diffusion of oxygen.

The experimental activation energies for diffusion of oxygen in dry oxides are not particularly reproducible from one experiment to another, and even from one sample to another by the same researchers. The extreme case of zinc oxide was mentioned previously, but there is a surprising lack of agreement between different experiments on different samples with close to the same composition, well outside experimental error. These results suggest an uncontrolled factor in the experiments, perhaps different concentrations of trace impurities in the samples; again, different trace water concentrations could be causing different effective diffusion coefficients in different samples. Synthesis of samples with carefully controlled amounts of impurities, especially water, is therefore highly desirable for diffusion experiments of oxygen in oxides.

In many experiments on diffusion of oxygen in crystalline oxides there are deviations from the expected mathematical treatment of the experimental data. These deviations are of two kinds. In experiments in which the amount of isotope in the gas phase is monitored as a function of time, the initial decrease of this amount is often less than expected for simple diffusion, as shown in Figs. 3 and 4 of ref. 2. This kind of deviation has almost always been considered to result from a slow surface exchange reaction (see Section 4.7). However, it can equally well result from nonequilibrium in the diffusion–exchange mechanism. The resulting dependence of isotope concentration in the gas phase on time is so similar in these two mechanisms that this dependence cannot distinguish between these mechanisms.

The second kind of deviation is incomplete exchange at the sample surface in a profile measurement. Again this result has usually been attributed to a slow surface exchange reaction; as shown in Section 4.7, the diffusion–exchange mechanism gives profiles so similar to this surface exchange mechanism that the two cannot be distinguished from the shapes of the profiles alone. An experimental method that can be used to distinguish these two mechanisms is the measurement of the surface concentration of exchanged isotope in the solid, as described in Section 4.7, as a function of time. I do not know of an example of this kind of measurement for oxygen diffusion in dry oxides.

The likelihood of a slow surface exchange reaction of oxygen on an oxide needs to be assessed. The rate of exchange of oxygen between an aqueous solution and oxygen on the surface of an aluminum polyoxocation $[AlO_4Al_{12}(OH)_{24}(H_2O)_{12}{}^{7+}]$ has been measured to be quite rapid at room temperature [59]; the exchange between solution and internal oxygen is much slower. This result suggests that the rate of exchange of oxygen atoms between adsorbed molecules on an oxide surface and surface oxygen atoms should be very fast at elevated temperatures. Thus I feel it is quite unlikely that slow surface exchange is the reason for the deviations in experiments on oxygen diffusion in oxides, and the diffusion–exchange mechanisms are a more likely cause for these deviations. Definition of the mechanisms of diffusion of oxygen in dry crystalline oxides awaits further experiments.

12.7 DIFFUSION OF OXYGEN IN NOMINALLY DRY SILICATE MELTS

Many measurements of oxygen diffusion in silicate melts have been made [83–88]. Experiments before 1982 are summarized by Dunn [83]. Effective diffusion coefficients D_e at 1400°C vary roughly over the range 10^{-9} to 10^{-5} cm^2/s, with activation energies from 100 to 400 kJ/mol. I have been unable to find any systematic variation of these quantities with melt composition. The ionic porosity Z or openness of these melts is difficult to calculate, because in most melts densities are not available.

In most reports on oxygen diffusion in nominally dry silicate melts, the mechanism of diffusion is considered to be diffusion of oxygen ions. However, there is no direct evidence for this mechanism. In one study of oxygen diffusion in $Na_2O{\cdot}4SiO_2$ melts [86] the diffusion coefficient was directly proportional to the pressure applied from 2.5 to 10 GPa; with close to zero D_e extrapolated to low pressures. This result is expected for molecular diffusion but not for ionic diffusion.

There is some evidence for a correlation between fluidity (reciprocal viscosity) of the melt and diffusion coefficients of oxygen in them [84–86]. This correlation is sometimes claimed to be consistent with oxygen ion diffusion, but it is also consistent with molecular diffusion. In molecular diffusion the molecules lodge in interstices (open regions) in the melt and jump from one interstice to another through "doorways" in the silicate network, as described in Section 5.3. An equation for the activation energy Q of diffusion derived for this mechanism is

$$Q = 8\pi G r_D (r - r_D)^2 \tag{12.13}$$

in which G is the shear modulus, r_D the doorway radius, and r the radius of the diffusing atom. This formula was derived by Frenkel [89] for the expansion of a hole in a crystal and is applied here for expansion of the silicate network in a viscous liquid. In this application the shear modulus is related directly to the viscosity of the melt. Thus as the viscosity increases, the activation energy for diffusion coefficient decreases. Therefore, the correlation of the fluidity of the melt with the diffusion coefficient of oxygen is consistent with a mechanism of molecular diffusion.

Mott has proposed a theory of viscosity of a silicate liquid in which the concentration of broken Si–O bonds increases as the temperature increases, providing for easier flow [90]. A pressure increase can either increase or decrease the number of broken bonds, depending on the structure of the melt. An increase in the number of broken bonds makes deformation of the doorways easier, and leads to a higher diffusion coefficient. These influences are consistent with a mechanism of molecular diffusion.

Lesher *et al.* [87] measured the diffusion coefficients of oxygen and silicon in a basaltic liquid, whose composition is given in Table 12.7. Measurements were made at temperatures from 1320 to 1600°C and pressures of 1 and 2 GPa with diffusion couples and isotopic exchange (^{18}O and ^{30}Si). The striking result was that the diffusion coefficients of oxygen and silicon were almost the same under all conditions. This result strongly implies that the same diffusing entity transports both oxygen and silicon. Since the oxygen and silicon are bonded very differently in the melt network, this result rules out the possibility of oxygen ion diffusion. I suggest that the diffusing entity in these experiments is silicon monoxide, SiO. This is the simplest and smallest molecular grouping of these atoms. When silicon dioxide vaporizes, the molecules in the gas phase are SiO. See Section 5.5 for a discussion of monoxides. The molecular size of SiO has apparently not been measured directly; it should be about the same as carbon monoxide, because both carbon and silicon have small atomic sizes in covalent compounds (see Chapter 5). Carbon monoxide molecules are about the same size as water molecules from measurements of gaseous viscosity [91] and the considerations described in Chapter 5, so silicon monoxide should diffuse rapidly in silicate melts because it has about the same size as these molecules. Silicon monoxide is also proposed as the carrier of oxygen and silicon in crystalline and amorphous silicon in Chapter 16.

The concentration of silicon monoxide in the gas phase in equilibrium with silicon dioxide depends on the oxidation state of the gas [92]. In pure oxygen the concentration is low, but in a one-to-one CO/CO_2 mixture the SiO pressure at 1500°C is $1.7(10)^{-6}$ atm, or $7(10)^{18}$ atoms/cm^3. If the solubility ratio of dissolved to gaseous SiO is 0.01 (see Chapter 5), the dissolved SiO concentration is about $7(10)^{16}$ molecules/cm^3, which is enough to provide rapid diffusion of oxygen and silicon.

More experiments are needed to clarify the mechanisms of diffusion of oxygen in silicate melts. Residual water and silicon monoxide molecules dissolved in the melt are the most likely carriers of oxygen in silicate melts. The oxygen in these dissolved molecules can exchange with oxygen in the silicon–oxygen network of the silicate melt, giving a profile of isotopic tracer during diffusion measurements. In this situation the diffusion coefficients measured from these profiles are effective diffusion coefficients, because the actual transport of oxygen involves the dissolved molecules. Experiments on carefully dried melts should demonstrate the role of

Table 12.7 Composition of a Basaltic Melt

Oxide	wt %	Oxide	wt %
SiO_2	48.5	MgO	7.6
Al_2O_3	13.8	CaO	10.9
TiO_2	2.7	Na_2O	2.5
FeO	12.6	K_2O	0.4

water in oxygen diffusion. Detection of a low concentration of SiO molecules in a silicate melt will be a challenge to experimentalists.

REFERENCES

1. F. J. Norton, *Nature* **191**, 701 (1961).
2. R. Haul and G. Dumbgen, *Z. Electrochem.* **66**, 636 (1962).
3. E. W. Sucov, *J. Am. Ceram. Soc.* **46**, 14 (1963).
4. E. L. Williams, *J. Am. Ceram. Soc.* **48**, 191 (1965).
5. L. D. Major, M. S. thesis, Case Western Reserve University, Cleveland OH, 1977.
6. E. Rosencher, A. Straboni, S. Rigo, and G. Amsel, *Appl. Phys. Lett.* **34**, 254 (1979).
7. F. Rochet, B. Agius, and S. Rigo, *J. Electrochem. Soc.* **131**, 914 (1984).
8. C. J. Han and C. R. Helms, *J. Appl. Phys.* **135**, 1824 (1988).
9. J. D. Cawley, J. W. Haloran, and A. R. Cooper, *Oxid. Met.* **28**, 1 (1987).
10. J. D. Kalen, R. S. Boyce, and J. D. Cawley, *J. Am. Ceram. Soc.* **74**, 203 (1991).
11. J. C. Mikkelson, *Appl. Phys. Lett.* **45**, 1187 (1984).
12. S. Rigo, F. Rochet, B. Agius, and A. Straboni, *J. Electrochem. Soc.* **129**, 867 (1982).
13. F. Rochet, S. Rigo, M. Froment, C. d'Anterroches, C. Maillot, H. Roulet, and G. Dufour, *Adv. Phys.* **35**, 237 (1986).
14. J. A. Costello and R. E. Tressler, *J. Electrochem. Soc.* **131**, 1944 (1984).
15. J. J. Ganem, I. Trimaille, P. Andreé, S. Rigo, F. C. Stedile, and I. J. R. Baumvol, *J. Appl. Phys.* **81**, 8109 (1997).
16. T. Akermark, J. J. Ganem, I. Trimaille, I. Vickridge, and S. Rigo, *J. Phys. Chem.* **B103**, 9910 (1999).
17. R. H. Doremus, *J. Mater. Res.* **10**, 2379 (1995).
18. R. Pfeffer and M. Ohring, *J. Appl. Phys.* **52**, 777 (1981).
19. R. Pfeffer and M. Ohring, in *The Physics of MOS Insulators*, G. Lucovsky, S. T. Panlelides, and F. L. Galeener, eds., Pergamon Press, New York, 1980, p. 162.
20. E. P. Gusev, H. C. Lu, T. Gustafsson, and E. Garfunkel, *Phys. Rev.* **B52**, 1759 (1995).
21. T. C. Labotka, D. R. Cole, and L. R. Nicikuti, *Am. Mineral.* **85**, 488 (2000).
22. L. A. Simpson and R. E. Carter, *J. Am. Ceram. Soc.* **49**, 139 (1966).
23. T. H. Etsell and S. N. Flengas, *Chem. Rev.* **70**, 339 (1970).
24. A. B. Lidiard, in *Encyclopedia of Physics*, S. Flügge, ed., Springer-Verlag, Berlin, 1957, p. 246.
25. Y. M. Chiang, D. P. Birnie, W. D. Kingery, *Physical Ceramics*, Wiley, New York, 1997, p. 105ff.
26. A. J. Moulson and J. P. Roberts, *Trans. Faraday Soc.* **57**, 1208 (1961).
27. Y. Zhang, E. M. Stolper, and P. D. Ihinger, *Am. Mineral.* **80**, 593 (1995).
28. P. J. Harrop, *J. Mater. Sci.* **3**, 206 (1968).
29. R. Freer, *J. Mater. Sci.* **15**, 803 (1980).
30. B. Gautason and K. Muehlenbachs, *Science* **260**, 518 (1993).

31. J. E. Shelby, *Handbook of Gas Diffusion in Solids and Melts*, ASM International Materials Park, OH, 1996, pp. 84, 98.
32. J. B. Brady, in *Mineral Physics and Crystallography*, T. J. Ahrens, eds., American Geophysical Union, Washington, DC, 1995, p. 269.
33. C. Connolly and K. Muehlenbachs, *Geochim. Cosmochim. Acta* **52**, 1585 (1988).
34. T. Hayashi and K. Muehlenbachs, *Geochim. Cosmochim. Acta* **50**, 585 (1986).
35. P. F. Dennis, in *Progress in Experimental Petrology,* C. M. B. Henderson, ed., Cambridge University Press, Cambridge, 1984, p. 260.
36. S. C. Elphick, C. M. Graham, and P. F. Dennis, *Contrit, Mineral. Petrol.* **100**, 490 (1988).
37. K. P. R. Reddy and A. R. Cooper, *J. Am. Ceram. Soc.* **66**, 664 (1983).
38. K. Ando and Y. Oishi, *J. Chem. Phys.* **61**, 625 (1974).
39. K. P. R. Reddy and A. R. Cooper, *J. Am. Ceram. Soc.* **64**, 368 (1981).
40. K. P. R. Reddy and A. R. Cooper, *J. Am. Ceram.* **65**, 634 (1982).
41. B. A. Thompson and R. L. Strong, *J. Phys. Chem.* **67**, 594 (1963).
42. B. K. Kim, S. J. Park, and H. Hamaguchi, *J. Am. Ceram. Soc.* **77**, 2648 (1994).
43. Z. D. Sharp, B. J. Giletti, and H. S. Yoder, *Earth Planet. Sci. Lett.* **107**, 339 (1991).
44. R. L. Deaton and C. J. Wiedenheft, *J. Inorg. Nucl. Chem.* **35**, 649 (1973).
45. K. Ando, Y. Oishi, and T. Hikada, *J. Chem. Phys.* **65**, 275 (1976).
46. M. O'Keeffe and W. J. Moore, *J. Phys. Chem.* **65**, 1438 (1961).
47. A. W. Smith, F. W. Meszaros, and C. D. Amata, *J. Chem. Phys.* **49**, 24 (1966).
48. J. Belle and A. B. Auskern, *J. Chem. Phys.* **28**, 171 (1958).
49. J. F. Marin and P. Contamin, *J. Nucl. Mater.* **30**, 161 (1969).
50. P. Contamin, J. J. Baemann, and J. F. Marin, *J. Nucl. Mater.* **42** (1972).
51. F. J. Ryerson and K. D. Keegan, *Geochim. Cosmochim. Acta* **58**, 3713 (1994).
52. B. J. Giletti, M. P. Semet, and R. A. Yund, *Geochim. Cosmochim. Acta* **42**, 45 (1978).
53. J. R. Farver, *Earth Planet. Sci. Lett.* **121**, 575 (1994).
54. J. R. Farver, *Earth Planet. Sci. Lett.* **92**, 386 (1989).
55. E. B. Watson and D. J. Cherniak, *Earth Planet. Sci. Lett.* **148**, 527 (1997).
56. R. H. Doremus, *Earth Planet. Sci. Lett.* **163**, 43 (1998).
57. D. K. Moore, D. J. Cherniak, and E. B. Watson, *Am. Mineral.* **83**, 700 (1998).
58. R. H. Doremus, *Glass Science*, Wiley, New York, 1994, p. 269ff.
59. B. L. Phillips, W. H. Casey, and M. Karlsson, *Nature* **404**, 369 (2000).
60. K. Ando, H. Kurokawa, and Y. Oishi, *J. Am. Ceram. Soc.* **64**, C30 (1981).
61. O. Jaoul, B. Houlier, and F. Abel, *J. Geophys. Res.* **88**, 613 (1983).
62. F. J. Ryerson, W. B. Durham, D. J. Cherniak, and W. A. Lanford, *J. Geophys. Res.* **94**, 4105 (1989).
63. D. Derdau, R. Freer, and K. Wright, *Contrib. Mineral. Petrol.* **199** (1998).
64. R. L. Haul, D. Just, and G. Dumbgen, in *Reactivity of Solids, Proceedings,* J. A. DeBoer, ed., Elsevier, Amsterdam, 1961, p. 65.
65. W. K. Chen and R. A. Jackson, *J. Phys. Chem. Solids* **30**, 1309 (1969).
66. W. C. Hagel, *J. Am. Ceram. Soc.* **48**, 70 (1965).

67. D. J. Derry, D. G. Lees, and J. M. Calrert, *J. Phys. Chem. Solids* **42**, 57 (1981).
68. W. J. Moore, Y. Ebisuzaki, and J. A. Sluss, *J. Phys. Chem.* **62**, 1438 (1958).
69. Y. Kiessling, J. Claus, G. Bouchardt, S. Weber, and S. Scheuer, *J. Am. Ceram. Soc.* **77**, 2188 (1994).
70. J. Denk, W. Münch, and J. Maier, *J. Am. Ceram.* **78**, 3265 (1995).
71. W. J. Moore and E. L. Williams, in *Crystal Imperfections and the Chemical Reactivity of Solids,* Faraday Society, Aberdeen, Scotland, 1959, p. 86.
72. J. W. Hoffman and I. Lauder, *Trans. Faraday Soc.* **66**, 2346 (1970).
73. R. Robin, A. R. Cooper, and A. H. Heuer, *J. Appl. Phys.* **44**, 3770 (1973).
74. G. W. Tomlins, J. L. Rutbort, and T. O. Mason, *J. Am. Ceram. Soc.* **81**, 869 (1998).
75. J. B. Goodenough, J. E. Ruiz-Diuz, and Y. S. Zhen, *Solid State Ion.* **44**, 21 (1990).
76. J. B. Goodenough, *Nature* **404**, 821 (2000).
77. T. Takahashi and H. Iwara, *J. Appl. Electrochem.* **3**, 65 (1973).
78. H. A. Harwig and A. G. Gerards, *J. Solid State Chem.* **26**, 265 (1978).
79. T. Ishihara, H. Matsuda, and Y. Takita, *J. Am. Chem. Soc.* **116**, 3802 (1994).
80. F. Abraham, J. C. Bovin, G. M.Vesse, and G. Nowogrocki, *Solid State Ion.* **40–41**, 934 (1990).
81. S. A. Kramer and H. L. Tuller, *Solid State Ion.* **82**, 15 (1995).
82. P. Lacorre, F. Goutenoire, O. Bohnke, R. Retoox, and X. Laligant, *Nature* **404**, 856 (2000).
83. T. Dunn, *Geochim. Cosmochim. Acta* **46**, 2293 (1982).
84. N. Shimizu and I. Kushiro, *Geochim. Cosmochim. Acta* **48**, 1295 (1984).
85. D. Canil and K. Muehlenbachs, *Geochim. Cosmochim. Acta* **54**, 2947 (1990).
86. D. C. Rubie, C. R. Ross, M. R. Carroll, and S. C. Elphick, *Am. Mineral.* **78**, 574 (1993).
87. C. E. Lesher , R. L. Herrig, and D. Tinker, *Geochim. Cosmochim. Acta* **60**, 405 (1996).
88. S. Chakraborty, *Rev. Mineral.* **32**, 441 (1995).
89. J. Frenkel, *Kinetic Theory of Liquids*, Oxford University Press, Oxford, 1946, p. 7.
90. N. F. Mott, *Philos. Mag.* **B56**, 257 (1987).
91. J. Jeans, *An Introduction to the Kinetic Theory of Gases*, Cambridge University Press, Cambridge, 1948, p. 183.
92. H. L. Schick, *Chem. Rev.* 331 (1960).

13

OXIDATION STATE

13.1 INTRODUCTION

The oxidation state of a solid or melt is important in a number of properties, such as chemical reactivity, defect formation and removal, optical absorption of ions with multiple valence states, and removal of bubbles from glass melts. The final stage of sintering a solid from a powder involves removal of voids or bubbles; the gases in them depend on the oxidation state. Formally, the oxidation state of a solid or melt depends on the chemical potential of oxygen in them (see Section 2.2 for a discussion of chemical potential and diffusion). However, control of this chemical potential is highly complex and depends on the history of the solid or melt, such as times, temperatures, atmospheres, and processing or reaction details. There have been few experiments in which these factors have been carefully controlled or studied; thus much of the discussion of this chapter is rather speculative; more definite conclusions await controlled experiments.

Equilibrium of oxidation–reduction reactions in solids in melts is rarely verified. It is often assumed that a solid or melt treated at high temperature (e.g., above 1000°C) reaches equilibrium with the surrounding atmosphere rapidly. However, many oxidizing and reducing constituents of gases diffuse slowly in solids and melts even at these temperatures, for example, molecular oxygen, carbon dioxide, and oxygen ions in the structure of the solid or melt. Nonuniformities in color resulting from different oxidation states of ions are frequently seen in minerals, glasses, and synthetic crystals, showing a slow approach to equilibrium. Minerals and synthetic crystals often cool rapidly from their crystallization or sintering temperatures, trapping gases and stopping oxidation–reduction reactions from reaching equilibrium at lower temperatures. Thus the oxidation state of a solid or melt is usually not at equilibrium at low temperatures, but instead, results from conditions at higher temperatures.

Despite these uncertainties, some likely controlling factors for oxidation states can be identified. These factors lead to ideas for controlled experiments to elucidate the constituents that control the oxidation state. In this book it is shown that gases, especially water and hydrogen, dissolve and diffuse in a wide variety of solids and melts. Thus one can consider the dissolved molecular oxygen as an indicator of the oxidation state of a solid or melt. Thermodynamically, a chemical reaction can be described in terms of minor constituents or compounds present at low concentrations. Thus the oxidation state of a solid is often discussed in terms of bare oxygen ions $O^{=}$ not bonded to other ions. In my opinion the concentration of such ions is negligibly small in oxides; therefore, I prefer to describe the oxidation state in terms of dissolved molecular oxygen, which is always present, even if in low concentration and can be directly related to other dissolved gases and the composition of the surrounding atmosphere. Admittedly, this preference is a matter of taste; it has the advantage of being directly related to practical processing parameters. Thus the chemical potential of oxygen μ_x is related to the concentration (more properly, activity) of dissolved oxygen C_x:

$$\mu_x = \mu_x^0 + R_g T \ln C_x \tag{13.1}$$

in which μ_x^0 is a constant dependent only on temperature, R_g the gas constant, and T the absolute temperature. See books on thermodynamics for discussion of reference states for thermodynamic activity (e.g., Ref. 1). This concentration C_x of dissolved molecular oxygen therefore determines the oxidation state of a solid or melt.

In Section 13.2 the influence of dissolved gases on the oxidation state is discussed; then in Section 13.3, oxidation and reduction of metals is treated. In Section 13.4 the changes of valence states of ions in solids and melts are related to their oxidation states, and in Section 13.5 the thermodynamic phase rule is related to constituents and oxidation states in solids. Finally, in Section 13.6 an unusual mechanism for oxidation of melts by transport of divalent ions is described.

13.2 DISSOLVED GASES

Some dissolved gases that can react in oxidation and reduction reactions are water, hydrogen, oxygen, carbon monoxide, and carbon dioxide. In this section the transport and reactions of these gases are considered. In previous chapters we have shown that water molecules dissolve, diffuse, and react in many different oxides. Natural and synthetic oxides contain from 0.01 to 1 wt % water unless very special efforts are made to dry them. Thus water is a natural candidate to consider first in oxidation–reduction reactions in solids. Water also diffuses rapidly compared to the other oxidizing and reducing gases, except for hydrogen; hydrogen will be considered after water.

The water in oxides is mainly held in reacted form as OH groups. At equilibrium these groups are balanced with dissolved molecular water; for example, the reaction in silicates can be written

$$\text{Si—O—Si} + H_2O + 2\text{SiOH} \tag{13.2}$$

with an equilibrium constant K:

$$K_2 = \frac{S^2}{C_W R} \tag{13.3}$$

in which S is the concentration of OH groups in the solid, C_W the concentration of dissolved molecular water, and R the concentration of unreacted Si–O–Si links. Often, R is large with respect to the other concentrations, and then it remains nearly constant. In Eq. 13.3, K is unitless. At temperatures above about 700°C in crystalline quartz, silica glass, and rhyolite glass, Eq. 13.2 goes to equilibrium rapidly (see Chapters 6 to 8). Thus reacted water (OH groups) in these oxides and most others leads to dissolved molecular water that can diffuse rapidly in the oxide. This dissolved molecular water can control the oxidation state of an oxide. In Chapters 6 to 9 it is shown that in many oxides, oxygen is transported most rapidly by molecular water as compared to transport by diffusion of molecular oxygen, oxygen atoms, or oxygen ions as interstitials or aided by vacancies. In these oxides changes in the oxidation state will therefore be controlled by diffusion of molecular water, and/or molecular hydrogen, as described next.

The relationship between concentrations of dissolved molecular hydrogen, oxygen, and water in a solid is controlled by the reaction

$$2H_2 + O_2 = 2H_2O \tag{13.4}$$

This reaction is governed by an equilibrium constant K_4 with units of 1/concentration:

$$K_4 = \frac{{C_W}^2}{C_X {C_H}^2} \tag{13.5}$$

The C's are concentrations of dissolved molecules of W (water), X (oxygen), and H (hydrogen). The approach to equilibrium for Eq. 13.4 can be slow if the concentrations of dissolved gases are low. From direct measurements of diffusion coefficients D_i in fused silica and the relative sizes of these molecules, their relative rates of diffusion in most oxides are probably $D_H \gg D_W \gg D_X$. Thus one can consider the dissolved molecular oxygen to be relatively immobile and the rate of approach to equilibrium to be controlled by the diffusion of molecular water.

In this paragraph I describe some estimates of the mean diffusion distance x of water molecules in oxides, using the approximate equation derived by Einstein [2] (see Ref. 3):

$$x^2 = 2Dt \tag{13.6}$$

In this equation D is the diffusion coefficient and t is the time. In silica glass the diffusion coefficient of molecular water has been calculated to be $1.7(10)^{-7}\,cm^2/s$ at 1000°C [4] and has been measured directly in rhyolite glass to be $8.6(10)^{-7}\,cm^2/s$ [5–8] and in crystalline pyrope (garnet) to be $6.7(10)^{-7}\,cm^2/s$. In crystalline quartz a value in this range can also be deduced [9]. If D is approximately $2(10)^{-7}\,cm^2/s$, the mean diffusion distance x at 1000°C for molecular water is about 0.04 cm in 1 h and 0.2 cm in 24 h. Thus if a sample or powder has thickness or diameter about 1 mm, it should reach an equilibrium oxidation state after a few hours; a sample of thickness or diameter 1 cm will take several days to reach an equilibrium oxidation state. There are few direct data to test these crude estimates. In a study of a sodium borosilicate melt ($Na_2O{\cdot}1.2SiO_2{\cdot}1.2B_2O_3$) it was found that oxidation–reduction equilibrium was reached after about 1 h at 1000°C [10]. In a sodium borate melt ($3Na_2O{\cdot}7B_2O_3$) at 1100°C, cerous–ceric equilibrium was reached after about 8 h [11]. The dimensions of these samples were uncertain, and there were no measurements of diffusion coefficients of water in them. Wilke and Behrens [12] found that in a melt of approximate base composition 70 wt % SiO_2, 19% Al_2O_3, 7.4% CaO, and 3.5% Na_2O, water had to be added to nominally dry components to reach equilibrium at 1200°C. It is clear that a careful study of times to reach oxidation–reduction equilibrium in crystalline and amorphous materials with known diffusion coefficients of molecular water would be highly desirable to test the present hypothesis that transport of water often controls the state of oxidation of an oxide.

Let us return to Eq. 13.4 and the equilibrium expression, Eq. 13.5. In the gas phase this reaction shows that the partial pressure of oxygen (more properly, fugacity) is controlled by the ratio of partial pressures of water and hydrogen. This well-known property of these gases is often used to control the partial pressure of oxygen; the Ellingham diagram (Fig. 5.6) shows the relationship between $p(H_2O)/p(H_2)p(O_2)$, where the p's are partial pressures. The standard Gibbs free energy at 1000°C for Eq. 13.4 is 352 kJ/mol; from this value $K_4 = 2.645\,(10)^{14}\,atm^{-1}$, from which the values of $p(O_2)$ in Table 13.1 were calculated. I assume that these values for partial pressures in the gas phase, where equilibrium is

Table 13.1 Comparison of Ratios of Partial Pressures of Water to Hydrogen and Carbon Dioxide to Carbon Monoxide with Partial Pressures of Oxygen at 1000°C

$p(H_2O)/p(H_2)$	$p(O_2)$	$p(CO_2)/p(CO)$
$1.63(10)^7$	1	$1.14(10)^7$
$1.63(10)^5$	10^{-4}	$1.14(10)^5$
$1.63(10)^3$	10^{-8}	$1.14(10)^3$
1	$3.78(10)^{-15}$	0.701
10^{-2}	$3.78(10)^{-19}$	$7.01(10)^{-3}$
10^{-4}	$3.78(10)^{-23}$	$7.01(10)^{-5}$

rapidly reached, also apply to concentrations of these gases in solids and melts of oxides above about 700°C. This assumption means that Henry's law is valid for water and hydrogen, so that the molecular concentrations of these gases in the solid or melt are proportional to their partial pressures in the gas phase. The concentration of the relatively immobile molecular oxygen in the solid or melt reaches the equilibrium value of Eqs. 13.4 and 13.5 as a result of diffusion and reaction of hydrogen and water with each other and with the dissolved molecular oxygen.

Another way to view the development of the oxidation-state equilibrium is to consider an initial state consisting of only OH groups in the solid or melt as a result of rapidly cooling or heating from different conditions. At a particular temperature these groups then react to form dissolved molecular water, from Eq. 13.2. To calculate these water concentrations it is necessary to estimate the value of K_2 in Eq. 13.3. In silica glass a value of about 1.1 is found [4], and for a rhyolite melt the value of K_2 extrapolated to 1000°C is about 0.4 [5], so K_2 is of order unity. If silica glass contains 0.2 wt % H_2O, all in the form of SiOH groups, their concentration is about $2.9(10)^{19}\,\mathrm{cm}^{-3}$. The value of R in Eq. 13.3 is $4.41(10)^{22}$O atoms in silica glass, so with $K = 1.1$, the concentration C of dissolved water molecules is about $1.8(10)^{16}\,\mathrm{cm}^{-3}$.

To calculate the equilibrium constant K_4 of Eq. 13.5 for a solid or melt, I assume that at equilibrium and a ratio of $C_W/C_H = 1$, the concentration of dissolved oxygen molecules in silica glass corresponds to the partial pressure of $3.78(10)^{-15}$ atm in the gas phase (see Table 13.1). This partial pressure equals a concentration of $2.2(10)^4$ molecules/cm^3. Since the ratio (Ostwald solubility) of gas to solid concentration for oxygen in silica is about 0.01, this partial pressure corresponds to 220 oxygen molecules/cm^3 dissolved in the silica glass. Therefore, $K_4 = 4.60\,(10)^{-3}\mathrm{cm}^{-3}$ at 1000°C from Eq. 13.5.

With this value of K_4, $C = 1.8(10)^{16}\,\mathrm{cm}^{-3}$, and the assumption that $C_X = C_H/2$ for Eq. 13.4, the concentration of dissolved molecular hydrogen is $5.2(10)^{11}\,\mathrm{cm}^{-3}$ and of dissolved molecular oxygen is $2.6(10)^{11}\,\mathrm{cm}^{-3}$. Thus the ratio of water to hydrogen concentrations is about $3.5(10)^4$, and this concentration of dissolved oxygen molecules corresponds to a partial pressure of oxygen in the gas phase of about $4.6(10)^{-6}$ atm. Thus the presence of water in a solid or melt leads to definition of the oxidation state as a result of Eqs. 13.2 and 13.4.

In a real solid or melt, the situation is usually more complicated. There is often additional dissolved molecular oxygen, which can come from entrapped air during processing, crystallization, or glass formation, or from oxidation–reduction reactions. Thus this dissolved oxygen concentration is strongly dependent on the history of a material. If the concentration of dissolved oxygen is less than 10^{11} molecules/cm^3 in the example above, it will not have much influence on the oxidation state, but if this oxygen concentration is greater, it will change the oxidation state as follows. Assume that the dissolved oxygen concentration is $1.2(10)^{16}$ molecules/cm^3, which corresponds to a partial pressure of about 0.21 atm (air) in the gas phase. Then the main effect of this oxygen is to react with the dissolved hydrogen and reduce its concentration. Since this concentration is quite low, this reaction does not change the oxygen concentration appreciably. From Eq. 13.5 with $K = 4.6(10)^{-3}$, $C_W = 1.2(10)^{16}$, then $C_H = 2.4(10)^9$ molecules/cm^3. The ratio of water to hydrogen concentration is increased to about $7.5(10)^6$ as compared to $3.5(10)^4$ when there is initially no dissolved oxygen, so the former condition is more oxidizing.

These examples show how important sample history is in determining the oxidation state of a solid or melt. Even in materials in which ionic diffusion of oxygen is fast, such as zirconia–yttria or calcia solid solutions, the concentration of dissolved gases is still the main factor controlling the oxidation state of a solid or melt.

Often, there are organic materials and carbonates in real materials, either natural or synthetic. The organic matter reacts with oxygen or water to form carbon dioxide and possibly carbon monoxide and hydrogen. A solid organic compound such as sugar (sucrose, $C_{12}O_{11}H_{22}$) can be added during processing to make a solid or melt more reducing. A possible reaction of sucrose with oxygen is

$$C_{12}O_{11}H_{22} + 12O_2 = 12CO_2 + 11H_2O \tag{13.7}$$

Thus the carbon monoxide and carbon dioxide formed by reactions of organic material or carbonates can also contribute to the oxidation state of a material.

In Table 13.1 the ratio of partial pressures of carbon dioxide to carbon monoxide is compared to the partial pressure of oxygen and the water–hydrogen ratio. The values in the table were calculated from the reaction

$$2CO + O_2 = 2CO_2 \tag{13.8}$$

with equilibrium constant

$$K_8 + \frac{[CO_2]^2}{[O_2][CO]^2} \tag{13.9}$$

in which the brackets represent dissolved concentrations of the components. More exactly, the brackets represent thermodynamic activities; concentrations are used here because the dissolved concentrations are low, resulting in little intermolecular interaction. The equilibrium constant for Eqs. 13.8 and 13.9 is $K_8 = 1.3(10)^{14}$ atm^{-1}, from which the values in the table were calculated. If there are appreciable concentrations of dissolved carbon dioxide or carbon monoxide, they will also influence the oxidation state of a melt or solid. The diffusion coefficient of carbon dioxide in an oxide should be small because of its large size. The carbon monoxide molecule probably has about the same size as a water molecule, so it could diffuse fairly rapidly in a solid or melt. In a study of the diffusion of water and "carbon" in calcite, the effective diffusion coefficient of water (oxygen) was about 100 times greater than that of carbon [13]. Perhaps this result shows that the diffusion coefficient of dissolved carbon monoxide in calcite is appreciable.

The discussions in this section show how the oxidation state of a solid or melt can be controlled by dissolved gases. Thus the role of these gases in oxidation–reduction reactions in solids and melts should always be considered. This

Table 13.2 Standard Free Energy of Some Oxidation–Reduction Reactions

Reaction[a]	$\Delta G_T{}^0$ (cal)[b] A	B	C	Temperature Range (K)
$4\langle Ag\rangle + (O_2) = 2\langle Ag_2O\rangle$	−14,600	—	31.60	298–?
$\frac{4}{3}\langle Al\rangle + (O_2) = \frac{2}{3}\langle Al_2O_3\rangle$	−267,207	−2.653	58.427	298–923
$\frac{4}{3}\{Al\} + (O_2) = \frac{2}{3}\langle Al_2O_3\rangle$	−270,507	−2.500	61.480	923–1800
$\frac{4}{3}\langle B\rangle + (O_2) = \frac{2}{3}\langle B_2O_3\rangle$	−204,070	—	42.3	298–?
$2\langle Ba\rangle + (O_2) = 2\langle BaO\rangle$	−271,600	—	46.400	298–983
$2\{Ba\} + (O_2) = 2\langle BaO\rangle$	−278,000	—	53.200	983–1600
$2\langle Be\rangle + (O_2) = 2\langle BeO\rangle$	−286,900	−3.32	56.100	298–1557
$2\{Be\} + (O_2) = 2\langle BeO\rangle$	−291,900	−3.32	59.320	1557–2000
$\frac{4}{3}\langle Bi\rangle + (O_2) = \frac{2}{3}\langle Bi_2O_3\rangle$	−91,450	—	42.95	298–544
$2\langle C\rangle + (O_2) = 2(CO)$	−53,400	—	41.900	298–2500
$\langle C\rangle + (O_2) = (CO_2)$	−94,200	—	−0.20	298–2000
$2\langle Ca\rangle + (O_2) = 2\langle CaO\rangle$	−302,650	—	47.32	298–1124
$2\{Ca\} + (O_2) = 2\langle CaO\rangle$	−307,100	—	51.28	1124–1760
$2(Ca) + (O_2) = 2CaO$	−380,200	—	93.24	1760–2500
$2(Cd) + (O_2) = 2\langle CdO\rangle$	−174,500	—	99.9	1241–1379
$\langle Ce\rangle + (O_2) = \langle CeO_2\rangle$	−260,900	−5.00	65.25	298–1000
$\frac{4}{3}\langle Ce\rangle + (O_2) = \frac{2}{3}\langle Ce_2O_3\rangle$	−290,000	—	47.13	298–?
$2\langle Co\rangle + (O_2) = \langle CoO\rangle$	−111,800	—	33.8	298–1400
$6\langle CoO\rangle + (O_2) = 2\langle Co_3O_4\rangle$	−87,600	—	33.8	298–1400
$\frac{4}{3}\langle Cr\rangle + (O_2) = \frac{2}{3}\langle Cr_2O_3\rangle$	−178,500	—	41.1	298–2100
$4\langle Cu\rangle + (O_2) = 2\langle Cu_2O\rangle$	−81,000	−7.84	59.0	298–1356
$4\{Cu\} + (O_2) = 2\langle Cu_2O\rangle$	−93,400	−7.84	68.2	1356–1503
$2\langle Cu_2O\rangle + (O_2) = 4\langle CuO\rangle$	−69,900	−12.20	88.6	298–1300
$2\langle Fe\rangle + (O_2) = 2\langle FeO\rangle$	−124,100	—	29.9	298–1642
$2\{Fe\} + (O_2) = 2\{FeO\}$	−111,240	—	21.66	1808–2000
$6\langle FeO\rangle + (O_2) = 2\langle Fe_3O_4\rangle$	−149,240	—	59.80	298–1642
$4\langle Fe_3O_4\rangle + (O_2) = 6\langle Fe_2O_3\rangle$	−119,240	—	67.24	298–1460
$2(GeO) = \langle Ge\rangle + \langle GeO_2\rangle$	−109,200	−13.8	124	—
$2(H_2) + (O_2) = 2(H_2O)$	−114,500	8.96	−4.42	298–2500
$4\langle K\rangle + (O_2) = 2\langle K_2O\rangle$	−173,200	−9.2	89.2	298–?
$4\langle Li\rangle + (O_2) = 2\langle Li_2O\rangle$	−285,200	—	58.68	298–454
$2\langle Mg\rangle + (O_2) = 2\langle MgO\rangle$	−288,700	−5.9	67.9	298–923
$2\{Mg\} + (O_2) = 2\langle MgO\rangle$	−290,700	−0.48	53.9	923–1380
$2(Mg) + (O_2) = 2\langle MgO\rangle$	−363,200	−14.74	151.4	1380–2500
$2\langle Mn\rangle + (O_2) = 2\langle MnO\rangle$	−183,900	—	34.8	298–1500
$2\{Mn\} + (O_2) = 2\langle MnO\rangle$	−190,800	—	39.4	1500–2050
$4\langle Na\rangle + (O_2) = 2\langle Na_2O\rangle$	−200,000	—	64.0	298–1193
$2\langle Ni\rangle + (O_2) = 2\langle NiO\rangle$	−116,900	—	47.10	298–1725
$\langle P\rangle_{white} + (O_2) = \frac{2}{5}\langle P_2O_5\rangle$	−142,600	—	46.50	298–317
$2\{Ni\} + (O_2) = 2\langle NiO\rangle$	−125,300	—	51.96	1725–2200
$2\langle Pb\rangle + (O_2) = 2\langle PbO\rangle$	−105,700	−6.9	68.86	298–600
$2\{Pb\} + (O_2) = 2\langle PbO\rangle$	−107,500	—	52.6	600–760
	−106,600	—	51.4	760–1150

(Continued)

Table 13.2 (Continued)

Reaction[a]	$\Delta G_T{}^0$ (cal)[b] A	B	C	Temperature Range (K)
$2\langle Pd\rangle + (O_2) = 2\langle PdO\rangle$	−44,800	−11.50	73.8	298–1133
$\langle Pt\rangle + (O_2) = (PtO_2)$	−39,270	—	0.93	1373–1823
$\langle Rh\rangle + (O_2) = (RhO_2)$	−45,140	—	4.94	1473–1773
$\langle Ru\rangle + (O_2) = (RuO_2)$	−73,600	−6.9	62.4	298–1850
$\frac{1}{2}(S_2) + (O_2) = (SO_2)$	−86,620	—	17.31	298–2000
$\frac{1}{3}(S_2) + (O_2) = \frac{2}{3}(SO_3)$	−72,813	—	25.78	318–1800
$\langle Si\rangle + (O_2) = \langle SiO_2\rangle$	−215,600	—	41.5	700–1700
$\{Si\} + (O_2) = \langle SiO_2\rangle$	−227,700	—	48.7	1700–2000
$2(SiO) = \langle SiO_2\rangle + \langle Si\rangle$	−169,600	−12.9	124.3	298–1700
$\{Sn\} + (O_2) = \langle SnO_2\rangle$	−140,180	—	51.52	770–980
$2\langle Ti\rangle + (O_2) = 2\langle TiO\rangle$	−244,600	—	42.6	600–2000
$4\langle TiO\rangle + (O_2) = 2\langle TiO_3\rangle$	−228,300	—	38.1	298–2000
$6\langle Ti_2O_3\rangle + (O_2) = 4\langle Ti_3O_5\rangle$	−177,000	—	39.4	700–2000
$2\langle Ti_3O_5\rangle + (O_2) = 6\langle TiO_2\rangle$	−146,000	—	46.0	298–2123
$2\langle V\rangle + (O_2) = 2\langle VO\rangle$	−411,800	—	71.8	900–1800
$4\langle VO\rangle + (O_2) = 2\langle V_2O_3\rangle$	−175,300	—	40.5	823–1385
$2\langle V_2O_3\rangle + (O_2) = 4\langle VO_2\rangle$	−102,800	—	33.5	1020–1180
$4\langle VO_2\rangle + (O_2) = 2\langle V_2O_5\rangle$	−64,480	—	45.6	298–943
$4\langle VO_2\rangle + (O_2) = 2\{V_2O_5\}$	−84,300	—	40.3	943–1800
$2\langle Zn\rangle + (O_2) = 2\langle ZnO\rangle$	−168,200	−13.8	88.28	298–693
$2(Zn) + (O_2) = 2\langle ZnO\rangle$	−230,840	−20.7	164.76	1170–2000
$\langle Zr\rangle_\alpha + (O_2) = \langle ZrO_2\rangle_\alpha$	−259,940	−4.33	59.12	298–1143
$\langle Zr\rangle_\beta + (O_2) = \langle ZrO_2\rangle_\beta$	−260,200	−6.44	65.99	1478–2138

[a]⟨ ⟩, Solid; { }, liquid; ( ), gas.
[b]$\Delta G_T{}^0 = A + BT\log T + CT$ (calories). T, °K.
Source: Ref. 11.

discussion also suggests that the best way to control the oxidation state of a solid or melt is with a gaseous mixture of water and hydrogen. These are the fastest diffusing gases and are the most likely to reach equilibrium above about 600°C. At lower temperatures water usually diffuses too slowly to reach equilibrium in usual experimental times (up to a few days). Hydrogen alone can still be used to reduce the oxidation state of a solid at temperatures below 600°C, as shown by the study by Cronemeyer [14] of the reduction of rutile (TiO_2). He found that hydrogen reduced rutile at 350°C in 24 h, but the reaction was not observable at 300°C. Sometimes the oxidation rate is controlled with oxygen alone or a mixture of carbon dioxide and carbon monoxide. However, molecular oxygen and carbon dioxide diffuse slowly in oxides even at high temperatures (above 1000°C), so often equilibrium is difficult to react with these gases. They do react with residual water in the gas phase, but such reactions result in uncertain concentrations of the oxidizing and reducing gases. Thus the best mixture of gases for controlling the oxidation state of a solid is of water and hydrogen . The use of a combination of metal and oxide in the gas phase to control the oxidation state is discussed in the next section.

13.3 METALS AND OXYGEN

A listing of the Gibbs free energies of oxidation reactions of metals as a function of temperature is given in Table 13.2. The list is more extensive than the reactions given in the Ellingham diagram (Fig. 5.6). These reactions show the relative reactivity of metals with oxygen and therefore also show their relative ease of being reduced in a solid or melt. These data are useful for calculating free energies of reactions, as done in Section 13.2.

Sometimes a "buffer" of a metal plus its oxide is used to attempt to control the oxidation state in a gaseous mixture. The combination of nickel and nickel oxide is a favorite pair. At 1000°C the partial pressure of oxygen in equilibrium with Ni–NiO is $4(10)^{-11}$ atm, and the ratio of partial pressure of water vapor to hydrogen is about 100. I contend that it is better to use a carefully controlled mixture of gases (e.g., water and hydrogen) than a metal–oxide pair to control the oxidation state of the gas and thus also to control the oxidation state of the solid or melt of interest. The reason is that if there is not sufficient metal or oxide, the oxidation or reduction reaction will not go to the equilibrium state, leaving the oxidation state of the gas uncertain. Furthermore, the rate of oxidation or reduction may not be fast enough to reach equilibrium in experimental time; the total amount of reaction depends on temperature, time, and sample size (geometry). Thus the more certain and reliable method is to control the composition of the gas mixture.

13.4 IONS WITH MULTIPLE VALENCE STATES AND THE PHASE RULE

The oxidation state of a solid or melt determines the concentrations of different valence states of certain ions dissolved in the condensed phase. The relative ability to oxidize various ions in solids and melts can be estimated from the thermodynamic properties of the pure oxides. A list of the relative ability to oxidize an ion from one state to another is given in Table 13.3. Thus of the pairs shown in Table 13.3 chromium is the easiest to reduce and cerium the easiest to oxidize.

Oxidation–reduction reactions in solids and melts can be examined by considering the equilibrium between different oxidation states of ions in the glass. For example, consider the equilibrium between solid phases of ferrous and ferric oxide:

$$4\text{FeO} + \text{O}_2 = 2\text{Fe}_2\text{O}_3 \tag{13.10}$$

The equilibrium constant K for this reaction is given by

$$K = \frac{[\text{Fe}_2\text{O}_3]^2}{[\text{FeO}]^4 P} \tag{13.11}$$

Table 13.3 Standard Free Energies at 25°C and 1 atm of Ions with Different Valence States

Reaction[a]	ΔG (kJ/mol O_2)
$\frac{2}{3}Cr_2O_3+O_2=\frac{4}{3}CrO_3$	+29
$As_2O_3+O_2=As_2O_5$	−197
$Sb_2O_3+O_2=Sb_2O_5$	−220
$4MnO+O_2=2Mn_2O_3$	−311
$4FeO+O_2=2Fe_2O_3$	−504
$2SnO+O_2=2SnO_2$	−525
$2Ce_2O_3+O_2=4CeO_2$	−686

[a]Oxidation reactions.

Source: Refs. 15 and 16.

where the brackets denote thermodynamic activities with the pure oxide as the standard state and P is the pressure of oxygen in equilibrium with the solid phase. Oxygen fugacity must be used if the gas phase is nonideal.

The phase rule for this system is

$$P_h + F = C + 2 \tag{13.12}$$

where P_h is the number of phases, F the number of degrees of freedom in specifying the system, and C the number of components. The number 2 results from two environmental variables, pressure and temperature. In complex systems, C is often difficult to determine. It can be defined [1] as "the minimum number of pure chemical substances that are required for the preparation of arbitrary amounts of all phases of the system." For the system of Eq. 13.10, there are two components and three phases, two solids and the oxygen gas. Thus from Eq. 13.12 there is one degree of freedom. This means that if the temperature is fixed, the oxygen pressure has a particular value. This result can also be deduced from Eq. 13.11. In the pure form, the activities of the solid oxides are unity, so $P = 1/K$.

If the oxides are dissolved in a melt, there are three components because the melt adds an additional one, but only two phases, melt and gas. The result is three degrees of freedom, for example, temperature, oxygen pressure, and activity of one oxide in the melt; then the activity of the other oxide is fixed. Equation 13.10 still is valid with the understanding that the oxides are components dissolved in the melt, and Eq. 13.11 represents the reaction equilibrium with the activities of the oxides in the melt. If the temperature and oxygen pressure are fixed, the activity ratio (or the concentration ratio of an ideal solution) of iron plus three to iron plus two is fixed, and the total amount of iron in the melt is the third variable.

A more complicated situation arises if the ion has three possible oxidation states. Copper can occur in melts in the +2, +1, and atomic state. Reaction equations for the pure oxides are

$$4Cu + O_2 = 2Cu_2O \tag{13.13}$$

$$Cu_2O + O_2 = 2CuO \tag{13.14}$$

Equilibrium constants K are

$$K_{13} = \frac{[Cu_2O]^2}{[Cu]^4 P} \tag{13.15}$$

$$K_{14} = \frac{[CuO]^2}{[Cu_2O]P} \tag{13.16}$$

For Eq. 13.13 there are three phases (gas, copper metal, and Cu_2O) and two components, so there is one degree of freedom, and the same is true for Eq. 13.14. If all three pure solid phases are present, there are four phases and three components and still one degree of freedom. If these constituents are dissolved in the melt as copper +1 and copper +2 ions, there are two phases and four components, so there are four degrees of freedom. If the copper atom concentration is greater than the solubility of bulk copper, copper metal particles will form. Then there are three phases and four components, giving three degrees of freedom; the concentration of dissolved copper atoms is fixed by the solubility of copper metal in the melt. Thus, for example, if temperature and oxygen pressure are fixed, there is one more degree of freedom, the total concentration of copper in the melt.

Douglas [17] discussed the thermodynamics of oxidation–reduction reactions in silicate glass and their influence on ionic ratios and colors of ions. For example, he showed how the amount of different oxidation states of copper vary with the oxygen pressure, as shown in Fig. 13.1. At low oxygen pressure (concentration in the glass), copper metal is present and its solubility determines the concentration of dissolved copper atoms, which is independent of the oxygen concentration. Above a certain oxygen concentration, the concentration of dissolved copper atoms begins

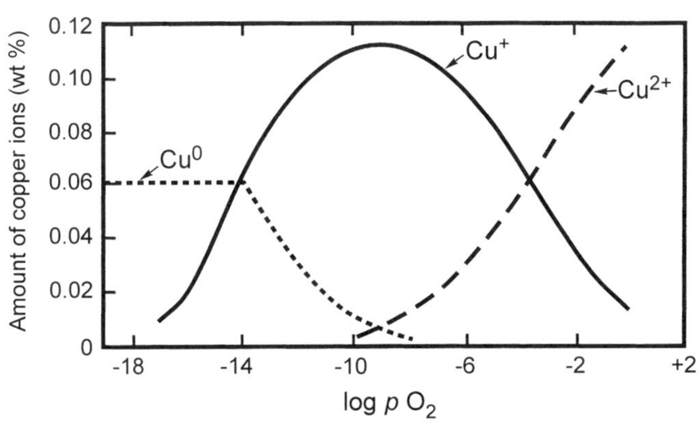

Figure 13.1 Amount of copper in different oxidation states as a function of log oxygen pressure at 1000°C in a $0.3Na_2O{\cdot}0.7B_2O_3$ sample with 0.12 wt % copper. (From Ref. 17.)

to decrease, and there is no copper metal present; the amount of monovalent copper increases to a maximum and then decreases. At high oxygen pressure, only divalent copper is present in appreciable amounts.

Let us now consider the situation in which there are two different ions in a solid or melt with two valence states. As an example, consider iron (Eq. 13.10) and cerium:

$$2Ce_2O_3 + O_2 = 4CeO_2 \tag{13.17}$$

For pure oxides there are now five phases present (FeO, Fe_2O_3, Ce_2O_3, CeO_2, and oxygen gas) and three components, giving zero degrees of freedom. This result means that there is only one temperature and oxygen pressure at which both Eqs. 13.10 and 13.17 are at equilibrium; this condition is determined by the intersection of the lines for these two reactions on the Ellingham diagram.

If these four ion species are dissolved in the melt, there are two phases and four components, giving four degrees of freedom. Two of these are normally the total amount of cerium and iron in the melt. If the temperature and oxygen pressure are fixed, the ratios of the two valent states of the two ions are determined independently by Eqs. 13.11 and 13.18:

$$K_{17} = \frac{[CeO_2]^4}{[Ce_2O_3]^2 P} \tag{13.18}$$

This conclusion is valid only if there is equilibrium between the melt oxygen concentration and the gas phase. If the melt is cooled to some intermediate temperature, for example 500°C, the oxygen concentration in the melt will be fixed by the melt condition ("frozen in"), and will not be in equilibrium with the gas phase. In this situation, the oxygen concentration in the melt will change as the ions in it are oxidized and reduced. The final oxygen concentration will depend on the initial conditions from the melt, such as the amounts of the ions in different oxidation states, and cannot be predicted from thermodynamics. The oxidation–reduction reactions can be combined into one reaction by eliminating oxygen from Eqs. 13.10 and 13.17.

$$Ce_2O_3 + Fe_2O_3 = 2CeO_2 + 2FeO \tag{13.19}$$

$$Ce^{3+} + Fe^{3+} = Ce^{4+} + Fe^{2+} \tag{13.20}$$

The equilibrium constant of Eq. 13.20 is

$$K_{16} = \frac{[Ce^{4+}][Fe^{2+}]}{[Ce^{3+}][Fe^{3+}]} \tag{13.21}$$

Thus the ratios of the different states are interrelated, but their absolute values depend on the oxygen concentration in the melt. At lower temperatures still (e.g., room temperature), the diffusion of water or oxygen is too slow to achieve equilibrium even between neighboring ions, and their oxidation states are determined by conditions at higher temperatures and quenching rates.

13.5 OXIDATION WITH TRANSPORT OF CATIONS

An unusual mechanism of oxidation in electronically conducting oxides has been demonstrated by Cooper and his co-workers [18–20]. When a basaltic glass of composition given in Table 13.4 was heated in air at 550 or 600°C, calcium and magnesium ions diffused to the glass surface, and their motion was compensated by the diffusion of electron holes into the glass. These electron holes oxidized iron +2 ions in the region depleted of Ca^{2+} and Mg^{2+} to iron +3 ions, and sodium ions diffused into the depleted zone to compensate the iron +3 ions tetrahedrally bonded to four oxygen ions. The schematic sketch in Fig. 13.2 shows the flows of holes and ions. Thus the net effect is the formation of CaO and MgO on the glass surface, the oxidation of iron +2 to iron +3, and the flow of sodium ions into the depleted regions, compensated by electron holes, without any diffusion of oxidizing or

Table 13.4 Composition of Basaltic Glass for Oxidation Studies

Oxide	Mol %	Oxide	Mol %
SiO_2	45.92	MgO	14.95
TiO_2	2.13	CaO	12.78
Al_2O_3	8.85	Na_2O	3.40
MnO	0.17	P_2O_5	0.32

Source: Ref. 18.

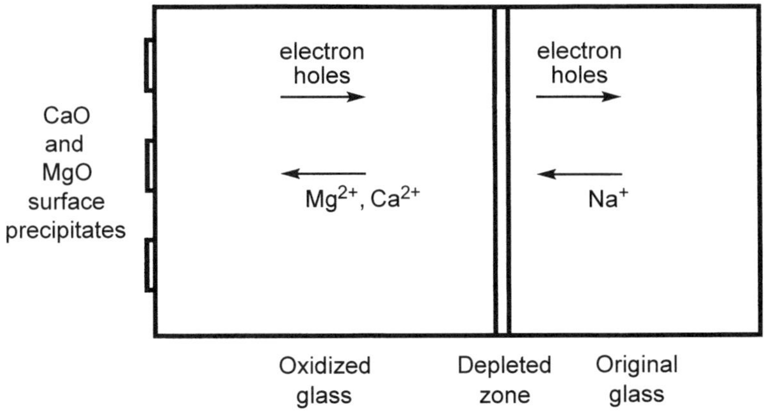

Figure 13.2 Schematic diagram of the transport of ions during oxidation of an electronically conducting glass that contains iron. Fe^{2+} is oxidized to Fe^{3+} in the depleted zone, which moves into the glass.

reducing components. This mechanism is, of course, limited to solids and melts that are electronic conductors and contain mobile ions such as alkaline earths or alkalis. Whether or not this mechanism involving ionic transport is dominant compared to gaseous diffusion depends on the relative rates of ionic and gaseous diffusion. Faster ionic diffusion in an electronic conductor gives this mechanism; faster gaseous diffusion (water) gives control by water diffusion.

REFERENCES

1. F. T. Wall, *Chemical Thermodynamics*, W. H. Freeman, San Francisco, 1974, p. 400ff.
2. A. Einstein, *Ann. Phys.* **17**, 549 (1905).
3. R. H. Doremus, *Rates of Phase Transformations*, Academic Press, San Diego, CA, 1985, p. 26ff.
4. R. H. Doremus, *J. Mater. Res.* **10**, 2379 (1995).
5. R. Zhang, E. M. Stolper, and G. J. Wasserburg, *Geochim. Cosmochim. Acta* **55**, 441 (1991).
6. J. R. Delaney and J. L. Karsten, *Earth Planet. Sci. Lett.* **55**, 441 (1991).
7. K. E. Lapham, J. R. Holloway, and J. R. Delaney, *J. Non-Cryst. Solids* **67**, 179 (1984).
8. R. H. Doremus, *J. Non-Cryst. Solids* **261**, 101 (2000).
9. R. H. Doremus, *Earth Planet. Sci. Lett.* **163**, 43 (1998).
10. J. Y. Tilquin, P. Duveiller, J. Gilbert, and P. Claes, *J. Non-Cryst. Solids* **224**, 216 (1998).
11. A. Paul, *Chemistry of Glasses*, Chapman & Hall, London, 1982, p. 157.
12. M. Wilke and H. Behrens, *Contrib. Mineral. Petrol.* **85**, 102 (1999).
13. T. C. Labotka, D. R. Cole, and L. R. Riciputi, *Am. Mineral.* **85**, 488 (2000).
14. D. C. Cronemeyer, *Phys. Rev.* **87**, 876 (1952).
15. H. J. Tress, *Phys. Chem. Glasses* **1**, 196 (1960).
16. D. D. Wagman *et al.*, *NBS Tables of Chemical Thermodynamics Properties*, National Bureau of Standards, Washington, DC, (1982).
17. R. W. Douglas, in *Proceedings of the 10th International Congress on Glass*, Ceramic Society of Japan, Tokyo, 1974.
18. R. F. Cooper, J. B. Fanselow, and D. B. Poker, *Geochim. Cosmochim. Acta* **60**, 3253 (1996).
19. R. F. Cooper, J. B. Fanselow, J. K. R. Weber, D. R. Merkley, and D. B. Poker, *Science* **274**, 1173 (1996).
20. G. B. Cook and R. F. Cooper, *Am. Mineral.* **85**, 397 (2000).

14

DIFFUSION IN SILICON: FAST AND INTERMEDIATE DIFFUSERS

14.1 INTRODUCTION

Silicon is one of the most important materials of modern technology. Electronic devices, computers, sensors, and controls depend on silicon as an essential component. To make these devices requires control of impurity concentrations on an increasingly small scale. Diffusion of impurities is involved in making junctions and devices. Thus there is a huge literature on diffusion in silicon. In this chapter certain aspects of diffusion in silicon are discussed, including some applications of the diffusion reaction model.

For the information in this chapter and in Chapters 15 and 16, I have relied on a few review articles [1–5] and a large number of individual papers. I have also surveyed several years of chemical abstracts for pertinent articles. Nevertheless, the discussion is based on a selected set of papers that were chosen to demonstrate certain aspects of diffusion in silicon. There is no pretense of a complete survey of data, theories, and commentary on diffusion in silicon. Some important and pertinent papers are undoubtedly neglected.

In Section 14.2 experimental results for diffusion in silicon are compared and diffusing atoms classified into different groups. Some comments on experimental results are included. Then mechanisms of diffusion in fast and intermediate diffusers are discussed, and the role of defects considered. The exchange or kick-out mechanism of diffusion is treated in a separate section because of its importance in silicon. Then the influence of dislocations in silicon on intermediate diffusion is discussed, and diffusion in amorphous silicon. Finally, these discussions are summarized and subsequent chapters on other aspects of diffusion in silicon described.

14.2 CLASSIFICATION OF DIFFUSING ATOMS

In Fig. 14.1 a comparison between experimental measurements of diffusion of different atoms in silicon is shown. In addition, diffusion coefficients of different atoms in silicon at 1156°C are compared in Table 14.1, roughly in order of their values. The activation energies for diffusion in silicon scale roughly with the D values; a low activation energy corresponds with a high diffusion coefficient, and a high activation energy with a low diffusion coefficient. The diffusion coefficients in Fig. 14.1 and Table 14.1 lead to a classification of diffusing substances in silicon, as also given in Table 14.1. There are some chemical characteristics of the various groups, mainly with regard to the valence of the diffusing atoms. The fast diffusers are all of valence +1 or +2. The intermediate diffusers all have multiple valence states, two or more of which are customary. The acceptors have +3 valence almost exclusively, and the donors usually have +5 valence and belong to group V of the periodic table; both acceptors and donors can substitute for silicon in the silicon network structure. The group IV atoms have +4 valence exclusively, except for tin, and can also substitute for silicon. They have the diamond structure in the pure

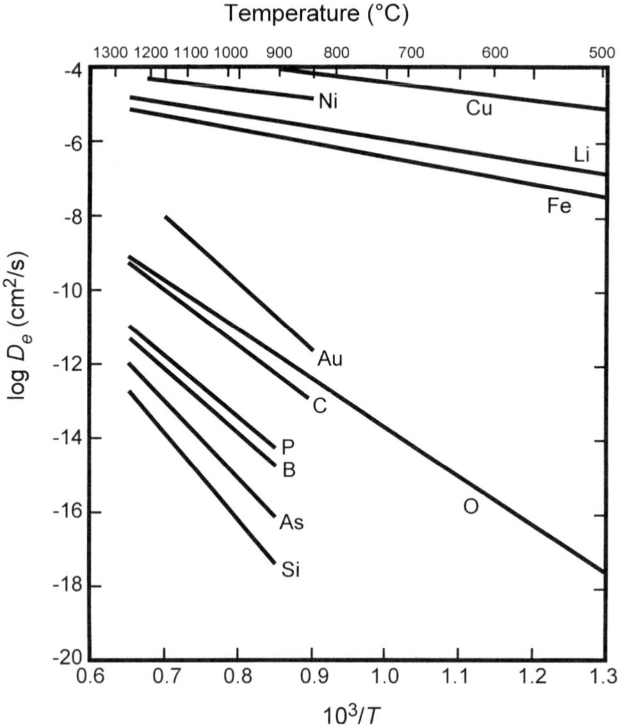

Figure 14.1 Log effective diffusion coefficients of various elements in silicon as a function of reciprocal temperature. Cu, Ni, Li, Fe, and Au, from Frank [1, 2]. O, see Chapter 16. P, B, As, and Si mean values from Chapter 17. (After Ref. 2.)

Table 14.1 Diffusion Coefficients Measured in Silicon at 1156°C and Classification of Elements

Element	D (cm^2/s)	Review Refs.
Fast diffusers		
Chromium	$3(10)^{-6}$	5
Cobalt	$8(10)^{-6}$	5
Copper	$1(10)^{-5}$	2, 3
Iridium	$1(10)^{-6}$	5
Iron	$5(10)^{-6}$	2
Manganese	$4(10)^{-6}$	5
Nickel	$4(10)^{-5}$	2
Palladium	$5(10)^{-5}$	3
Silver	$5(10)^{-5}$	3
Zinc (?)	$2(10)^{-7}$	3
Acceptors		
Aluminum	$7(10)^{-13}$	Chap. 17
Boron	$7(10)^{-13}$	Chap. 17
Gallium	$1(10)^{-12}$	Chap. 17
Indium	$2(10)^{-13}$	Chap. 17
Scadium	$5(10)^{-13}$	Chap. 17
Thallium	$2(10)^{-13}$	Chap. 17
Intermediate diffusers		
Gold	$1(10)^{-8}$	2
Platinum	$1(10)^{-6}$	3
Titanium	$1(10)^{-6}$	3
Vanadium	$3(10)^{-8}$	3
Group VI		
Oxygen	$2(10)^{-10}$	2
Selenium	$3(10)^{-10}$	5
Sulfur	$2(10)^{-8}$	3
Telurium	$8(10)^{-13}$	5
Donors		
Antimony	$1(10)^{-13}$	Chap. 17
Arsenic	$1(10)^{-13}$	Chap. 17
Bismuth	$1(10)^{-13}$	Chap. 17
Phosphorus	$2(10)^{-12}$	Chap. 17
Unclassified		
Cadmium	$4(10)^{-13}$	5
Hydrogen	?	
Nitrogen	$7(10)^{-15}$	5
	$3(10)^{-7}$	3
Potassium	$2(10)^{-11}$	5
Sodium	$4(10)^{-3}$	5
	$5(10)^{-8}$	5
	$8(10)^{-11}$	5
Diamond structure		
Carbon	$1(10)^{-10}$	Chap. 17

(*Continued*)

Table 14.1 (Continued)

Element	D (cm^2/s)	Review Refs.
Germanium	$2(10)^{-14}$	Chap. 17
Silicon	$1(10)^{-14}$	Chap. 17
Tin	$2(10)^{-14}$	Chap. 17

state. Within each grouping the diffusion coefficients at 1156°C vary no more than about two orders of magnitude except for the intermediate diffusers and nonmetals. These groupings are thus significant compared to the overall difference of about 11 orders of magnitude between copper, the fastest atom in silicon, and self-diffusion of silicon. These groupings suggest that mechanisms of diffusion within each group are similar, except for the nonmetals.

A striking aspect of the experimental diffusion data in silicon is that of the large discrepancies between measurements by different experimenters on the same diffusing material. The most extreme example of this spread is for the hydrogen diffusion coefficients in Ref. 3, which vary by more than 10 orders of magnitude at about 300°C. Some of this variation probably results from structural features such as single-crystal and amorphous structures and dislocation content; also, dopant concentrations are often different. Nevertheless, even in samples of the same structure and impurity concentrations, there are large differences between measurements by different investigators. These results suggest that there are uncontrolled factors in many experiments that have substantial influence on diffusion coefficients. This situation makes generalizations such as the groupings and conclusions about mechanisms in subsequent sections rather uncertain.

These differences provide a challenge for future experiments to compare single-crystal silicon samples with low dislocation density and low impurity content. In recent years most experiments on diffusion in silicon were for samples with variable impurity and defect levels, for example from irradiation. It seems premature to deduce mechanistic effects in these samples when other variable factors have not been isolated and controlled.

14.3 INTERSTITIALS, VACANCIES, AND FAST DIFFUSERS IN SILICON

Just after World War II there began an explosive increase in the knowledge of solid-state behavior. It was established that diffusion in metals and alkali halides usually involved vacant lattice sites. The presence of vacancies was proven by experiments such as measurements of bulk volume and lattice parameters as a function of temperature, and by the influence of cooling rates and quenching from high temperatures on diffusion coefficients. These experiments led to a convincing and detailed description of the vacancy method of diffusion in these materials. Both the predominant and impurity atoms that are in the crystal structure can jump into lattice vacancies and therefore diffuse through the crystal. This self or substitutional diffusion of lattice elements is consequently directly related to the concentration of vacant lattices sites. In metals small atoms such as carbon, nitrogen, and hydrogen can dissolve and diffuse as interstitial atoms even in close-packed structures such as face-centered cubic and hexagonal close packed structures and in less closely packed body-centered cubic structures. In alkali halides lattice vacancies of both cations and anions are similarly related to diffusion of these ions.

The satisfying success of the vacancy mechanism in explaining diffusion in these materials naturally led to its application in other materials. Thus diffusion in silicon is almost always described exclusively in terms of interstitial or vacancy diffusion (see Ref. 4 for an extensive review). However, there are at least two structural factors of silicon that are significantly different from those of metals and that influence the mechanisms of diffusion in silicon. First, the silicon structure is much more open than those of metals. If the metal atoms are considered to be spheres, the face-centered cubic and hexagonal close-packed structures contain 26% open space, and the body-centered cubic structure has 32% open space. The hard-sphere model for calculating open space breaks down in silicon because of the strong covalent bonding (see Chapters 5 and 16 for further discussion). If silicon atoms are considered to be 1 Å in radius and the volume of cylindrical bonds is included, a free space of 47% is found for silicon (see Section 5.2). In Fig. 14.2 the (110) plane in silicon is shown; a comparatively large hole exists in the middle of the unit cell. This hole is connected to four equivalent holes by "doorways," so

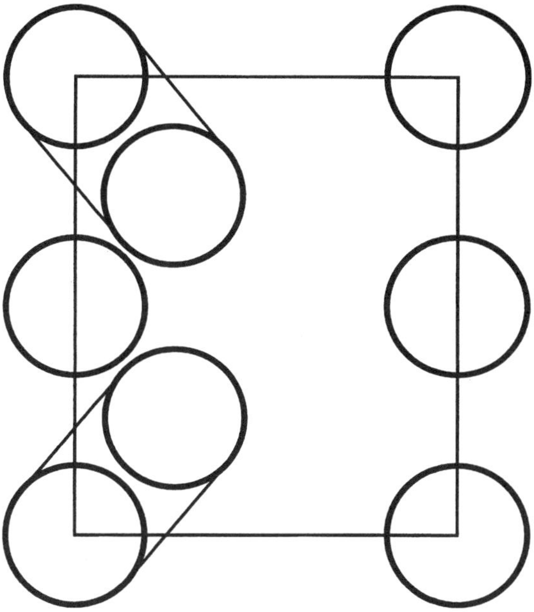

Figure 14.2 Section through a (110) plane in silicon, showing the hole in the middle of the unit cell.

atoms lodged in this hole can easily jump from one hole to another. Thus these interstitial atoms, either silicon or impurity atoms, should diffuse rapidly in the silicon structure. There is experimental evidence for such rapid interstitial diffusion, described in later sections and Chapters 15 and 16.

The second structural feature is the strong covalent bonding between silicon atoms in the diamond structure. The energy of the silicon–silicon bond was estimated by Pauling to be about 255 kJ/mol (2.6 eV). To form a vacancy–interstitial pair in the interior of bulk silicon requires breaking four of these energetic bonds. Thus the equilibrium concentration of these pairs in a silicon crystal should be extremely low.

The strong bonding between silicon atoms in silicon also means that self-diffusion by a vacancy mechanism (or diffusion of vacancies) should be exceedingly slow. For a silicon atom neighboring a vacancy to jump into the vacancy requires breaking three strong silicon–silicon bonds. Thus vacancies in silicon should be highly immobile. Since interstitital atoms can diffuse rapidly in silicon, they should annihilate any excess vacancies in the silicon. The participation of vacancies in diffusion in pure single-crystal silicon with low dislocation density requires their production at the surface of the silicon. However, there is no direct evidence for such production. One would expect the surface to be a sink for the highly energetic silicon vacancies, rather than a source. The silicon surface has also been suggested as a source of silicon interstitials, especially to explain effects of oxidation on diffusion in silicon and other phenomena. However, the experiments on diffusion of gold in silicon produce good evidence that the silicon surface is a sink, not a source, for silicon interstitial atoms. There is no direct evidence for generation of silicon interstitials at the silicon surface.

My conclusion from the discussion above is that the concentration of silicon vacancies in silicon is very low, and any such vacancies in the silicon are immobile. Thus in discussions of mechanisms of diffusion in silicon, I will assume that motion of lattice vacancies does not play a role. Alternatively, the diffusion of interstitial atoms, either of silicon or other elements, should be rapid. The equilibrium concentration of self-interstitials should be very low, but generation of self-interstitials by, for example, the exchange or kick-out mechanism, can lead to significant concentrations of these self-interstitials, and even to their control of diffusion of some elements, as described in Section 14.4.

In view of the considerations above, it is natural to consider the mechanism of diffusion of the fast diffusers in Table 14.1 to be interstitial solution in the open space in the silicon structure (e.g., Fig. 14.2) and rapid jumping from one interstitial site to another. This simple picture should be a good first approximation, but there are some complicating experimental results even for these fast diffusers that are not easy to explain. The diffusion of copper is the fastest of the fast diffusers and has been studied extensively [3, 6–8], so it will be used as an example of fast diffusion in silicon. There is evidence that the copper moves in an electrical field, so it is possibly in the form of positive ions [6]. There is also some evidence of interactions of copper with acceptor atoms in silicon [7, 8]. These interactions are possibly the source of differences between diffusion coefficients measured by different

Table 14.2 Some Approximate Atomic Radii

Atom	Radius (Å)
Aluminum	1.43
Antimony	1.5
Arsenic	1.25
Boron	0.97
Carbon	0.77
Copper	1.28
Gallium	1.3
Germanium	1.22
Gold	1.44
Indium	1.6
Iron	1.24
Nickel	1.24
Nitrogen	0.71
Palladium	1.37
Phosphorus	1.09
Platinum	1.38
Silicon	1.20
Tin	1.6
Titantium	1.4
Vanadium	1.3

investigators. They emphasize again that experiments should be compared on silicons of exactly the same doping compositions. It would be highly desirable to have a series of measurements of important elements in highly pure, intrinsic, and dislocation-free silicon. Nevertheless, it is generally accepted that fast diffusers in silicon diffuse as interstitial atoms or ions and that they do not substitute appreciably for silicon in the silicon structure.

The size of the diffusing atom does not seem to influence the rate of diffusion of fast diffusers very much. Some atomic sizes are given in Table 14.2, and they do not appear to correlate with the diffusion coefficients in Table 14.1. The reason is probably that the holes and doorways in the silicon structure are large enough so that the atomic size is not a critical factor in their rate of jumping from one site to another.

14.4 EXCHANGE OR KICK-OUT MECHANISM AND DIFFUSION OF INTERMEDIATE DIFFUSERS IN SINGLE-CRYSTAL, DISLOCATION-FREE SILICON

Atoms diffusing interstitially in silicon can exchange with silicon atoms in the silicon structure, described by the following equation:

$$\begin{matrix} M_i & + & \mathrm{Si}_n & = & M_n & + & \mathrm{Si}_i \\ C & & N & & S & & I \end{matrix} \tag{14.1}$$

In this equation M is the diffusing atom and subscript i is interstitial and n is network. The symbols below these are those used for the concentrations of the atoms in this section. An equilibrium constant K for Eq. 14.1 is

$$K = \frac{SI}{CN} \tag{14.2}$$

There are two possible diffusing atoms in Eq. 14.1, the interstitial impurity atom and the self-interstitial silicon atom. Thus for a complete treatment of the diffusion process one should solve two simultaneous partial differential equations. Such a general solution is difficult to achieve, so simplifying assumptions are necessary. In their treatment of this problem, Frank and his co-authors [1, 9] have introduced a number of limiting assumptions and special conditions. Different dislocation densities and amorphous structure are considered in subsequent sections. In this section single crystals of low ($< 10^4$ lines/cm^3) dislocation density are considered, so that the only significant sinks for silicon self-interstitials are the sample surfaces.

The equations for diffusion plus exchange from Chapter 3 and Eq. 14.1 are

$$\frac{\partial I}{\partial t} = \frac{\partial}{\partial X}\left(D_I \frac{\partial I}{\partial X}\right) + \frac{\partial S}{\partial t} \tag{14.3}$$

$$\frac{\partial C}{\partial t} = \frac{\partial}{\partial X}\left(D_C \frac{\partial C}{\partial X}\right) - \frac{\partial S}{\partial X} \tag{14.4}$$

These equations describe the diffusion of self-interstitials (I) and interstitials of a diffusing impurity atom (C). A complete solution of these equations is difficult and requires extensive information on reaction 1 and values of diffusion coefficients that is not available. To provide useful solutions that can be compared with experimental data, some simplifying assumptions are necessary. Frank and his co-authors have assumed that there is local equilibrium of Eq. 14.1 in silicon, that is, that Eq. 14.2 is valid for all times, positions, and concentrations. At temperatures above about 600°C this same assumption was valid for diffusion in oxides (see Chapters 6 to 10). Thus it seems to be reasonable that local equilibrium is also established in silicon at temperatures above about 600°C. At lower temperatures it is possible that equilibrium is not reached, and then treatments similar to those in Chapter 4 are required. The possibility of nonequilibrium of Eqs. 14.1 and 14.2 will be ignored in this discussion.

Three limiting conditions for Eqs. 14.3 and 14.4 will be considered here. They seem to be consistent with experimental data on diffusion in silicon, but of course many other conditions are possible. These three different conditions each lead to a different functional dependence of the effective diffusion coefficient D_e, so in principle they can be identified by comparison with experimental data. These three conditions can be briefly described as constant C, constant I, and I proportional to S, with reference to Eqs. 14.1 and 14.2. These three conditions lead to the following functional dependence of the effective diffusion coefficient D_e on exchanged concentration S in the silicon structure: constant C, D_e proportional to $1/S^2$; constant I, constant D_e; I proportional to S, D_e directly proportional to exchanged concentration. In all these conditions the actual diffusing atoms are either silicon interstitials (concentration I) or impurity interstitials (concentration C), not exchanged atoms, which are immobile. In all these conditions it is assumed that the total number of network silicon atoms N available is constant; this assumption means that other concentrations must be below about $5(10)^{21}$ atoms/cm^3, because $N = 5.0(10)^{22}\,\text{cm}^{-3}$.

These three conditions (I, II, and III) are described as follows.

Condition I. In this condition it is assumed that the concentration C of an impurity or dopant atom is constant. This condition requires that the diffusion coefficient of these atoms is high, so that they become uniform rapidly, before other reactions and diffusions are appreciable. This condition was suggested and worked out by Frank and associates, for example in Section 2.4 of Ref. 9. Its validity was confirmed dramatically by experiments on diffusion of gold in silicon [2, 9, 11–15]. The profiles of reacted gold (gold substituted for silicon) are the same when gold is deposited on one surface of a slab or on both surfaces. An example of these U-shaped profiles is shown in Fig. 14.3. These profiles can only be understood if the gold interstitially diffuses rapidly across the entire wafer thickness, and then the dissolved gold concentration remains constant at its equilibrium value. The concentration of *reacted* gold S, which is the quantity measured experimentally, is a function of time and distance (x and t). The profile of S is determined by the diffusion of the silicon self-interstitials formed in reaction 1 to the wafer surfaces. Thus the profiles are the same at the two surfaces, even though gold was deposited only on one surface.

In Eq. 14.2 the product KCN is constant:

$$\text{KCN} = K_I = \text{SI} \tag{14.5}$$

Equation 14.4 is not needed because C is constant. From Eq. 14.5,

$$\frac{\partial I}{\partial X} = -\frac{K_I}{S^2}\frac{\partial S}{\partial X} \tag{14.6}$$

With the assumption that $\partial S/\partial t \gg \partial I/\partial t$, Eq. 14.3 becomes

$$\frac{\partial S}{\partial t} = \frac{\partial}{\partial X}\left(\frac{K_I D_I}{S^2}\frac{\partial S}{\partial X}\right) \tag{14.7}$$

This equation is just Fick's second law with a concentration-dependent effective diffusion coefficient D_e:

$$D_e = \frac{K_I D_I}{S^2} \tag{14.8}$$

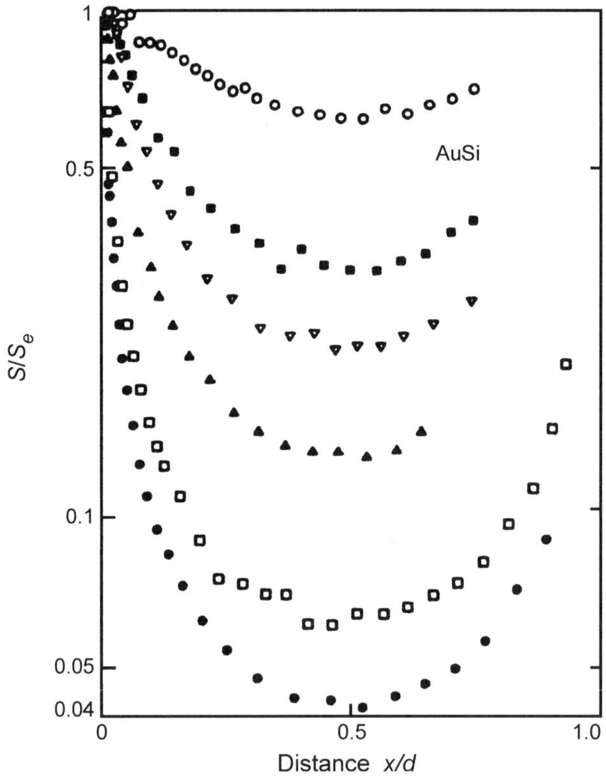

Figure 14.3 Profiles of gold in a dislocation-free single crystal silicon wafer of thickness 0.5 mm. Gold was deposited on one surface only ($x/d = 0$) and the wafer heated at 964°C for different times: •, 0.467 h; □, 1.03 h; ▲, 4.27 h; ▽, 4.27 h, 0.3-mm-thick wafer; ■, 26.8 h; ○, 100.6 h. (From Refs. 2 and 12.) The ordinate is the log of the ratio of gold concentration to the equilibrium gold concentration.

The boundary conditions for the reacted concentration $S(x, t)$ for diffusion from a layer of gold deposited on one or both surfaces of a silicon wafer are

$$S(x, t) = 0 \qquad \text{at } t = 0 \tag{14.9}$$

$$S(0, t) = S(d, t) = S_e \qquad \text{at } x = 0, d \tag{14.10}$$

in which S_e is the concentration of substitutional gold in silicon in equilibrium with the dissolved interstitial gold in equilibrium with the deposited gold layer. Gösele *et al.* [9] show that an approximate solution to Eq. 14.7 for these conditions is

$$S = g(x)\sqrt{D_I t} \tag{14.11}$$

in which g is a function of x only. Then the approximate solution of Eq. 14.7 for boundary conditions 14.9 and 14.10 is [2, 9, 17]

$$\pm \operatorname{erf}\left(\ln \frac{S}{S_m} \right)^{1/2} = \frac{2x}{d} - 1 \qquad (14.12)$$

in which S_m is the reacted concentration at the middle of the slab ($x = d/2$). This solution fails at the slab surfaces ($x = 0, d$) because the calculated concentration becomes large there, but the region in which this excessive concentration occurs is small. Some profiles of gold diffusion of Fig. 14.3 are compared to Eq. 14.12 in Fig. 14.4; the agreement between calculated profiles and experimental ones is excellent.

Equation 14.11 shows that at a particular value of x, such as $x = d/2$ at the center of the wafer, the reacted concentration for the concentration-dependent diffusion coefficient of Eq. 14.8 should be directly proportional to the square root of time. This dependence is quite different from that expected for a diffusion coefficient not dependent on concentration, as shown in Fig. 14.5. This concentration of reacted atoms at the center of the wafer S_m was found to follow the square-root dependence of time over a wide range of time [9, Fig. 2], consistent with the diffusion coefficient of Eq. 14.8.

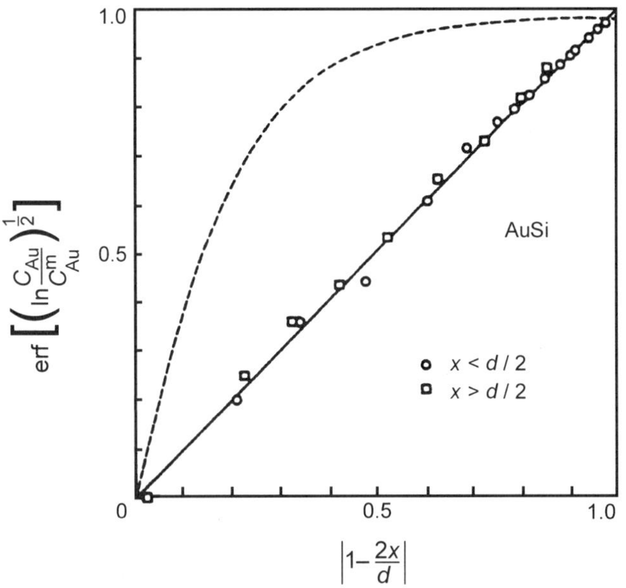

Figure 14.4 The left-hand side of Eq. 14.12 is plotted on the ordinate as a function of distance on the abscissa, for the profiles of gold diffusing in silicon at 964°C for 1.03 h as shown in Fig. 14.3 (points); the solid line is expected from Eq. 14.12, and the dashed line from the solution for a diffusion coefficient independent of concentration. (From Refs. 1 and 2.)

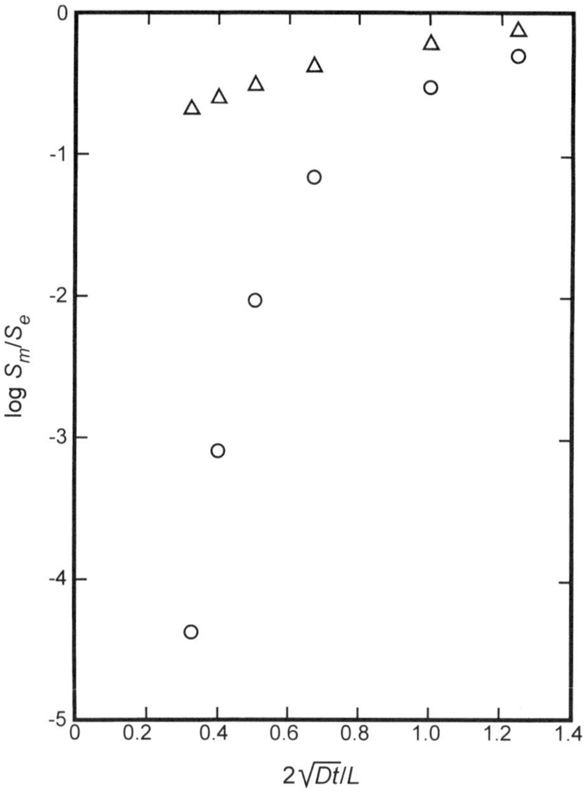

Figure 14.5 Logarithm of relative concentration S_m/S_e at the center of a slab as a function of square root normalized time; for diffusion coefficient inversely proportional to S^2 (△) from Eq. 14.11 and independent of concentration (○). (From Eq. 4.19 of Ref. 14.)

These comparisons of measured concentration profiles and concentrations at the center of the wafer provide good evidence that gold in silicon diffuses by the exchange or kick-out mechanism, with the condition that the interstitial gold diffuses rapidly across the wafer, so that its concentration C is constant with distance and time. This dissolved gold exchanges with silicon atoms in the silicon structure by Eq. 14.1, forming substitutional gold atoms in the silicon structure. The diffusion rate is controlled by the diffusion of interstitial silicon atoms formed by Eq. 14.1 from the interior to the silicon surface, which is a sink for silicon interstitials.

Condition II. A second possibility is that the concentration of silicon interstitials is constant throughout the silicon, and the diffusion is controlled by the impurity interstitials M in Eq. 14.1. In this case the equilibrium expression of Eq. 14.2 contains the constant factor

$$K_{\mathrm{II}} = \frac{KN}{I} \tag{14.13}$$

Then

$$\frac{\partial C}{\partial X} = \frac{1}{K_{\mathrm{II}}} \frac{\partial S}{\partial X} \tag{14.14}$$

and from Eq. 14.4,

$$\frac{\partial S}{\partial t} = \frac{\partial}{\partial X}\left(\frac{D_c}{K_{\mathrm{II}}} \frac{\partial S}{\partial X}\right) \tag{14.15}$$

if $\partial S/\partial t \gg \partial C/\partial t$. In this case the effective diffusion coefficient D_e is

$$D_e = \frac{D_C}{K_{\mathrm{II}}} \tag{14.16}$$

and is not a function of concentration. Then the usual solutions of the diffusion equations for constant diffusion coefficients apply. It is shown in Chapter 17 that this case probably applies to interstitial diffusion and Eq. 14.1 of certain donor atoms in silicon such as arsenic and antimony at high concentrations.

One condition that leads to this case is that the reacted concentration S is relatively high, as compared to condition I, where it is low. For gold in silicon the reacted concentration in equilibrium with solid gold on the silicon surface is about $6(10)^{14}$ gold atoms/cm^3 at 970°C [15]. For donors and acceptors the reacted concentrations in equilibrium with the solid elements are much higher, at about 10^{20} atoms/cm^3 or more at 970°C. I suggest that this difference in reacted concentrations can lead to the different concentration dependencies of the diffusion coefficients for different substitutional atoms in silicon. One reason that the reacted or exchanged concentration of gold in silicon is so much smaller than for group III and V donors and acceptors is possibly that gold has two valence states, +1 and +3, so that a smaller fraction of gold atoms are accommodated in the silicon structure than for the traditional donor and acceptor atoms.

Another condition for which the diffusion coefficient is not a function of concentration is in a tracer diffusion experiment in which all the concentrations C, I, and S are uniform spatially and with time and are equilibrium values. In this situation there is no net reaction, and the effective diffusion coefficient is

$$D_e = \frac{C_T D_c}{S_T} \tag{14.17}$$

in which C_T and S_T are the total (equilibrium) concentrations of interstitial and substituted gold, respectively (see Section 3.3). The diffusion coefficient measured in this experiment is that of interstitial gold, not of interstitial silicon. Some unpublished experiments apparently fit this situation [2].

Condition III. Another possible condition for the exchange reaction 14.1 is that the concentration of self-interstitials I is proportional to the exchanged concentration S. One might expect this condition to occur when the concentration S of reacted atoms is less than for conditon II. A constant factor K_{III} in Eq. 14.2 is therefore

$$K_{\mathrm{III}} = \frac{KN}{\alpha} \tag{14.18}$$

in which α is the proportionality factor between I and S. Then

$$K_{\mathrm{III}} = \frac{S^2}{C} \tag{14.19}$$

and

$$\frac{\partial C}{\partial X} = \frac{2S}{K_{\mathrm{III}}}\frac{\partial S}{\partial X} \tag{14.20}$$

so that Eq. 14.4 becomes

$$\frac{\partial S}{\partial t} = \frac{\partial}{\partial X}\left(\frac{2SD_C}{K_{\mathrm{III}}}\frac{\partial S}{\partial X}\right) \tag{14.21}$$

if $\partial S/\partial t \gg \partial C/\partial t$. D_C is the diffusion coefficient of interstitial impurity atoms, which is assumed to be independent of concentration C. Thus the effective diffusion coefficient D_e is

$$D_e = \frac{2SD_C}{K_{\mathrm{III}}} \tag{14.22}$$

and is directly proportional to the exchanged concentration S. Solutions of the diffusion equation for a diffusion coefficient proportional to concentration are given in Ref. 16, p. 119ff., and in Section 2.6 and Chapters 6 and 8. This dependence is found for certain donor atoms in silicon, as described in Chapter 17.

The detailed experiments on gold diffusion in single-crystal dislocation-free silicon show that they fit the exchange (kick-out) mechanism condition I, as described above. The concentration S_m of platinum at the center of a wafer is proportional to the square root of time [2, Fig. 6], so platinum apparently also diffuses with the effective diffusion coefficient of Eq. 14.8. However, in Ref. 18 a profile of platinum diffusing into silicon shows a region in the middle of the wafer that is flat and deviates from the profile expected from the exchange mechanism, although the profiles near the silicon surfaces are those expected from this mechanism (see Fig. 14.6). The reason for this deviation is apparently that the supersaturation of silicon interstititals away from the surface sinks leads to precipitation of these interstitials to form stacking faults. A light micrograph in the center region of higher platinum concentration shows short streaks attributed to the stacking faults [18]. In Eq. 14.2 a smaller value of I with constant C and N leads to a higher value of S (platinum substituted for silicon in the silicon structure).

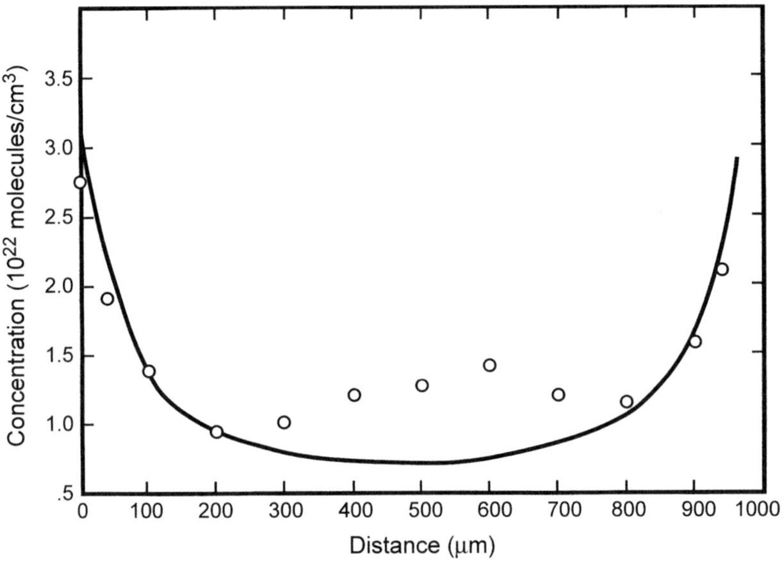

Figure 14.6 Profile of exchanged platinum in dislocation-free single-crystal silicon after 6 h at 1086°C. Points, experimental data (not all points are shown); line, expected from Eq. 14.12. (From Ref. 18.)

Nevertheless, it appears that the diffusion of platinum in silicon follows the exchange mechanism. The reason for the precipitation of interstitials during platinum diffusion and not in most experiments on gold diffusion is probably that the concentration of reacted platinum is about an order of magnitude greater than that of gold [15, 18, 19].

Profiles of gold diffusion with a *W* shape [2, Fig. 19; 13] apparently result in a similar way from formation of stacking faults from condensation of silicon interstitials. These *W* profiles are measured at higher temperatures (1200°C), where the concentration of substitutional gold is high, giving a high supersaturation of silicon interstitials.

Vanadium and titanium have effective diffusion coefficients about two orders of magnitude lower than other transition metals (iron, nickel, cobalt, chromium, and manganese) in silicon, so in Table 14.2 they are classified as intermediate diffusers. However, their diffusion apparently does not follow the exchange mechanism [20–22]. Profiles of diffusing titanium and vanadium are those for a diffusion coefficient independent of concentration, and the dissolved concentrations of these atoms are low (10^{12} to $10^{14}/\text{cm}^3$ at temperatures from 950 to 1200°C), suggesting that these elements diffuse directly as interstitials, without any appreciable exchange with silicon in the silicon structure. These elements do show deep donor and acceptor levels, so there is some interaction of these atoms with the silicon. Perhaps it is this interaction that leads to lower diffusion coefficients of titanium and vanadium in silicon. These elements have more stable higher valance states than those of the other transition metals, which is possibly the reason for their interaction with silicon.

14.5 INFLUENCE OF DISLOCATIONS IN SILICON ON INTERMEDIATE DIFFUSERS

As the dislocation density increases above about $10^4\,\mathrm{cm}^{-2}$, the concentration of gold that replaces silicon in the silicon structure increases. Figure 14.7 shows the exchanged gold concentration as a function of dislocation density [23, Fig. 1] at 960°C. Below about 10^4 dislocation lines/cm^2 the substitutional gold concentration is constant at about $6.3(10)^{14}\,\mathrm{cm}^{-2}$ and increases proportional to the square root of dislocation density at higher densities up to about $2(10)^6$ lines/cm^2, above which density the substitutional concentration levels off at about $4(10)^{15}\,\mathrm{cm}^{-2}$. This increase of gold concentration with dislocation density is interpreted to result from the segregation of silicon interstitial atoms, generated from reaction 1, to the dislocation lines. From Eq. 14.2 with C and N constant, if I decreases, S increases. In Ref. 23 it is assumed that the decrease in concentration I of silicon interstitials results from their removal at the line dislocations resulting from plastic deformation

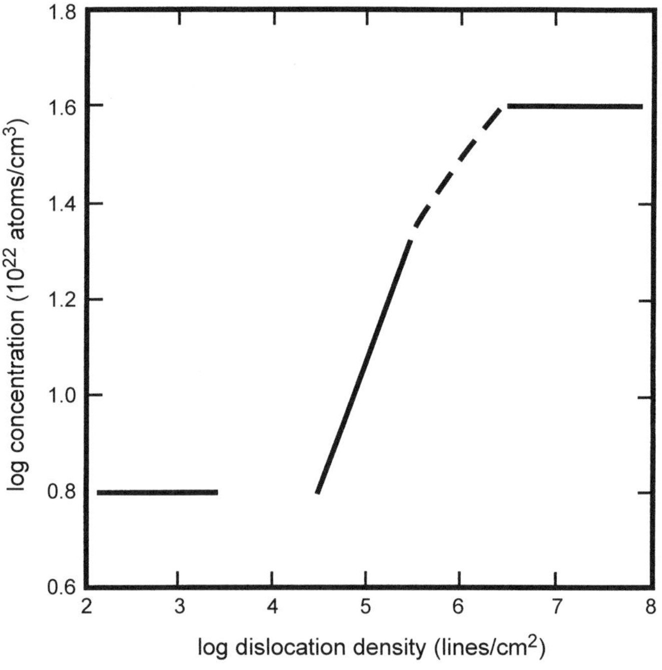

Figure 14.7 Substitutional gold concentration in silicon as a function of dislocation density. The slope of the middle line corresponds to $N_D^{1/2}$. (From Ref. 23.)

of the silicon. However, the usual methods for removal of point defects in a crystal are formation of stacking faults or dislocation climb. There is evidence [2, 23] that stacking faults form in regions of high dislocation density, so they may be the actual mechanism for the removal of silicon interstitials.

At dislocation densities below about $10^4\,\text{cm}^{-2}$ the effective diffusion coefficient of gold is still influenced by the dislocation density, even though the concentration of substitutional gold is the same as for zero dislocation density. From Fig. 2 of Ref. 9, the effective diffusion coefficient of silicon interstitials is about a factor of 14 greater at 10^4 dislocations lines/cm^2 than at zero lines, and the exchange or kick-out mechanism is valid. Thus these dislocation lines act to speed up diffusion of silicon interstitials, but do not reduce their concentration. This pipe or enhanced diffusion at or near a dislocation line increases the mean-square displacement of an atom without changing the form of the diffusion profiles [24, 25].

At high dislocation densities (e.g., 10^7 to $10^9\,\text{cm}^{-2}$) in silicon, the shapes of the profile of substitutional gold or platinum change to those for a diffusion coefficient independent of concentration [13, 18]. In this regime, therefore, profiles of substitutional gold are controlled by the diffusion of gold insterstitials instead of silicon interstitials (condition II in Section 14.4), and the effective diffusion coefficient of gold is about two orders of magnitude higher for a high dislocation density than for a low one. There are apparently two separate reasons for this difference. The diffusion coefficient of silicon interstitials is increased substantially by pipe diffusion along the dislocation lines, and their concentration is reduced by formation of stacking faults caused by the higher supersaturation of silicon interstitials near the dislocation lines. These lines can possibly also remove silicon insterstitials by climb. There is a very similar increase in effective diffusion coefficient of platinum in silicon with a high dislocation density, and a change of mechanism from condition I to condition II as well [18].

14.6 DIFFUSION IN AMORPHOUS SILICON

Diffusion coefficients of several of the fast and intermediate diffusers in amorphous silicon are given in Fig. 14.8. The concentrations of these elements in Fig. 14.8 were much greater than those of the interstitial or substitutional atoms in crystalline silicon. The diffusion coefficients for copper and palladium in crystalline silicon extrapolated from higher temperatures are many orders of magnitude higher than those in amorphous silicon. For gold the diffusion coefficients in dislocation-free crystalline and amorphous silicon are about the same at 1000°C, but the activation energy in amorphous silicon is lower than in crystalline [26]. The diffusion profiles for experiments in Fig.14.8 are those for a diffusion coefficient independent of concentration. The authors in Ref. 26 suggest that the elements in Fig.14.8 diffuse interstitially in amorphous silicon, and are "retarded by temporary trapping at defects intrinsic to the α-Si structure." It is also possible that this "trapping" could be described as exchange with substitutional silicon (Eq. 14.1). Exchange should be easier with the silicon atoms in amorphous silicon; the structure of amorphous

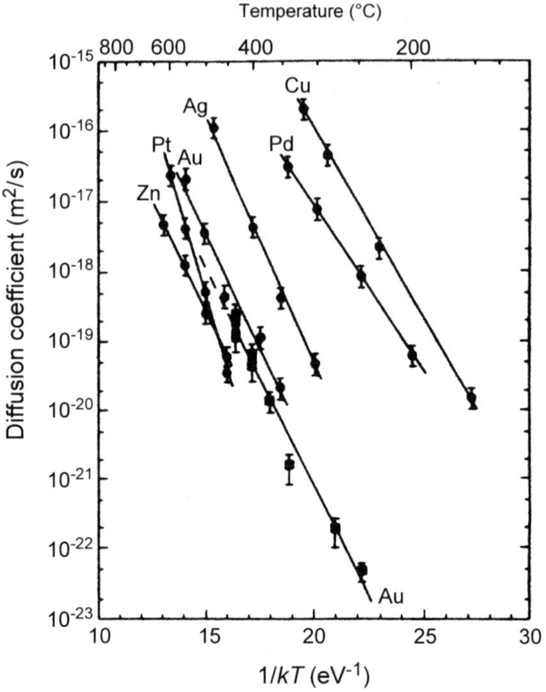

Figure 14.8 Logarithm of diffusion coefficients in amorphous silicon as a function of reciprocal temperature. ●, From Rutherford backscattering; ■, From tracer diffusion. (From Ref. 26.)

silicon is quite similar to that of crystalline silicon, with the difference possibly being some broken and "bent" silicon–silicon bonds. The densities of crystalline and amorphous silicon are about the same. The resultant higher reactivity in the amorphous silicon could lead to higher substitutional concentration of atoms, consistent with condition II of Section 14.3 with an effective diffusion coefficient independent of concentration. More experiments in amorphous silicon are needed to test this possibility for all the elements in Fig. 14.8.

14.7 SUMMARY AND SUBSEQUENT CHAPTERS

In this chapter fast diffusion of impurity atoms in silicon is described as interstitial, with possible minor interactions with other impurities and silicon itself. Two elements, gold and platinum, diffuse interstitially in silicon but can exchange to an appreciable extent with silicon and become substitutional. The concentration profiles of these substitutional gold and platinum atoms in single-crystal dislocation-free silicon are described by the exchange or kick-out mechanism. Dislocations in the silicon can serve as short-circuit paths for gold and platinum and also cause removal of exchanged silicon-self-interstitial atoms, probably by formation of stacking faults.

In subsequent chapters diffusion of other elements in silicon are considered. In Chapter 15, diffusion of hydrogen in crystalline and amorphous silicon is discussed. In Chapter 16, diffusion of oxygen is suggested to result from transport of silicon monoxide, in contrast to the more usual model of diffusion of bonded oxygen. Also in this chapter possible mechanisms of silicon self-diffusion are discussed. In Chapter 17 diffusion of other elements in silicon is discussed, with special emphasis on donor and acceptor atoms.

REFERENCES

1. W. Frank, U. Gösele, H. Mehrer, and A. Seeger, in *Diffusion in Crystalline Solids*, G. E. Murch and A. S. Nowick, eds., Academic Press, San Diego, CA, 1984, p. 63.
2. W. Frank, *Defect Diffusion Forum* **75**, 121 (1991).
3. D. J. Fisher, ed., Diffusion in Silicon: 10 Years of Research, *Defect Diffusion Forum* **153–155** (1998).
4. P. M. Fahey, P. B. Griffin, and J. D. Plummer, *Rev. Mod. Phys.* **61**, 289 (1989).
5. *Diffusion Defect Data* **47** (1986).

Copper

6. C. J. Gallagher, *J. Phys. Chem. Solids* **3**, 82 (1957).
7. C. J. Mesli, T. Heiser, and E. Mulheim, *Mater. Sci. Eng.* **B25**, 141 (1994).
8. A. Zamouche, T. Heiser, and A. Mesli, *Appl. Phys. Lett.* **66**, 631 (1995).

Gold

9. U. Gösele, W. Frank, and A. Seeger, *Appl. Phys.* **A23**, 361 (1980).
10. F. A. Huntley and A. F. W. Willoughby, *Philos. Mag.* **28**, 1319 (1973).
11. J. Hauber, N. A. Stolwijk, L. Tapfer, H. Mehrer, and W. Frank, *J. Phys.* **C19**, 5817 (1986).
12. N. A. Stolwijk, B. Schuster, J. Hölzl, H. Mehrer, and W. Frank, *Physica* **B116**, 335 (1983).
13. N. A. Stokwijk, J. Hölzl, W. Frank, E. R. Weber, and H. Mehrer, *Appl. Phys.* **A39**, 37 (1986).
14. T. K. Monson, J. A. Van Vechten, and Q. S. Zhang, *J. Electrochem. Soc.* **142**, 2077 (1955).
15. S. Coffa, L. Calcagno, S. Y. Campisano, and G. Feria, *J. Appl. Phys.* **69**, 1350 (1991).

16. J. Crank, *Mathematics of Diffusion,* 2nd ed., Oxford University Press, Oxford, 1975, p. 48.
17. G. Bluman and S. Kumei, *J. Math. Phys.* **21**, 1019 (1980).

Platinum

18. W. Lerch, N. A. Stokwijk, H. Mehrer, and Ch. Poisson, *Semicond. Sci. Technol.* **10**, 1257 (1995).
19. S. Coffa, V. Privitera, F. Frisina, and F. Priolo, *J. Appl. Phys.* **74**, 195 (1993).

Titanium

20. S. Hocine and D. Mathiol, *Appl. Phys. Lett.* **53**, 1269 (1988).
21. S. Kuge and H. Nakashima, *Jpn. J. Appl. Phys.* **11A**, 2659 (1991).

Vanadium

22. T. Sadoh and H. Nakashima, *Appl. Phys. Lett.* **58**, 1653 (1991).

23. B. Pichaud, G. Mariami, W. J. Taylor, and W. S. Yang, *Phys. Status. Solidi.* **138**, 465 (1993).
24. E. W. Hart, *Acta Metall.* **5**, 597 (1957).
25. J. W. Christian, *The Theory of Transformation of Metals and Alloys*, Pergamon Press, Oxford, 1965, p. 372ff.
26. S. Coffa, J. M. Poate, D. C. Jacobson, W. Frank, and G. Gustin, *Phys. Rev.* **B45**, 8355 (1992).

15

DIFFUSION OF HYDROGEN IN SILICON

15.1 INTRODUCTION

Hydrogen in silicon improves its electronic properties. Hydrogen reacts with dangling silicon bonds to "passivate" the silicon [1]. In amorphous silicon these dangling bonds are a major defect, and treatment in hydrogen makes the amorphous silicon useful for applications such as solar cells. Polycrystalline silicon has possible applications as active material in photovoltaic devices and large-area thin-film transistors [2]. Again, hydrogen reacts with defects in grain boundaries and on the surface to passivate them. However, in some treatments hydrogen can degrade the electronic properties of the silicon by adding defects, including platelets and dislocations [1]. Therefore, it is important to understand the mechanisms of transport and reaction of hydrogen in amorphous and crystalline silicon.

There is an enormous literature on the diffusion of hydrogen in silicon. There are many short papers in two series of reports of conferences [3, 4], and there are several books on the subject [5–8]. In Ref. 8, which is a summary of papers on diffusion in silicon for the period 1987–1998, there are abstracts of 113 papers on diffusion of hydrogen in silicon. A summary plot in this reference of "quantitative data" from 13 of these studies shows a variation in the diffusion coefficient of hydrogen or deuterium of 11 orders of magnitude at about 300°C. A thorough review, explanation, and reconciliation of the experimental results in this vast literature would be a daunting task. In this chapter I attempt a more modest goal: I have selected and described certain experimental results that provide good evidence that they follow the diffusion–reaction mechanisms described in this book.

Many of the papers on diffusion of hydrogen in silicon demonstrate pernicious aspects of present-day publishing practice. Often, short papers in conference proceedings and letters journals are so fragmentary as to be of little value.

Experimental methods are sketchily described, one or two profiles only are shown in tiny graphs, and there is no attempt to reconcile the data with similar studies. The authors boast of the value of their work, and move quickly to describe a small variation of previous interpretations of data. There is much evidence of lax reviewing. It would seem that future papers on this subject should (1) describe experimental methods carefully and thoroughly; (2) list *all* raw data, preferably in tabular form; and (3) try to compare results with previous ones. These strictures will lead to longer papers, preferably published in full in regular journals.

In the next section, mechanisms for the diffusion and reaction of hydrogen in silicon are discussed. The application of the diffusion–reaction mechanism to hydrogen in silicon is described and contrasted with mechanisms suggested previously. In the following section some equations of the diffusion–reaction model that are pertinent to diffusion of hydrogen in silicon are presented. Next, experimental measurements of diffusion of hydrogen in silicon are compared to the diffusion–reaction model, and finally, the subject is briefly summarized and discussed.

15.2 MECHANISMS OF HYDROGEN DIFFUSION IN SILICON

The usual mechanism proposed for hydrogen transport in silicon is that the hydrogen bonded to silicon breaks away from the silicon and is recaptured at other trapping sites [1–11]. One problem with this mechanism is that the silicon–hydrogen bond energy is about 323 kJ/mol (3.36 eV), which is much greater than the activation energy for the hydrogen diffusion found in most studies. Furthermore, to explain the concentration dependence of the diffusion coefficient of hydrogen in silicon and hydrogen or deuterium profiles, a number of different energies and distributions of trapping sites have been invoked. Nevertheless, the measured profiles, and their time dependence, have not been reproduced especially well.

In this chapter I propose an alternative diffusion–reaction mechanism for the diffusion of hydrogen in silicon. In this mechanism molecular hydrogen dissolves in the silicon and diffuses by jumping between interstitial sites. The hydrogen can react with the silicon to form silicon hydride groups:

$$H_2 + 2Si = 2SiH \tag{15.1}$$

Deuterium as a hydrogen tracer can exchange between dissolved and reacted species:

$$HD + SiH = SiD + H_2 \tag{15.2}$$

The mobile hydrogen species in silica are usually considered to be hydrogen atoms H or protons H^+, with molecular hydrogen being immobile [10]. However, there is no direct experimental evidence for hydrogen atoms or protons. Raman spectroscopy [11] reveals the 4158-cm^{-1} vibrational line for molecular hydrogen H_2, and the Si–H stretching vibration at 2100 cm^{-1}; NMR experiments also confirm Si–H groups [12]. Thus H_2 and Si–H are the only hydrogen-related species in silicon for which there is direct experimental evidence. The 4158-cm^{-1} line is very close to the frequency of the most intense Raman line of gaseous hydrogen of 4155 cm^{-1}. Hydrogen molecules dissolved in crystalline silicon also show a rotational line at 590 cm^{-1} [11]. Thus the dissolved hydrogen molecules do not interact much with the surrounding silicon matrix and are free to jump from one interstice to another in the open silicon structure with low activation energy.

In Chapter 14 the fast diffusion of many different interstitial atoms in silicon is described. For example, copper atoms diffuse very rapidly in silicon. These atoms, of radius 1.28 Å, probably dissolve in the large interstitial holes in the silicon structure and diffuse by jumping through "doorways" between these holes. The hydrogen molecule, with an effective radius of 1.25 Å, should easily dissolve in these hole sites in silicon and diffuse interstitially in the same way as do many metal atoms of the same size. There is no need to postulate a high energy jumping of bound hydrogen atoms in the silicon structure. These bound atoms should be immobile compared to the hydrogen molecules.

At temperatures below about 600°C, hydrogen is often introduced in silicon from a hydrogen plasma. The concentration of hydrogen dissolved in amorphous silicon at its surface is greater in the plasma than in the same pressure of hydrogen without the plasma (e.g., Ref. 13, Fig. 2). The reason for this difference can be understood as follows. Normally, there is an oxide layer on the silicon. The concentration of hydrogen dissolved in the amorphous silicon dioxide is about 0.03 of the concentration of hydrogen molecules in the gas phase [Section 6.3]. The very reactive hydrogen atoms of the plasma react with any surface oxide and remove it as water, so the concentration of hydrogen that can be dissolved in the bare silicon is about 30 times greater than that in silicon covered with oxide. A similar result was found on polycrystalline silicon [2, Fig. 6]; when there was 10 nm of oxide on the silicon, the amount of deuterium that dissolved in the surface of the silicon was much less than when the surface with a very thin oxide layer was treated in a deuterium plasma. See also Ref. 14.

Hydrogen atoms are extremely reactive. The Gibbs free energy for the reaction:

$$H_2(g) = 2H(g) \tag{15.3}$$

is about 360 kJ/mol at 500°C, as deduced from standard thermodynamic quantities [15]. Thus the ratio of molecular hydrogen to atomic is about 10^{13} at 500°C. There is ample time for hydrogen atoms on the silicon surface to react to molecular hydrogen by this reaction. If the diffusion coefficient of hydrogen in silicon is attributed to hydrogen atoms, their mean diffusion distance in 1h is about 400 nm for a diffusion coefficient of about $2(10)^{-13}$ cm^2/s at 500°C; a typical concentration of 10^{18} hydrogen atoms/cm^3 in silicon gives an atomic spacing of about 10 nm. Thus both thermodynamically and kinetically, the concentration of hydrogen atoms in silicon should be negligibly small, even when the surface is in a hydrogen plasma.

Proton diffusion in solids is often postulated. However, a bare proton is also extremely reactive, and there is no reliable experimental evidence for it in solids. Protons are always strongly bonded in solids. Ernsberger has considered proton transport in solids [16, 17] and concludes that "the bare proton cannot exist even momentarily in any solid" and that it is transported only when it is bonded to other constituents. Von Hippel showed that ice does not conduct by proton transport and that ionic defects in ice cannot follow the electric field in an ice crystal [18]. Johnson and Herring [10] diffused hydrogen through a *p-n* junction and deduced that protons were the mobile species and that molecular hydrogen was immobile. These experiments can equally well be interpreted as resulting from transport of molecular hydrogen with immobile Si–H groups, as described in a subsequent section.

The permeation of hydrogen [19] and tritium [20] through the walls of a silicon tube has been studied. The diffusion coefficient and solubility were calculated from the initial and final transients in permeation. However, the presence of a chemical reaction such as in Eq. 15.1 means that these experiments cannot be interpreted as these workers did. For example, the data of Fig. 5 of Ref. 20, of the quantity of tritium in the silicon as a function of time do not extrapolate to the correct value at zero time. It is shown in a later section that these data result from the slow reverse reaction of Eq. 15.1, and not diffusion at all. Original data are not provided in Ref. 19, so it is not possible to interpret these results, and they cannot be used to deduce mechanisms of hydrogen transport in silicon.

The discussions above bring into question previous mechanisms for transport of hydrogen in silicon by hydrogen atoms or protons. Theoretical models for positions of hydrogen atoms and protons give reasonable estimates for bonding energies and configurations [21] but no estimates of the concentrations of these particles. It is likely that these concentrations are extremely small. Thus it seems to be worthwhile to consider the alternative mechanism of transport by molecular hydrogen and reaction to Si–H, since these are the only species that have direct experimental confirmation.

The apparent diffusion coefficient of hydrogen in single-crystal silicon [22] and polysilicon [2] are similar to, although not identical to, those in amorphous silicon, so it is likely that the same mechanism of hydrogen transport is valid for both amorphous and crystalline silicon. Of course, this conclusion is valid only for some studies that are comparable. The wide variation in diffusion coefficients of hydrogen in silicon [8], mentioned previously, shows no trend with the structure of the silicon (single crystalline, polycrystalline or amorphous), suggesting that uncontrolled factors other than structure are causing these wide variations.

The diffusion–reaction model agrees well with experimental results of water diffusion in many different oxides. Inert gases, hydrogen, oxygen, and nitrogen all dissolve and diffuse molecularly in oxides (Chapter 5). Thus it is not surprising that the small hydrogen molecule should dissolve and diffuse easily in the relatively open structure of amorphous silicon, and also in crystalline silicon, which has about the same density as amorphous silicon.

15.3 EQUATIONS

In this section some equations of the diffusion–reaction model, derived in Chapters 3 and 4, are summarized as they apply to diffusion of hydrogen in silicon. An equilibrium constant K_2 for Eq. 15.1 is

$$K_2 = \frac{S^2}{C} \tag{15.4}$$

in which S is the concentration of immobile reacted SiH groups and C is the concentration of dissolved molecular hydrogen. The form of Eq. 15.4 assumes that the concentration of silicon available for Eq. 15.1 is large and constant. In one kind of experiment, hydrogen diffuses into unhydrogenated silicon or out of hydrogenated silicon. The concentration S of reacted hydrogen is much greater than the concentration C of dissolved molecular hydrogen and the concentration S is actually measured in the profile studies. Thus the diffusion coefficient D_e calculated from the profiles of S is an effective diffusion coefficient, because the diffusing material is actually the dissolved interstitial hydrogen. The effective diffusion coefficient for diffusion of hydrogen into unhydrogenated silicon or out of hydrogenated silicon with local equilibrium is

$$D_e = \frac{2SD}{K_2} \tag{15.5}$$

for the equilibrium constant of Eq. 15.4; D is the diffusion coefficient of dissolved molecular hydrogen. This diffusion coefficient is derived in Section 3.2 (Eq. 3.12). From Eq. 15.5 the measured effective diffusion coefficient is proportional to the reacted concentration S, although the diffusion coefficient of the molecular hydrogen D is independent of concentration.

Profiles for the diffusion coefficient of Eq. 15.5 are given in Ref. 24, Figs. 7.5 to 7.7, Section 2.6, and in Figs. 15.1 and 15.2, as described in more detail in the following section. These profiles are quite different from those for a constant diffusion coefficient (error function or conjugate error function for infinite and semi-infinite geometries).

15.3.1 Diffusion of Hydrogen in Silicon, No Local Equilibrium

Consider the situation in which there is a uniform concentration C_T of dissolved molecular hydrogen in amorphous silicon. At equilibrium, from Eq. 15.4, $K_2 = S_T{}^2/C_T$ throughout the silicon, in which S_T is the total concentration of reacted hydrogen. If the system is not at equilibrium, the concentration of reacted hydrogen is less than S_T.

In diffusion experiments into unhydrogenated silicon there is a surface source of hydrogen, either from hydrogen gas or from a solid, usually hydrogenated amorphous silicon. In recent years the source has often been a hydrogen plasma. Any hydrogen atoms in the plasma adsorb on the silicon (actually silica, SiO_2)

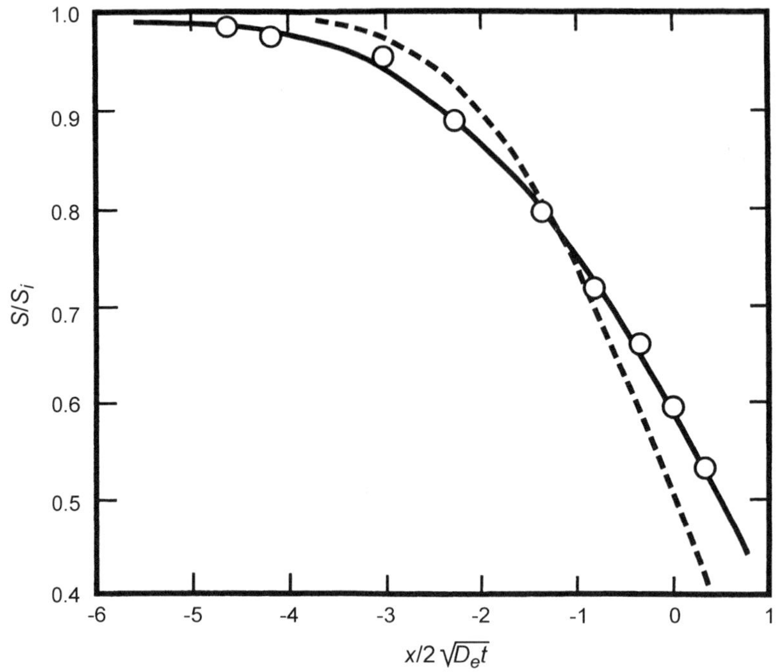

Figure 15.1 Diffusion of hydrogen from initially hydrogenated silicon ($t = 0$, $x < 0$) into unhydrogenated silicon ($t = 0$, $x > 0$). S is the concentration of reacted hydrogen (SiH) as a function of distance x. Line from profile with D_e of Eq. 15.5, and Fig. 2.2; from Fig. 7.6 of Ref. 24; points from data of Song *et al.* [25] after 4h at 300°C; dashed line, erfc profile [14].

surface and recombine rapidly to form molecular hydrogen, as shown by the considerations in Section 15.2. Thus the dissolving species is molecular hydrogen. These sources lead to a certain concentration of dissolved molecular hydrogen in the unhydrogenated silicon at its surface. If the source is hydrogen gas that is not depleted with time, the surface concentration C_T will be constant with time and proportional to the pressure of hydrogen (Henry's law). If the source is some silicon containing hydrogen, with concentrations C_1 and S_1 of dissolved and reacted hydrogen, assumed to be at equilibrium, the relation $K_2 = S_1{}^2/C_1$ holds. At the interface between the hydrogenated and unhydrogenated silicon, the concentration of reacted SiH groups S_0 is $0.59S_1$ at equilibrium for the diffusion coefficient of Eq. 5 (see discussion in Ref. 24, pp. 119–125; see also Fig. 15.1). This situation is perhaps clarified in Fig. 15.3. The ratio of interface concentration C_0 of dissolved molecular hydrogen to C_1 in $C_0/C_1 = (S_0/S_1)^2 = 0.348$. If the forward reaction of Eq. 15.1 is slow, equilibrium may not be reached in the time of diffusion; in this case the surface concentrations S in regions 1 and 2 are different from $0.59S_1$. This difference gives a sharp drop in the experimentally measured hydrogen concentration at the interface (see Fig. 15.3) and is often found experimentally. This drop is sometimes described as *suppressed diffusion*; the present discussion

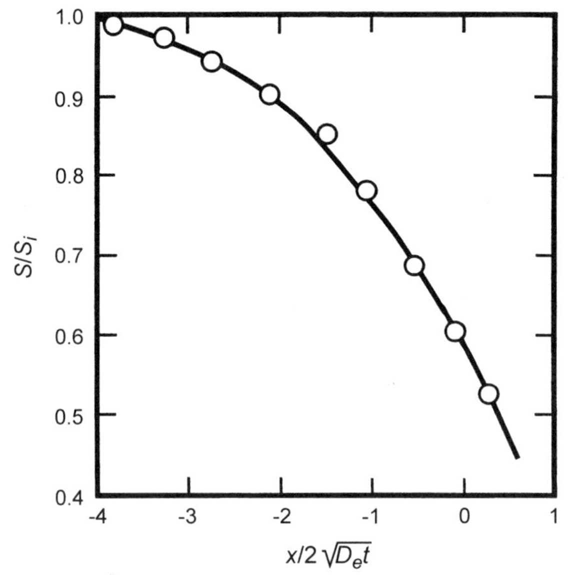

Figure 15.2 Diffusion of hydrogen from initially hydrogenated silicon ($t = 0$, $x < 0$) into unhydrogenated silicon ($t = 0$, $x > 0$). S is the concentration of reacted hydrogen (SiH) as a function of distance x. Line from profile with D_e of Eq. 15.5, and Fig. 2.2; from Fig. 7.6 of Ref. 24; points from data of Song *et al.* [25] after 4h at 250°C. (From Ref. 14.)

shows that it arises naturally in the diffusion–reaction model and does not require invocation of a complicated sequence of different trapping energies of sites. In the discussion above, it is assumed implicitly that the concentration of dissolved molecular hydrogen is always at equilibrium at the interface between the source and the unhydrogenated silicon. See also Sections 4.3 and 4.4.

Another possible experiment is the outgassing of hydrogen from hydrogenated silicon. In this case the concentration of dissolved molecular hydrogen is zero at the surface of the silicon; at equilibrium the concentration of reacted SiH groups (S_0) is also zero. However, at nonequilibrium this concentration can be greater than zero. The rate of reaction to form or remove SiH groups is, from Eq. 15.1 (see also Section 4.4),

$$\frac{\partial S}{\partial t} = \lambda C - \frac{\mu S^2}{S_T} \tag{15.6}$$

in which λ and μ are reaction rate coefficients with units of reciprocal time. S_T is the equilibrium concentration of reacted SiH groups and is included so that the reverse reaction rate coefficient μ has the same units as those of the forward coefficient.

Consider the outgassing of hydrogenated amorphous silicon with an initial concentration S_T of reacted SiH groups. From the discussion above, the

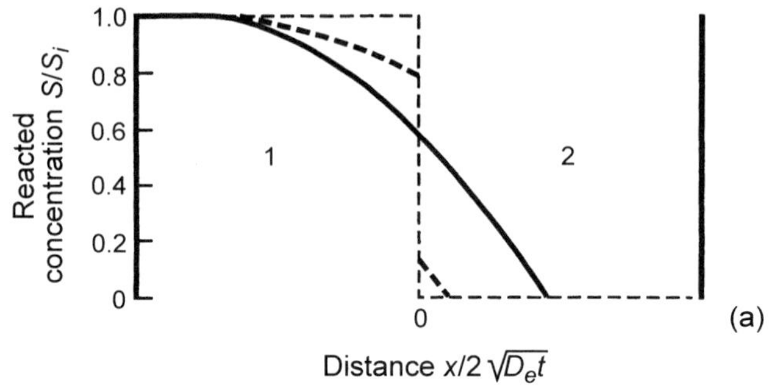

Figure 15.3 (*a*) Schematic diagram of diffusion from hydrogenated amorphous silicon (region 1) into unhydrogenated amorphous silicon (region 2). The initial concentration of reacted hydrogen is the line of light dashes; after some diffusion the profile of reacted hydrogen (SiH groups) is given by the solid line when there is local equilibrium (Eq. 15.5 and Fig. 7.6 of Ref. 24). At $x=0$, $S=0.59S_1$. The line of heavy dashes is the profile of reacted hydrogen with no equilibrium; the interfacial concentrations are given by Eqs. 15.12 and 15.13. (*b*) Schematic diagram of diffusion from deuterated silicon (region 1) into hydrogenated silicon (region 2). It is assumed that the total concentration of dissolved molecular deuterium plus dissolved molecular hydrogen is uniform and constant (constant $C_T = C^* + C$). It is also assumed that the concentration of reacted deuterium or hydrogen (S^* or S) is much greater than the dissolved concentration C_T. The solid line is the concentration profile of reacted deuterium after a time of diffusion, with local equilibrium. The line of heavy dashes is the profile of reacted deuterium with no local equilibrium (interfacial concentrations from Eqs. 15.19 and 15.20).

concentration C_0 of dissolved molecular hydrogen at the silicon surface is zero, so from Eq. 15.9 at the silicon surface,

$$\frac{\partial S_0}{\partial t} = -\frac{\mu S_0{}^2}{S_T} \tag{15.7}$$

in which S_0 is the concentration of reacted SiH groups at the silicon surface. Integration of Eq. 15.7 for limits $S_0 = S_T$ at $t=0$ and S_0, t gives

$$\frac{S_0}{S_T} = \frac{1}{1+\mu t} \tag{15.8}$$

At long times $\mu t \gg 1$, S_0 goes to zero, and at $t = 0$, $S_0 = S_T$, as required by the boundary conditions above.

Now consider the diffusion of hydrogen into unhydrogenated amorphous silicon from hydrogen gas that maintains a fixed concentration C_T of dissolved hydrogen in the silicon at its surface. Then integrating Eq. 15.6 from $S_0 = 0$ at $t = 0$ to S_0, t gives

$$\frac{S_0}{S_T} = \frac{1 - e^{-2\mu t}}{1 + e^{-2\mu t}} \tag{15.9}$$

using $C_T = S_T{}^2/K_2$. At long times $\mu t \gg 1$, $S_0 \rightarrow S_T$, and at short times $\mu t \ll 1$, $S_0 \rightarrow 0$. If the source is hydrogenated silicon with concentration S_1 of reacted hydrogen, Eq. 15.9 can be written as

$$\frac{\partial S}{\partial t} = \lambda C - \frac{\mu S^2}{S_1} \tag{15.10}$$

At equilibrium at the interface $x = 0$ between hydrogenated and unhydrogenated silicon, $S_0 = 0.59 S_1$. Let us assume that the concentration C_0 at the interface is constant with time and equal to $(.59)^2\ C_1 = 0.348 C_1$, which should be valid after a short initial transient. Then at $x = 0$, Eq. 15.10 becomes (see Section 4.4)

$$\frac{\partial S_0}{\partial t} = 0.348 S_1 \mu - \frac{\mu S_0{}^2}{S_1} \tag{15.11}$$

since $C_1 = S_1 \mu / \lambda$. Integrating Eq. 15.11 with the limit $S_0 = 0$ at $t = 0$ gives

$$\frac{S_0(2)}{0.59 S_1} = \frac{1 - e^{-1.18\mu t}}{1 + e^{-1.18\mu t}} \tag{15.12}$$

for the concentration of reacted SiH groups $S_0(2)$ in region 2 (the unhydrogenated silicon) at the interface $x = 0$. When $t = 0$, $S_0(2) = 0$, and as $1.18\,\mu t \gg 1$, $S_0(2) \rightarrow 0.59 S_1$, or equilibrium at the interface. For region 1 (hydrogenated silicon), Eq. 15.11 is integrated from $S_0 = S_1$ at $t = 0$, giving

$$\frac{S_0(1)}{0.59 S_1} = \frac{1 + 0.258\, e^{-1.18\mu t}}{1 - 0.258\, e^{-1.18\mu t}} \tag{15.13}$$

The factor 0.258 arises in the integration and equals $(1 - 0.59)/(1 + 0.59)$. As $t \rightarrow 0$, $S_0(1) \rightarrow S_1$, and as $\mu t \ll 1$, $S_0(1) \rightarrow 0.59 S_1$, or equilibrium at the interface. The difference $S_0(1) - S_0(2)$ gives the "step" at the interface that has been found in many experiments on diffusion of hydrogen from hydrogenated to unhydrogenated amorphous silicon.

15.3.2 Diffusion of Deuterium into Hydrogenated Silicon (Exchange)

In this kind of experiment, hydrogenated and deuterated silicon are placed in contact at time zero. For simplicity, assume that the total concentrations of hydrogen plus deuterium in the two regions are the same. Let us first assume that local equilibrium holds throughout the silicon. It is shown in Section 3.3 that with Eq. 15.2 and local equilibrium, diffusion is governed by an effective diffusion coefficient

$$D_e = \frac{DC_T}{S_T + C_T} \approx \frac{DC_T}{S_T} \tag{15.14}$$

For hydrogen in silicon $S_T \gg C_T$. From Eq. 15.14 the effective diffusion coefficient D_e is constant with time and space, so the solutions for profiles are just those for a constant diffusion coefficient. For semi-infinite geometry with a constant surface concentration $S_0{}^*$ the solution is

$$S^* = S_0{}^* \operatorname{erfc}\left(\frac{x}{2\sqrt{D_e t}}\right) \tag{15.15}$$

in which erfc is the conjugate error function (see Section 2.3) and $S_0{}^*$ is the concentration of reacted deuterium (SiD) at the interface between the deuterium and hydrogen containing silicons. S^* is the concentration of SiD groups. At equilibrium $S_0{}^* = S_1{}^*/2$, in which $S_1{}^*$ is the initial deuterium concentration in the deuterated silicon. If all the isotope in the deuterated region is initially deuterium, $S_1{}^* = S_T$; this condition is assumed in the following treatment.

If local equilibrium does not hold, the interface concentration $S_0{}^*$ of reacted deuterium will be smaller than $S_T/2$ in the hydrogenated region (region 2) and larger than $S_T/2$ in the deuterated region (region 1) at the interface. Thus again as above, there will be a step change in the deuterium concentration at the interface for nonequilibrium, and this step change does not represent suppressed diffusion. This step can be calculated as follows. From the exchange reaction (Eq. 15.2),

$$\frac{\partial S^*}{\partial t} = k^1(C^*S - CS^*) \tag{15.16}$$

in which k^1 is a reaction rate coefficient with units of reciprocal concentration and time. From Eqs. 3.15, 3.16, and 15.16,

$$\frac{\partial S^*}{\partial t} = k^1(C^*S_T - C_T S^*) \tag{15.17}$$

The molecularly dissolved hydrogen and deuterium interdiffuse, and at the interface between regions 1 and 2 their concentrations are equal, so that $C^* = C_1^*/2 = C_T/2$; then, from Eq. 15.17,

$$\frac{\partial S_0^*}{\partial t} = k\left(\frac{S_T}{2} - S_0^*\right) \tag{15.18}$$

in which $S_0{}^*$ is the interface concentration of deuterium in either region 1 or 2, and S_T is the total of the hydrogen and deuterium concentrations; it is assumed that initially only deuterium is in region 1 and only hydrogen in region 2. In Eq. 15.18, $k = k^1 C_T$.

To find the interface concentration in region 1, Eq. 15.18 is integrated with limits of S_T and $t = 0$ to $S_0{}^*(1)$ at t. Then

$$S_0{}^*(1) = (e^{-kt} + 1)\frac{S_T}{2} \tag{15.19}$$

Similarly, for $S_0{}^*(2)$ in region 2, Eq. 15.18 is integrated from 0 at $t = 0$ to $S_0{}^*(2)$:

$$S_0{}^*(2) = (1 - e^{-kt})\frac{S_T}{2} \tag{15.20}$$

Thus the step height is

$$S_0{}^*(1) - S_0{}^*(2) = S_T\, e^{-kt} \tag{15.21}$$

Initially, the step height is S_T, and at long times it is zero.

A complete solution for diffusion and exchange and nonequilibrium is given in Chapter 4. If the dimensionless time kt is less than about unity, there is a steady-state solution in which the profile of molecularly dissolved deuterium is exponential and constant with time (Section 4.2) [24, p. 334]:

$$C^* = C_0{}^* \exp\left(- x\sqrt{\frac{k}{D_e}} \right) \tag{15.22}$$

In Eq. 15.22, $k = k^1 C_T$ and D_e is from Eq. 15.14. The profile of SiD is

$$S^* = \frac{S_T\, kt}{2} \exp\left(- x\sqrt{\frac{k}{D_e}} \right) \tag{15.23}$$

since $C_0{}^* = C_T/2$. If one tries to fit these exponential profiles with an erfc solution such as Eq. 15.15, the resulting effective diffusion coefficient becomes time dependent.

15.4 COMPARISON WITH EXPERIMENTS

In most reports of hydrogen diffusion in silicon, only one or two profiles are given, so it is impossible to make detailed comparisons with the present model at different temperatures, times, and concentrations. I have chosen two studies for detailed comparison, because they report several profiles at different temperatures. Other diffusion results are also compared to the model.

15.4.1 Song et al.

Song *et al.* [25] measured profiles of hydrogen in a couple, one side hydrogenated and the other not. This boundary condition forms an *infinite couple*, as shown in Fig. 15.3. The couples were held for 4 h at temperatures from 150 to 300°C. From reaction 15.1 one expects an effective diffusion coefficient directly proportional to concentration (Eq. 15.5). Two measured profiles are given in Figs. 15.1 and 15.2, and fit very closely to the calculated profiles for an infinite couple with the diffusion coefficient of Eq. 15.5 [24, Fig. 7.6]. The measured profiles do not fit the error function solution. The time of diffusion was long enough that local equilibrium was established. Effective diffusion coefficients calculated from the profiles at different temperatures are given in Table 15.1. These diffusion coefficients fit an Arrhenius relation:

$$D = D_0 \exp\left(-\frac{Q}{RT}\right) \tag{15.24}$$

in which D_0 is and Q are constants, R the gas constant, and T the absolute temperature. From the data in Table 15.1, Q is about 15 kJ/mol (0.15 eV) and $D_0 = 5.6(10)^{-15}\,\mathrm{cm^2/s}$.

15.4.2 Zellama et al.

Zellama *et al.* [26] measured profiles of hydrogen with nuclear reactions after diffusion of hydrogen out of thin films of hydrogenated amorphous silicon, at temperatures from 300 to 640°C. The films were heated for 30 min in argon, so the surface concentration of dissolved molecular hydrogen was zero. At temperatures of 520°C and below, the surface concentration of reacted SiH groups was greater than zero, showing that local equilibrium was not reached at these temperatures. Values of the reaction coefficient μ were calculated from Eq. 15.8 for these different temperatures and surface concentrations, and the results are shown in Table 15.2 and Fig. 15.4. The activation energy of the least-squares line in Fig. 15.4 is 36 kJ/mol (0.37 eV).

The profiles of hydrogen measured by Zellama *et al.* [26, Fig. 9] (Fig. 15.5) were in a thin film of amorphous silicon on a substrate of fused silica. The profiles

Table 15.1 Effective Diffusion Coefficients[a]

T (°C)	D_e (cm^2/s $\times$ 10^{16})
150	0.87
200	1.66
250	1.83
300	2.63

[a]Calculated from the profiles of Song *et al.* [25] and the solution of Eq. 15.5 [24, Fig. 7.6].

Table 15.2 Reaction Coefficients μ and Effective Diffusion Coefficient D_e Calculated from Annealing Hydrogenated Amorphous Silicon for 30 min at Different Temperatures[a]

T (°C)	S_0/S_i	μ ($s^{-1} \times 10^4$)	D_e ($cm^2/s \times 10^{13}$)
300	0.68	2.6	
350	0.385	8.9	
400	0.30	13.0	0.67
440	0.22	19.7	3.3
460	0.18	25.3	5.8
480	0.155	30.3	11
520	0.065	73.8	28
540	—	—	36
560	—	—	58
580	—	—	69

[a]Calculated from the data of Zellama *et al.* [26, Fig. 9]. μ is calculated from surface concentrations of SiH groups and Eq. 15.8. D_e is calculated from fits in Fig. 15.5 to experimental hydrogen profiles. In Fig. 9 of Ref. 3, the curves for 350 and 300°C have been switched, as is clear from Table II in the paper.

showed that hydrogen did not diffuse into the substrate and that the hydrogen concentration at the amorphous silicon–substrate interface decreased with time. These conditions complicate the analysis of the profiles. Rigo *et al.* [27] solved the diffusion equation 3.3 with this thin-film geometry and the relation

$$\frac{\partial S}{\partial t} = \lambda C - \mu S \tag{15.25}$$

that is, for a first-order reverse reaction. The resulting profile of S is proportional to a summation of terms with a distance factor

$$\frac{\cos \pi(n + 1/2)(a - x)}{a} \tag{15.26}$$

in which $n = 0,1,2\ldots$, a is the film thickness, and x is the distance into the film from its surface (see Fig. 15.5). This profile does not fit the experimental profiles of Zellama *et al.* because the reverse reaction rate is actually proportional to S^2. A complete solution for this binary reverse reaction and the film geometry is apparently not available, so I have used the solution for semi-infinite geometry with the diffusion coefficient of Eq. 15.5 and the method of Jain [28] (see Section 2.4) to obtain an approximate solution.

Consider the following boundary conditions, which fit those of the experiments of Zellama *et al.* At $t = 0$, $S = S_T$ in the film of thickness a for all x; that is, initially, the concentration S_T of reacted SiH groups is uniform in the film. At the surface $x = 0$ the concentration of dissolved molecular hydrogen is zero for $t > 0$, but the concentration of reacted hydrogen S_0 at $x = 0$ can be greater than zero and follows

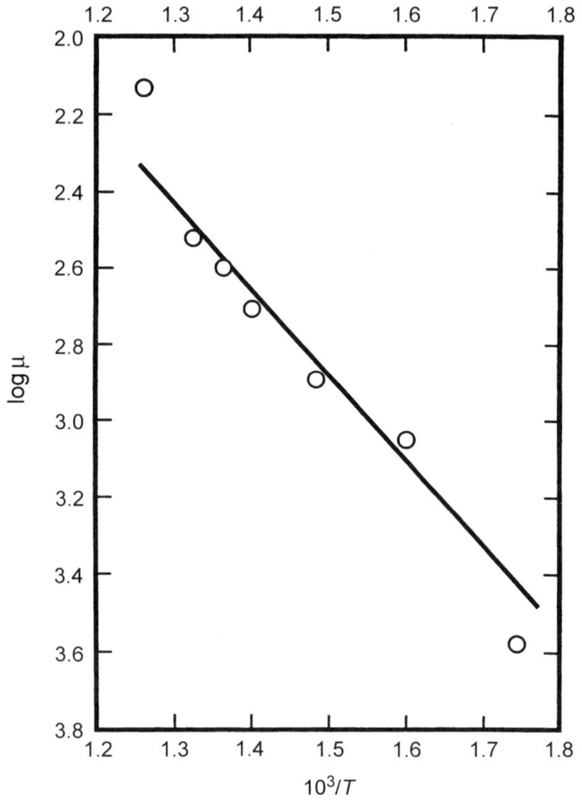

Figure 15.4 Log of the reaction rate coefficient μ as a function of $1/T$; μ calculated from Eq. 15.8 and the profiles of Zellama *et al.* [26, Fig. 9]; time, 30 min. (From Ref. 14.)

Eq. 15.8. The profile of hydrogen for semi-infinite geometry is a function $f(x/\sqrt{D_e t})$ of the distance x, the time t, and a constant effective diffusion coefficient $D_e = 2S_T D/K_2$, taken from Eq. 15.5 with $S = S_T$. This function f gives the profiles in Figs. 15.1 and 15.2 and comes from Fig. 7.6 of Ref. 24. In Figs.15.1 and 15.2 the profiles are $S/S_T = f(x/2\sqrt{D_e t})$, with $S_0 = 0$. In the method of Jain, two semi-infinite profiles for the two surfaces of a film or plane sheet are added to give an approximate solution for the profile in the sheet:

$$\frac{S}{S_0} = f\left(\frac{x}{2\sqrt{D_e t}}\right) + f\left(\frac{2a - x}{2\sqrt{D_e t}}\right) \tag{15.27}$$

Jain showed that this method gives a quite accurate solution for the profiles in the plane sheet, well beyond the reduction of S at $x = a$ below $S/S_T = 0.5$ (see Section 2.4). Profiles calculated from Eq. 15.27 are compared with the experimental data of Zellama in Fig. 15.5, and agree well. Apparent diffusion

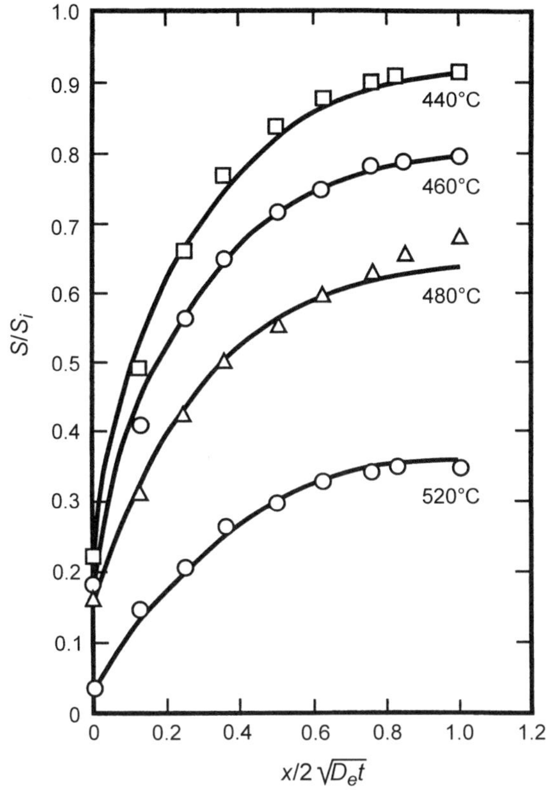

Figure 15.5 Profiles of hydrogen as a function of distance. Points from Fig. 9 of Ref. 26; lines from the diffusion–reaction model; see the text for details of the calculation of the profiles. Temperatures on the curve; time, 30 min. (From Ref. 14.)

coefficients calculated from these profiles and others are given in Table 15.2 and plotted in Fig. 15.6. The activation energy for this plot is about 122 kJ/mol (1.3/eV), with $D_0 = 2.8(10)^{-4}\,\text{cm}^2/\text{s}$.

15.4.3 Interface Concentration Steps and Exponential Profiles

A careful distinction must be made between two different kinds of interface steps between deuterated and hydrogenated silicon. In the first the step is caused because reaction 15.2 has not gone to local equilibrium at the interface. This step height decreases with time until it disappears when equilibrium between the deuterated and hydrogenated regions is reached. A direct example of this type of step is in Figs. 2 and 4 of Ref. 29. After 3 h there is a large interface step, which has disappeared after 16 days. There are many other experimental examples of these steps whose heights decrease with time and can be explained quantitatively by the

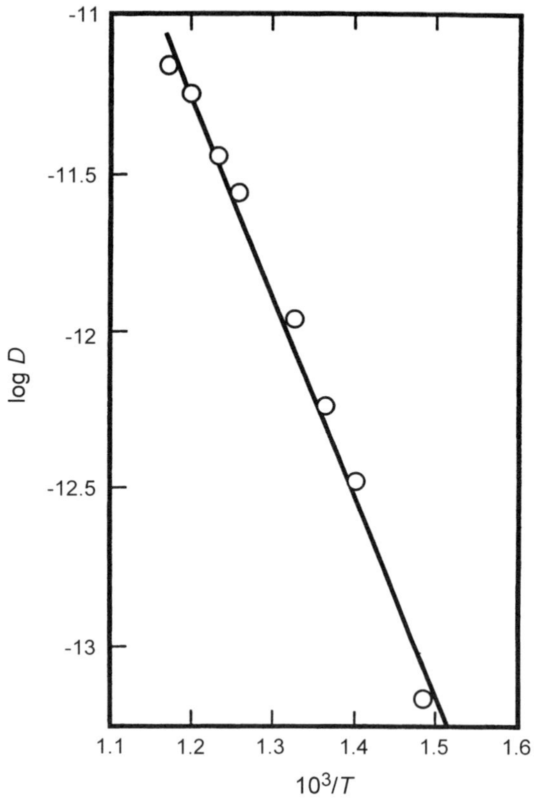

Figure 15.6 Log of the effective diffusion coefficient D_e as a function of $1/T$ for the profiles of Zellama *et al.* [26, Fig. 9] (see Table 15.2). (From Ref. 14.)

diffusion–reaction model, as described below, for example, in Refs. 26 and 30 to 32.

A second type of step results when two regions of silicon processed in different ways contain different total concentrations of deuterium or hydrogen at high concentration. An example is shown in Fig. 1b of Ref. 33. In this couple the deuterated region was prepared with a substrate temperature of 180°C, giving a deuterium concentration of about $5(10)^{21}$ atoms/cm^3, and the hydrogenated region prepared at a substrate temperature of 500°C had a concentration of about 10^{21} atoms/cm^3. These regions are essentially saturated with deuterium and hydrogen, respectively, so the difference in concentration between them does not change with time as they are annealed. Hydrogen–deuterium interdiffusion and exchange by Eq. 15.2 still occurs, but the step in total concentration at the interface does not change with time, as found experimentally.

For steps of the first kind, the measured concentration of deuterium in one medium and the diffusion time t a value of k can be calculated from Eq. 15.20, and then the concentration of deuterium in the other medium calculated from Eq. 15.19.

Good agreement with experimental measurements is found. These steps result because the reaction of Eq. 15.2 has not gone to local equilibrium at the interface in the time of heating and are not evidence for any particular combination of defects.

Exponential profiles of deuterium are often found at early times of diffusion, together with the concentration steps at the interface. Again these profiles are evidence that exchange reaction (Eq. 15.2) has not reached equilibrium. Such profiles are expected from the diffusion–reaction model at early times ($kt < 1$), see Eq. 15.23. At longer times after interfacial equilibrium is reached, the profiles follow the error function or conjugate error function (Eq. 15.15), as expected from the diffusion–reaction model.

The profiles of Branz *et al.* [31, Figs. 2 and 4] can be compared quantitatively with the diffusion–reaction model. After 3 h at 260°C [31, Fig. 4], the interface concentration $S_0{}^*(2)$ in region 2 is $1.1(10)^{19}$ deuterium atoms/cm^3, and S_T is $1.8(10)^{20}$ atoms/cm^3; then from Eq. 15.20, $k = 1.2(10)^{-5}\,\mathrm{s}^{-1}$. From the exponential profile at 3 h and Eq. 15.23, $D_e/k = 1.2(10)^{-12}\,\mathrm{cm}^2$, so $D_e = 1.4(10)^{-17}\,\mathrm{cm}^2/\mathrm{s}$. From the profile after 16 days, when local equilibrium has been established, from Eq. 15.15, $D_e = 1.3(10)^{-17}\,\mathrm{cm}^2/\mathrm{s}$. The good agreement between these two values of D_e shows that the diffusion–exchange equations explain the profiles of Branz *et al.* If the exponential profile is "fit" with Eq. 15.15 (erfc profile), a value of D_e of about $1.6(10)^{-16}\,\mathrm{cm}^2/\mathrm{s}$ is found. This result shows the apparent time dependence of D_e if one does not take into account the diffusion–exchange model (see below). From the deuterium profile in Fig. 2 of Ref. 31 after 30 days at 180°C, the interface concentration in region 2 is about $5.0(10)^{18}$ deuterium atoms/cm^3, and $S_T = 1.78(10)^{20}\,\mathrm{cm}^{-3}$, so from Eq. 15.20, $k = 2.2(10)^{-8}\,\mathrm{s}^{-1}$. From the exponential profile in this figure and Eq. 15.23, D_e at 180°C is about $1.3(10)^{-19}\,\mathrm{cm}^2/\mathrm{s}$. From these data at 260 and 180°C, the activation energy for k is about 150 kJ/mol (1.6 eV) and for D_e is about 110 kJ/mol (1.2 eV).

15.4.4 Time-Dependent Diffusion Coefficient

Several workers have reported that the effective diffusion coefficient of hydrogen in amorphous silicon from error function fits decreases as the time of diffusion increases [5, 30]. This dependence arises naturally in the diffusion–reaction model. This dependence is often approximated by a power law [30]. A detailed comparison between the time dependence of the effective diffusion coefficients of hydrogen in silicon and the diffusion–reaction model requires availability of the original profiles, or a study of this dependence in experiments with simplified boundary conditions, such as an infinite couple.

15.4.5 Diffusion in *p-n* Junctions

Johnson and Herring [10] diffused deuterium through a *p-n* junction and deduced from the resulting deuterium profiles with and without reverse bias that protons (H^+) were mobile and molecular hydrogen was formed and was immobile. Proton motion is questionable, as described in the introduction, and these results can equally well be interpreted as resulting from the diffusion of a low concentration of molecular hydrogen and its reaction to immobile Si–H or Si–D groups by reaction 15.1.

The profiles in Figs. 1 and 3 of Ref. 10 show a high concentration of deuterium below the silicon surface in the *n*-type region; this concentration decreases sharply as the depth of the depleted region of the junction is reached and then increases again beyond the depleted region in the *p*-type region. When reverse bias is applied, the depleted region increases in thickness, and the sharp increase in the *p* region shifts to the new edge of the depleted zone. Johnson and Herring took this shift as evidence that the deuterium (D^+) was moved in the bias field. However, it can also result if reaction 1 depends on the carrier concentration. If the equilibrium of Eq. 15.1 is shifted to the right (higher value of K_2 in Eq. 15.4) or the forward reaction rate ($H_2 \rightarrow 2H$) increases relative to the reverse rate ($2H \rightarrow H_2$) or both, the concentration of deuterium is greater in regions of higher carrier concentration (*n* and *p* regions) and lower in the depleted zone, just as found in the experiments. Thus the deuterium profiles in the *p-n* junction can just as well be interpreted as resulting from diffusion of dissolved molecular hydrogen and its reaction to immobile Si–H groups.

15.4.6 Permeation Measurements

The permeation of hydrogen through hollow cylinders of silicon was measured, and the diffusion coefficients and solubilities of hydrogen in silicon were deduced from the transients before a steady state was reached and after a change in hydrogen pressure [19, 20]. However, these deductions are valid only for diffusion without reaction and are invalid when reaction is present. This conclusion is confirmed by the extrapolations to zero time of M_t/M_0 in Fig. 5 of Ref. 20, in which M_t is the amount of tritium remaining in the sample at time t and M_0 is the initial amount. M_t/M_0 should extrapolate to $8/\pi^2 = 0.81$; the actual value is about 0.4, because of the slow release of hydrogen bound in Si–H groups. Exactly the same problem was found for hydrogen in vitreous silica, in which the rate of release of hydrogen is also slowed because of a chemical reaction [34, 35] (see Chapter 11).

The data in Ref. 20 can be clarified with the diffusion–reaction model. Integration of Eq. 15.6 gives the amount of hydrogen released from the silicon when the reaction is slow. The result is similar to Eq. 15.8:

$$\frac{M_0}{M_t} - 1 = \mu t \tag{15.28}$$

A plot of the data from Fig. 5 of Ref. 20 is given in Fig. 15.7; the plot is linear, as expected from Eq. 15.28, except for short times, where there is a small contribution from molecular hydrogen. Values of the reaction rate parameter μ (Eq. 15.6) are given in Table 15.3 for the different temperatures measured in Ref. 20. These values fit an Arrhenius equation with an activation energy of 53 kJ/mol (0.54 eV) and a preexponential factor of 0.48 s^{-1}. These results show that the release rate of hydrogen in these experiments is controlled by the slow reaction rate and not by

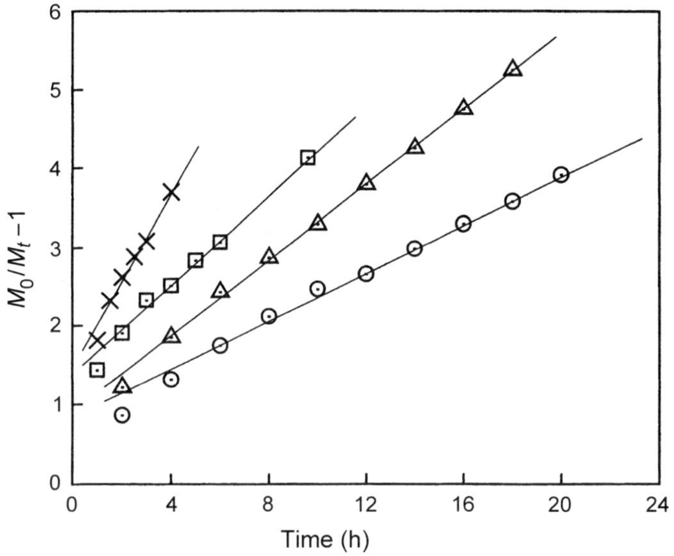

Figure 15.7 Plot of $M_o/M_t - 1$ versus time for data of and Furuichi and Ichimiya [14, 20].

Table 15.3 Reaction Rate Parameter μ (Eq. 15.28) for the Outgassing of Hydrogen from Crystalline Silicon

T (°C)	μ ($s^{-1} \times 10^5$)
400	4.5
420	6.8
450	8.4
500	16.5

Source: Ref. 14.

diffusion of hydrogen, except during the transient at short times. Thus diffusion coefficients or solubilities cannot be calculated from these measurements.

There are no raw data given for the permeation experiments reported in Ref. 19, so it is not possible to assess their validity; in particular, it is not possible to judge if a steady state was truly reached, despite the slow reaction rate. Thus the permeation results, including their pressure dependence, cannot be a basis for mechanistic deductions.

15.5 DISCUSSION

The excellent agreement between calculated and measured profiles in Figs. 15.1, 15.2, and 15.5, and the correspondence between the time dependence of interface steps, exponential profiles, and time dependence of D_e in model and experiment, provide strong evidence that the model of molecular diffusion and reaction of hydrogen in silicon is valid. The model is based on the reactions of Eqs. 15.1 and 15.2, with a single reacted species or "trap"; the various experimental features arise naturally in the model, with no need for a complicated system of different traps of different energies, and hydrogen states, that are sometimes invoked.

The activation energy for the diffusion of molecular hydrogen cannot be calculated from available results. Even when there is local equilibrium and no time dependence of the effective diffusion coefficient, Eqs. 15.5 and 15.14 show that the equilibrium coefficient K_2 or the ratio of reacted to dissolved molecular hydrogen (S_T/C_T) must be available to calculate the true diffusion coefficient D of molecular hydrogen. Some workers have found activation energies of about 140 kJ/mol (1.5 eV) for the apparent diffusion coefficient of hydrogen in amorphous silicon, but others have found activation energies as high as 290 kJ/mol (3 eV) [36]. In the present analysis the activation energies of Song *et al.* [25] (15 kJ/mol, 0.15 eV) and Zellama *et al.* [26] (122 kJ/mol, 1.3 eV) are smaller than those found by others. Without measurements of the equilibrium coefficient K_2, or λ and μ, the present model suggests that these various measured activation energies give no direct information about the diffusion and reaction energetics of molecular hydrogen in silicon.

The open structure of amorphous and crystalline silicon provides ample space for the molecular solution of the hydrogen molecule, whose gas kinetic diameter is about 0.25 nm. The ionic porosity Z of crystalline or amorphous silicon can be calculated from the formula

$$Z = 1 - \frac{V}{V_c} \tag{15.29}$$

where V is the total volume of atoms in the unit cell of volume V_c. This formula can be used for amorphous materials, where V is the volume of atoms in a "molecule" and V_c is the molecular volume $M/\rho A$, where ρ is the density, A Avogadro's number, and M the molecular weight. For silicon, Z is about 0.5 if the silicon atom has a diameter of about one Å, which may not be exact. Nevertheless, the silicon structure is quite open; vitreous silica, in which molecular hydrogen dissolves and diffuses easily, has an ionic porosity of about 0.51. Theoretical calculations find that the hydrogen molecule in crystalline silicon is stable in the tetrahedral site in the diamond structure (see Refs. 21, 37, and references therein). The vibrational frequency of hydrogen in this site is estimated to be 4470 cm^{-1} [37], in fair agreement with the Raman value of 4158 cm^{-1} [11]. The transport of molecular hydrogen can occur by jumps through a "doorway" between the tetrahedral sites in which it dissolves; a formula for the activation energy of diffusion Q for a doorway radius r_D is [38] (see Chapter 5)

$$Q = 8\pi G r_D (r - r_D)^2 \tag{15.30}$$

For silicon G is $8(10)^{10}$ Pa, and the radius of the hydrogen molecule is about 0.125 nm, so with a $Q = 110$ kJ/mol, r_D is 0.094 nm, which is possible. This doorway radius is much larger than that found for gases diffusing in silicate glasses (see Chapter 5), perhaps consistent with the rapid interstitial diffusion in slilicon.

REFERENCES

1. E. A. Davis, *J. Non-Cryst. Solids* **198–200**, 1 (1996).
2. N. H. Nickel, W. B. Jackson, and J. Walker, *Phys. Rev.* **B53**, 7750 (1996).
3. Amorphous Silicon Technology, *Mater. Res. Soc.* **118** (1988); **149** (1989); **192** (1990); **219** (1991); **258** (1992); **297** (1993); **336** (1994); **377** (1995); **420** (1996).
4. *J. Non-Cryst. Solids* **35**, **36** (1980); **59**, **60** (1983); **97**, **98** (1987); **114** (l989); **137**, **138** (1991); **164–166** (1993); **198–200** (1996); **227–230** (1998).
5. J. I. Pankove and N. M. Johnson, *Hydrogen in Semiconductors*, Academic Press, San Diego, CA, 1991.
6. R. A. Street, *Hydrogenated Amorphous Silicon*, Cambridge University Press, Cambridge, 1991.
7. M. Stutzmann and J. Chevallier, eds., *Hydrogen in Semiconductors*, North-Holland, Amsterdam, 1991.
8. D. J. Fisher, ed., *Diffusion in Silicon*, Scitec Publication's Zurich, Switzerland, 1998.
9. R. A. Street, C. C. Tsai, J. Kakalios, and W. B. Jackson, *Philos. Mag.* **B56**, 305 (1987).
10. N. M. Johnson and C. Herring, *Phys. Rev.* **B38**, 1581 (1988).
11. N. Fukata, S. Sasaki, K. Murakami, K. Ishioka, K. G. Nakamura, M. Kitajima, S. Fujimura, J. Kikuchi, and H. Haneda, *Phys. Rev.* **B56**, 6642 (1997).
12. R. E. Norberg, D. J. Leopold, and P. A. Fedders, *J. Non-Cryst. Solids* **227–230**, 124 (1998).
13. A. E. Widmer, R. Fehlmann, and C. W. Magee, *J. Non-Cryst. Solids* **54**, 199 (1983).
14. R. H. Doremus, *Mat. Res. Innovat.* **4**, 49 (2000).
15. D. D. Wagman *et al.*, *NBS Tables of Chemical Thermodynamic Properties*, National Burureau of Standards, Washington, DC, 1982.
16. F. M. Ernsberger, *J. Non-Cryst. Solids* **38–39**, 557 (1980).
17. F. M. Ernsberger, *J. Am. Ceram. Soc.* **66**, 747 (1983).
18. A. von Hippel, *J. Chem. Phys.* **54**, 145 (1971).
19. A. van Wieringen and N. Warmoltz, *Physica* **22**, 849 (1956).
20. T. Ichimiya and A. Furuichi, *Int. J. Appl. Radiat. Isot.* **19**, 573 (1968).
21. P. Deak, L. C. Snyder, and J. W. Corbett, *Phys. Rev.* **B37**, 6887 (1988).
22. W. Beyer and Y. Zastrow, *J. Non-Cryst. Solids* **227–300**, 880 (1998).
23. J. E. Shelby, *Handbook of Gas Diffusion in Solids and Melts,* ASM International, Materials Park, OH, 1996.
24. J. Crank, *Mathematics of Diffusion*, 2nd ed. Oxford University Press, Oxford, 1975.
25. Z. Song, F. Zhang, G. Yu, G. Kong, and G. Chen, *J. Non-Cryst. Solids* **164–166**, 305 (1993).

26. K. Zellama, P. Germain, S. Squelard, B. Bourdon, J. Fontenille, and R. Danielow, *Phys. Rev.* **B23**, 6648 (1981).
27. S. Rigo, F. Rocket, B. Agius, and A. Straboni, *J. Electrochem. Soc.* **129**, 867 (1982).
28. S. C. Jain, *Proc. R. Soc.* **A243**, 359 (1958).
29. H. M. Branz, S. Asher, B. P. Nelson, and M. Kemp, *J. Non-Cryst. Solids* **164–166**, 269 (1993).
30. J. Shinar, S. Mitra, X. L. Wu, and R. Shinar, in *Diffusion in Materials*, A. L. Laskar *et al.*, eds., Kluwer Academic, Amsterdam, 1990, p. 573; *Mater. Res. Soc. Symp. Proc.* **219**, 75 (1991).
31. D. E. Carlson and C. W. Magee, *Appl. Phys. Lett.* **33**, 81 (1978).
32. W. Beyer and H. Wagner, *Mater. Res. Soc. Symp. Proc.* **336**, 323 (1994).
33. W. Beyer, *Phys. Status Solidi* **A159**, 53 (1997).
34. R. W. Lee and D. L. Fry, *Phys. Chem. Glasses* **7**, 19 (1966).
35. J. E. Shelby, *J. Appl. Phys.* **48**, 3387 (1977).
36. A. J. Franz and J. L. Gland, *Mater. Res. Soc. Symp. Proc.* **420**, 539 (1996).
37. K. G. Nakamura, K. Ishioka, M. Kitajima, and K. Murakami, *Solid State. Commun.* **101**, 735 (1997).
38. O. L. Anderson and D. A. Stuart, *J. Am. Ceram. Soc.* **37**, 573 (1954).

16

DIFFUSION OF OXYGEN AND SILICON IN SILICON: SILICON MONOXIDE MODEL

16.1 INTRODUCTION

Silicon is one of the purest commercial materials; low concentrations of impurities that influence electronic properties in silicon are essential. However, in silicon for integrated circuits grown in a silica crucible there are usually about 10^{18} oxygen atoms/cm^3 [$2(10)^{-5}$ atom fraction, 10 ppm]. Electrically active oxygen and complexes and precipitation of oxygen can influence the electrical properties of silicon, so oxygen solubility and diffusion are processes of great practical importance.

Experiments on the diffusivity and solubility of oxygen in silicon have been summarized by Mikkelsen [1]. The diffusion coefficient fits on a single line on a log D versus $1/T$ plot, where D is the diffusion coefficient and T the absolute temperature, at temperatures from 350 to 1250°C. An activation energy of 244 kJ/mol (2.53 eV) and a preexponential factor of 0.13 cm^2/s are found. There is excellent agreement between different investigators and low and high temperature data, and the diffusion coefficients are independent of oxygen concentration, wafer orientation, dopant concentrations, and state of oxidation of the ambient atmospheres. Higher diffusion coefficients of oxygen in silicon at temperatures from 400 to 700° C than given by these results have been reported in several studies [2–6]. The solubility measurements of oxygen in silicon show more scatter than the diffusivity measurements; Mikkelsen [1] estimated the best fit to be an enthalpy of 146 kJ/mol (1.52 eV) and a preexponential factor of $9(10)^{22}$ oxygen atoms/cm^3.

Silicon containing oxygen has a strong infrared absorption band at about 1106 cm^{-1} (9.0 μm wavelength) and a less intense band at 515 cm^{-1} (19 μm wavelength) [7]. From isotopic shifts Hrostowski and Kaiser concluded that the

1106-cm^{-1} band results from a single Si–O vibration [7]. Kaiser *et al.* proposed a model for oxygen in silicon in which an oxygen atom breaks a silicon–oxygen bond and is bonded to the two silicon atoms from the broken bond [8]. The resulting silicon–oxygen–silicon bond is not linear and is bent away from the original bond, forming a silicon–oxygen–silicon angle of about 100°. The 1106-cm^{-1} band is assigned to ν_3, the antisymmetric stretching motion of this structure, and the 515-cm^{-1} band is assigned to the ν_2 bending motion of the bond [7].

The 1106-cm^{-1} band is dichroic under stress in a $\langle 111 \rangle$ direction in the crystal lattice, whereas with stress in a $\langle 100 \rangle$ direction there is no dichroism [9, 10]. This dichroism is consistent with the foregoing model of oxygen bonded to two silicon atoms in a $\langle 111 \rangle$ direction. There is also an internal friction peak in samples of $\langle 111 \rangle$ orientation and no peak in samples of $\langle 100 \rangle$ orientation [11]. Nowick and Berry have pointed out that these observations imply that the defect responsible for the internal friction peak has trigonal symmetry, whereas the model for oxygen above has monoclinic symmetry [12]. The model can be rescued by postulating that the oxygen atom has access to six equivalent sites, located in a circle around the $\langle 111 \rangle$ axis of the silicon atoms bonded to oxygen. It is assumed that the oxygen atom can jump from one of these sites to another with a low activation energy. This possible hopping changes the effective symmetry of the defect to trigonal [13,14], consistent with the experiment.

To explain diffusion of oxygen in silicon with the model above, it is assumed that the oxygen atom breaks its two silicon bonds, breaks a neighboring silicon–silicon bond, and forms another silicon–oxygen–silicon bond with the latter silicon atoms. The silicon–oxygen bond energy is about 466 kJ/mol (4.84 eV), and the silicon–silicon bond energy is about 228 kJ/mol (2.37 eV), so breaking two silicon–oxygen bonds and one silicon–silicon bond should require a high activation energy, much greater than the experimental value of 244 kJ/mol for diffusion. The high energy to form the silicon–oxygen–silicon defect that is covalently bonded suggests that the numbers of these defects will be quite small. The rapid motion of an oxygen atom from one Si–O–Si site to another around a circle is defended with drawings of the structure of this site with atoms drawn much smaller than their actual size; more realistic sizes show that the jumping of an oxygen atom from one circular site to another will encounter substantial steric hindrance from neighboring silicon atoms in the silicon structure (see Fig. 16.1). Thus it is doubtful that the symmetry of the Si–O–Si group is trigonal. For these reasons it seems to be worthwhile to consider an alternative model for oxygen in silicon.

At the surface of silicon there are always the chemical constituents silicon and silicon dioxide. These constituents can react to form silicon monoxide:

$$\mathrm{Si}(s) + \mathrm{SiO_2}(s) = 2\mathrm{SiO}(d) \tag{16.1}$$

where *d* means dissolved and *s* solid.

The purpose of this chapter is to propose a dissolution of SiO in silicon in the largest open site in the silicon structure. The SiO molecule is oriented in a $\langle 111 \rangle$ direction, with the oxygen lightly bonded to a network silicon atom (see Fig. 16.2).

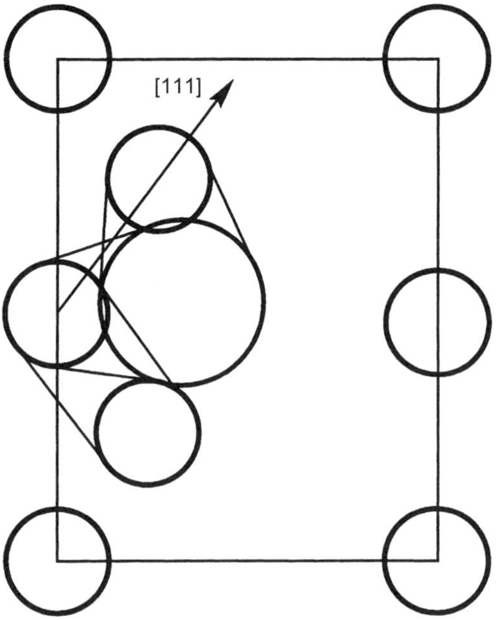

Figure 16.1 A (110) plane in the silicon (diamond) structure, showing an oxygen atom bonded to two silicon atoms.

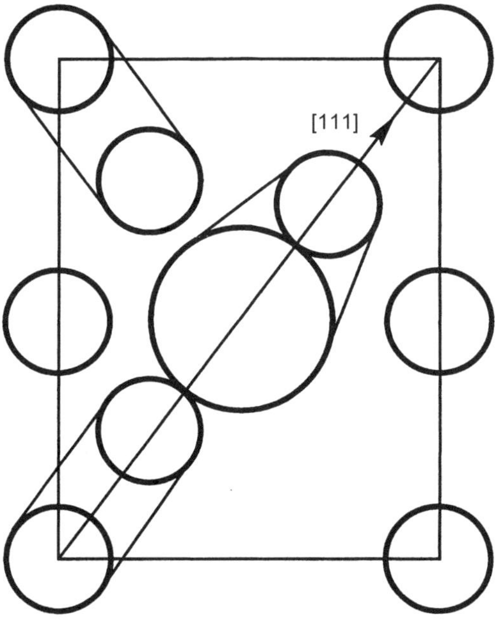

Figure 16.2 A (110) plane in the silicon (diamond) structure, showing an SiO molecule lightly bonded to a silicon atom and resting in a hole in the diamond structure.

This dissolved SiO molecule [SiO(d) in Eq. 16.1] is consistent with infrared spectra, including dichroism, internal friction, oxygen solubility, and diffusion in silicon. This dissolved SiO molecule can hop from one hole site to another.

The mechanism of the rapid diffusion of oxygen in silicon below about 700°C is uncertain. I suggest that a low concentration of dissolved molecular water should diffuse rapidly in silicon. There is no direct experimental evidence to support this suggestion, but the water molecule seems to be the most likely carrier of rapid diffusion of oxygen. Molecular oxygen is also a possibility.

Dissolved silicon monoxide also transports silicon. A diffusion–reaction model of SiO transport can perhaps explain silicon tracer diffusion in silicon.

In this chapter we discuss the position of dissolved SiO molecules in silicon, the infrared spectra and dichroism, diffusion and internal friction of SiO in silicon, fast diffusion of oxygen, and tracer diffusion of silicon, and provide conclusions.

16.2 STRUCTURE OF DISSOLVED SiO

The size of an SiO molecule dissolved in a solid should not be much greater than the size of oxygen ions in solids; the size of a water molecule dissolved in a solid is about the same as that of an oxygen ion [15]. Thus an SiO molecule can dissolve in the largest "space" between network silicon atoms; an example of this space is the center of the unit cell of the diamond structure, as shown in Fig. 16.2. The SiO molecule is oriented linearly along a $\langle 111 \rangle$ direction, with the silicon atom pointing in a $\langle 111 \rangle$ direction and the oxygen atom interacting lightly with a network silicon atom. The attack of water on silica is found from quantum mechanical calculations to start by interaction of the oxygen atom in the water with four-coordinated silica, forming a fifth lightly bonded atom [16], so this interaction of the oxygen atom in SiO with a network silicon atom is reasonable. There are four equivalent sites of this type around each silicon atom in the diamond structure.

In drawing Figs. 16.1 and 16.2, I have used a radius of the oxygen ion of 1.35 Å, which is the average value for oxygen coordinated to two other atoms in many different oxides [17]. The hard-sphere model of an atom is inadequate for a covalently bonded atom such as silicon in the diamond structure. As a compromise I have used a radius of 0.75 Å for the silicon atom in its crystal. This radius is larger than radii of silicon four-coordinated to oxygen in silicates [17] but smaller than the "touching" radius in the diamond structure. I have shown covalent bonds between atoms by cylindrical or conical connections between the bonded atoms. With these conventions the silicon monoxide molecule fits easily into a "hole" in the diamond structure. The oxygen ion bonded to two silicons with a 100° Si–O–Si angle does not fit without overlap, and jumping to another circular site around the [111] direction shows considerable steric interference. A Si–O–Si angle of 162° for this defect has been deduced from near- and far-infrared spectra of oxygen in silicon [18]. This nearly linear configuration would lead to substantial displacement of the two silicon atoms and large stress in the silicon network; it is improbable.

The SiO site could properly be called an interstitial site in the diamond structure, in analogy with interstitial sites in metals, for example, for the solution of carbon in iron. The term *interstitial site* has unfortunately been used for the covalently bonded site for oxygen in silicon proposed by Kaiser, so in this chapter I will not use the term *interstitial* for either site. I will refer to the Kaiser site as *bound oxygen* and to the dissolved SiO site as a *dissolution site.*

16.3 INFRARED SPECTRA AND DICHROISM

The infrared spectrum of oxygen in silicon shows two absorption bands at room temperature, one at 1106 cm^{-1} and another at 515 cm^{-1}. The band at 1106 cm^{-1} is close to that for the absorption of gaseous SiO at 1230 cm^{-1}. This 1230-cm^{-1} band can be assigned to the vibration of the diatomic SiO molecule; a force constant of 910 N/m is calculated for this vibration. The 1106-cm^{-1} band for SiO dissolved in silicon is also assigned to this SiO stretching vibration; the force constant is lower at 736 N/m because of the weak attraction of the oxygen atom to a lattice silicon atom. The 515-cm^{-1} band can be assigned to the vibration of this weakly bonded SiO pair; a force constant of 160 N/m is calculated for the vibration. See below for another possible assignment of this band.

As the silicon is cooled below room temperature, the 1106-cm^{-1} band shifts to higher frequencies, and below about 100 K it splits into two and possible three separate bands, each with a different temperature dependence [7]. There is little change in the 515-cm^{-1} band as the silicon is cooled; an additional weak band at 1205 cm^{-1} develops below about 80 K. Hrostowski and Kaiser [7] suggest that these shifts and changes upon cooling are related to interactions of the Si-O-Si group with the silicon network, without giving a detailed calculation of these effects. A similar explanation should be valid for the dissolved Si–O molecule. As silicon is cooled below room temperature, it contracts until about −125°C, below which it expands. The contraction presumably leads to the shift in the band position, and the expansion to the splittings and new band. A complete explanation of these effects awaits further work.

The isotopic shifts of the 1106 cm^{-1} and 515 cm^{-1} bands have been measured and interpreted by Pajot and his co-workers [19], as well as the shifts of a weak band at 1748 cm^{-1}. These authors based their interpretation on a linear Si-O-Si configuration of oxygen between two network silicon atoms, resulting from a calculation in ref. 20. They found agreement with the isotopic shifts with the following assignments of band positions at liquid helium temperatures: 1136 cm^{-1}, asymmetric ν_3 vibration; 518 cm^{-1} ν_2 bending vibration; 596 cm^{-1} ν_1 symmetrical vibration (not observed experimentally); and 1748 cm^{-1} $\nu_1+\nu_3$ combination.

Almost the same isotopic shifts observed by Pajot *et al.* are found for the present proposed configuration of an Si-O-Si group, resulting from the bonding of an Si-O molecule to a silicon atom, as calculated from equations II 198 and II 199 on p. 173 of ref. 21. Thus it is not possible to distinguish between these two models for oxygen in silicon with the spectroscopic bands and isotopic shifts. The linear Si-O-Si configuration of oxygen between two network silicon atoms leads to high distortions and strains in the silicon structure, and is improbable. The disssolved SiO molecule can easily dissolve in the center of a unit cell in silicon with minimum strain and distortion, so it is much more likely than the linear Si-O-Si configuration.

Far-infrared bands related to oxygen in silicon, especially one at 29.3 cm^{-1}, have been assigned to the ν_2 bending mode of the Si–O–Si groups [18]. It is possible that the dissolved SiO molecule lightly bonded to silicon has a low-energy bending vibration at far-infrared frequencies, but more work is needed to clarify these far-infrared bands. Such bands are usually associated with rotational motions, which are unlikely for molecules dissolved in a solid.

16.4 DIFFUSION OF OXYGEN, INFRARED DICHROISM, AND INTERNAL FRICTION

At temperatures from about 700 to 1230°C, diffusion coefficients of oxygen in silicon, measured by different investigators with different profiling techniques, agree closely, as summarized by Mikkelsen [1]. Diffusion coefficients of oxygen, calculated from the internal friction measurements of Southgate at about 1000°C with the equations of Haas [22], fit closely with those measured by the profiling techniques. Diffusion coefficients of oxygen in silicon have also been calculated for much lower temperatures (300 to 400°C) from the decay of infrared optical dichroism of the 1106-cm^{-1} band [23,24], again using the equations of Haas [22]. The low- and high-temperature measurements agree within experimental error over a range of 13 orders of magnitude. Both the internal friction and dichroism methods depend on the jumping of the oxygen constituent from its site to a preferred site with lowered stress. Since the effects occur with a stress applied in a $\langle 111 \rangle$ but not in a $\langle 100 \rangle$ direction, the sites must have trigonal symmetry in the $\langle 111 \rangle$ directions. The SiO site proposed in this paper fulfills these conditions as well as does the bound oxygen site. The jump distance between dissolved SiO sites is calculated to be $\left(\sqrt{3/4}\right)a$, where a is the lattice parameter of the silicon diamond lattice. This distance is the same as found by Haas for jumps of the bound oxygen, so all the calculations for this defect for internal friction and dichroism are directly applicable to jumps of the dissolved SiO molecule.

The activation energy Q for jumping of a dissolved molecule from one site in a solid to another can be described by a "doorway" equation (see Chapter 5):

$$Q = 8\pi G r_D (r - r_D)^2 \tag{16.2}$$

which is the elastic energy to dilate a spherical cavity from radius r_D to r [25]. This equation can be applied to diffusion in silicon by assuming that the dissolved SiO molecule passes through a "doorway" of radius r_D in jumping from one site to another; r is then the radius of the diffusing molecule and G is the shear modulus.

For silicon, G is 79.6 GPa, and if the effective radius of the SiO molecule is about 0.15 nm, the doorway radius is about 0.1 nm, which is about the same as found for molecular hydrogen in silicon in section 15.5.

16.5 FAST DIFFUSION OF OXYGEN

Between temperatures of about 400 and 700°C, some measured diffusion coefficients of oxygen in silicon [2–6] are much higher than those that agree from 350 to 1250°C. These higher results show a number of variants and special influences not found in the agreeing measurements. These "fast" diffusion coefficients are sometimes found to depend on the time of diffusion [6,26]. Often, the surface concentration of oxygen in diffusion out of silicon is not zero and decreases as the diffusion time increases [2,6]. There are substantial differences (up to an order of magnitude) between diffusion coefficients calculated for different wafers at the same temperature [2], with no apparent reason for these differences. Thus the mechanism of this fast diffusion appears to be quite different from that for the lower values, which all fit on the same Arrhenius curve. The findings of nonzero surface concentrations in out-diffusion, time-dependent diffusion coefficients and strong history dependence suggest a diffusion–reaction mechanism [27].

The fast diffusing species for transporting oxygen is uncertain. I suggest that dissolved water molecules are a likely candidate. When dissolved in solids, water has an effective radius of about 1.38 Å, which is close to the size of an oxygen atom in solids. Thus it should diffuse rapidly in the open silicon structure (see the discussion in Section 16.4). Oxygen in SiO should exchange to H_2O:

$$SiO + H_2 = H_2O + Si \tag{16.3}$$

The SiO can be considered to be immobile compared to the faster diffusing water, giving a diffusion–exchange mechanism for the water. There is no direct spectroscopic evidence for water in silicon, probably because the concentration is below detectable limits. However, many metal atoms diffuse very rapidly in silicon [28], so it is reasonable to postulate rapid diffusion of water, thus requiring only a small water concentration to transport oxygen. More work is needed to test this possibility.

Gösele and Tan [29] have reviewed the experiments on fast diffusion of oxygen carried out before 1982 and also explanations for these results. They proposed that the rapid transport of oxygen in silicon resulted from the diffusion of molecular oxygen. Diffusion of molecular oxygen is a possible explanation of fast diffusion of oxygen in silicon, although the oxygen molecule is larger (3.2 Å in diameter) than water (2.8 Å in diameter). Direct measurements of the concentrations of these gases dissolved in silicon by infrared optical absorption (for water) or Raman spectroscopy might show whether or not they are responsible for the fast oxygen diffusion in silicon.

16.6 DIFFUSION OF SILICON

There have been a number of studies of the tracer diffusion of silicon in silicon [29–37]. The values of the diffusion coefficients at 1200°C agree well for intrinsic samples heated in low-oxygen atmospheres, as shown in Table 16.1. However, there is good evidence from several studies that in an oxidizing atmosphere the diffusion coefficient is a factor of 2 to 3 higher than that in an inert atmosphere. Annealing in argon in an ampule or silica tube furnace leads to a slightly oxidizing atmosphere because of outgassing of the silica ampule or tube. When special care is taken to exclude oxygen, as in Ref. 35, the diffusion coefficient of silicon is even lower than those measured in nominally inert conditions. There is also evidence that in extrinsic *n*-type silicon the diffusion is somewhat faster than in intrinsic silicon; for *p*-type material the trend is uncertain. Furthermore, the activation energies found vary from about 390 kJ/mol (4.1 eV) to 490 kJ/mol (5.1 eV), even for intrinsic material in inert atmosphere.

These results are in striking contrast to the different measurements of oxygen diffusion, which agree closely and do not depend on atmosphere or dopant or carrier concentration; activation energies also agree well in the oxygen studies. These differences suggest that the silicon tracer measurements involve an exchange reaction that is influenced by external factors. The dissolved silicon monoxide postulated for oxygen diffusion should also transport silicon. The exchange reaction between the dissolved tagged Si^*O and network silicon can be written

$$\mathrm{Si}^*\mathrm{O} + \mathrm{Si}(n) = \mathrm{SiO} + \mathrm{Si}^*(n) \tag{16.4}$$

where Si(n) is network silicon. For this kind of exchange reaction, the model of diffusion plus exchange (Chapters 3 and 4 and Ref. 27) gives an effective diffusion coefficient D_e of:

$$D_e = \frac{DC_T}{S_T} \tag{16.5}$$

Table 16.1 Tracer Diffusion of Silicon in Silicon

	Diffusion Coefficient (cm²/s)		
Conditions During Diffusion Anneal	1200°C ($\times 10^{14}$)	900°C ($\times 10^{18}$)	Ref.
Ar in SiO_2 ampule	8.6		31
Ar in SiO_2 ampule	8		30
HCl 5 torr	2.6		32
HCl	2.0	1.0	33
Ar in SiO_2 tube	7.5	20	34
Low oxygen	1.0		35
Implanted	4.3	3.2	36
CVD at 900°C, FZ, Si	3.0		37
CVD, CZ, Si, inert	3.0		37
CVD, CZ, oxidizing	6.5		38

in which D is the diffusion coefficient of dissolved diffusing substance (SiO), C_T its concentration, and S_T the total concentration of exchanging material (network silicon). Thus if the diffusion–exchange mechanism is valid for silicon, the diffusion coefficient of oxygen D_{Ox} should be related to that D_{Si} of silicon

$$\frac{D_{Ox}}{D_{Si}} = \frac{S_T}{C_T} \tag{16.6}$$

To test this possibility I have gathered some data that I consider most reliable, as listed in Table 16.2. The diffusion coefficients and solubilities of oxygen are from Mikkelsen [1], and the diffusion coefficients of silicon (isotope diffusion) are from Bracht *et al.* [37]. The temperature range is from about 800 to 1400°C, which is the range of reliable data for D_{Si} and C_T. The concentration S_T of silicon atoms is $5.0(10)^{22}\,\text{cm}^{-3}$. The ratio D_{Ox}/D_{Si} is close to that of S_T/C_T at 838°C, but the ratios deviate up to an order-of-magnitude difference at 1400°C. It is possible that the D_{Si} values are influenced by some of the factors mentioned above, such as oxidation state and dopant concentrations. More experimental work is needed to determine the applicability of silicon transport by SiO.

The influence of oxygen in the annealing atmosphere on silicon diffusion can be understood from the experiments and ideas of Mikkelsen [1,39]. He found that the solubility of oxygen in silicon increases by about an order of magnitude at 800°C over that in an inert atmosphere when the atmosphere is oxidizing [38]; at 1000°C the difference was about a factor of 7. Mikkelsen showed that this increase of oxygen solubility is metastable and returns to the equilibrium value when the atmosphere is inert. Thus when the external atmosphere is oxidizing, the effective oxygen pressure at the silica–silicon interface is increased somewhat over the equilibrium value. At 1000°C the oxygen pressure at equilibrium with Si and SiO_2 is about 10^{-27} atm (10^{-32} Pa), so the metastable increase results in an effective oxygen pressure that is still very small [39]. This increase of oxygen pressure is responsible for an increase in the dissolved SiO concentration C_T, and so causes an increase in the effective diffusion coefficient for tracer silicon from Eq. 16.5.

Tan and Gösele [41] and Celler and Trimble [42] have suggested that the diffusion of silicon monoxide in the silicon dioxide layer on silicon can explain the influence of oxidation on the transport of dopants such as arsenic in the SiO_2. In

Table 16.2 Comparison of Diffusion of Oxygen and Silicon in Silicon[a]

	Diffusion Coefficients (cm^2/s)			Solubility	
T (°C)	D_{Ox}	D_{Si}	D_{Ox}/D_{Si}	C_T (0 atoms/cm$^{3)}$	S_T/C_T
1394	$2.9(10)^{-9}$	$2.3(10)^{-12}$	$1.3(10)^3$	$2.6(10)^{18}$	$1.9(10)^4$
1156	$1.6(10)^{-10}$	$9.4(10)^{-15}$	$1.6(10)^4$	$3.6(10)^{17}$	$1.4(10)^5$
977	$8.2(10)^{-12}$	$3.8(10)^{-17}$	$2.2(10)^5$	$6.3(10)^{16}$	$7.9(10)^5$
838	$4.4(10)^{-13}$	$1.5(10)^{-19}$	$2.9(10)^6$	$1.4(10)^{16}$	$3.6(10)^6$

[a] D_{Ox} and solubilities from Ref. 1; D_{Si} from Ref. 37.

these models SiO forms at the oxide–silicon layer. If SiO is generated at this interface, it seems reasonable that it could also diffuse into the silicon as well as the silicon dioxide.

The usual mechanism for the tracer diffusion of silicon involves vacancies or interstitials; recently, interstitials have been favored [43,44]. At temperatures below ambient, radiation damage experiments suggest that the diffusion coefficient of interstitials is very high and has low activation energies, 32 kJ/mol (.33 eV) or less. Such values appear to be inconsistent with the results on diffusion of silicon in silicon [45]. One might expect that any appreciable concentrations of vacancies and interstitials should annihilate rapidly, leaving low concentrations of these defects (see Chapter 14). The participation of interstitials or vacancies in diffusion in dislocation-free and single-crystal silicon requires their production at the silicon surface; the mechanism of this production and why it should be influenced by oxidation state have not been explained satisfactorily. The tracer diffusion of silicon is apparently not influenced by the dislocation density in the silicon, although dislocations might be sources for interstitials and vacancies. These uncertainties suggest that it is worthwhile to consider alternative mechanisms for tracer diffusion of silicon, such as the SiO mechanisms presented here.

16.7 CONCLUSIONS

There are sufficient uncertainties about mechanisms proposed for the diffusion of oxygen and silicon in silicon to justify exploration of other possible mechanisms. Silicon monoxide dissolved in silicon should result from the reaction of SiO_2 and silicon at the silicon surface. This dissolved SiO can transport both oxygen and silicon in silicon. I propose that the SiO dissolves in silicon in a $\langle 111 \rangle$ orientation, consistent with internal friction and optical dichroism measurements. Calculations from internal friction and optical dichroism for a bonded Si–O–Si group all apply to the dissolved SiO molecule.

The fast diffusion of oxygen in silicon at lower temperatures may result from transport of dissolved molecular water or oxygen, but more evidence is needed to support these possibilities. Dissolved SiO transport can also be the mechanism for tracer diffusion of silicon in silicon. The ratio between diffusion coefficients of oxygen and silicon in silicon is not too different from that expected for a diffusion and reaction mechanism for tracer diffusion of silicon, but the temperature dependence of this ratio has an unexplained variation.

REFERENCES

1. J. C. Mikkelsen, in *Oxygen, Carbon, Hydrogen and Nitrogen in Crystalline Silicon*, J. C. Mikkelsen, S. J. Pearton, J. W. Corbett, and S. J. Pennycock, eds., *Mater. Res. Soc.* Symp. Proc. **59**, 19 (1986).
2. S. T. Lee, P. Fellinger, and S. Chen, *J. Appl. Phys.* **63**, 1924 (1988).

3. P. Gaworzewski and G. Ritter, *Phys. Status Solidi* **A67**, 511 (1981).
4. W. Berghoz, J. L. Hutchinson, and P. Pirouz, *J. Appl. Phys.* **58**, 3429 (1985).
5. H. Takeno, Y. Hayamizu, and K. Miki, *J. Appl. Phys.* **84**, 3113 (1998).
6. S. A. McQuaid, B. K. Johnson, D. Gambaro, R. Falster, M. J. Ashwin, and J. H. Tucker, *J. Appl. Phys.* **86**, 1878 (1999).
7. H. J. Hrostowski and R. H. Kaiser, *Phys. Rev.* **107**, 966 (1957).
8. R. H. Kaiser, P. H. Keck, and C. F. Lange, *Phys. Rev.* **101**, 1264 (1956).
9. J. W. Corbett and G. D. Watkins, *J. Phys. Chem. Solids* **20**, 319 (1961).
10. J. W. Corbett, R. S. McDonald, and G. D. Watkins, *J. Phys. Chem. Solids* **25**, 873 (1964).
11. P. D. Southgate, *Proc. Phys. Soc.* **B70**, 804 (1957); **B76**, 385 (1960).
12. A. S. Nowick and B. S. Berry, *Anelastic Relaxation in Crystalline Solids,* Academic Press, San Diego, CA, 1972, p. 292ff.
13. C. Haas, *J. Phys. Chem. Solids* **15**, 108 (1960).
14. A. S. Nowick and W. R. Heller, *Adv. Phys.* **12**, 101 (1965).
15. Y. Zhang and Z. Xu, *Am. Mineral.* **80**, 670 (1995).
16. H. H. Dunken, in *Treatise on Materials Science and Technology*, Vol. 22, M. Tomozawa and R. H. Doremus, eds., Academic Press, San Diego, CA, 1982, p.1.
17. R. D. Shannon, *Acta Crystallogr.* **A32**, 751 (1976).
18. D. R. Bosomworth, W. Hayes, A. R. L. Spray, and G. D. Watkins, *Proc. R. Soc.* **A317**, 133 (1970).
19. B. Pajot, E. Artacho, C. A. J. Ammerlaan, and J. M. Spaeth, *J. Phys. Condens. Matter* **7**, 7077 (1995).
20. E. Artacho, A. Lizón-Nordström, and F. Yndurian, *Phys. Rev.* **B51**, 7862 (1995).
21. G. Herzberg, Molecular Spectra and Molecular Structure, Vol. 2, Infrared and Raman Spectra of Polyatomic Molecules, D. Van Nustrand, New York, 1945.
22. C. Hass, *J. Phys. Chem. Solids* **15**, 108 (1960).
23. M. Stavola, J. R. Patel, L. C. Kimmerling, and P. E. Freeland, *Appl. Phys. Lett.* **42**, 73 (1983).
24. R. C. Newman, J. H. Tucker, and F. M. Livingstone, *J. Phys.* **C16**, L151 (1983).
25. J. Frenkel, *Kinetic Theory of Liquids*, Oxford University Press, Oxford, 1946, pp. 10–11.
26. Y. Tokuda, T. Shimokata, M. Katayama, and T. Hattori, *Mater. Res. Soc. Symp. Proc.* **262**, 75 (1992).
27. R. H. Doremus, *J. Mater. Res.* **10**, 2379 (1995); **14**, 3754 (1999).
28. W. Frank, *Defect Diffusion Forum* **75**, 121 (1991).
29. V. Gösele and T. Y. Tan, *Appl. Phys.* **A28**, 79 (1982).
30. R. F. Peart, *Phys. Status Solidi* **15**, K119 (1966).
31. R. N. Ghoshtagore, *Phys. Rev. Lett.* **16**, 890 (1966).
32. J. M. Fairfield and B. J. Masters, *J. Appl. Phys.* **38**, 3148 (1967).
33. L. Kalinowski and R. Seguin, *Appl. Phys. Lett.* **35**, 211 (1979).
34. J. Hirvonen and A. Anttila, *Appl. Phys. Lett.* **35**, 703 (1979).
35. H. J. Mayer, H. Mehrer, and K. Maier, in *Radiation Effects in Semiconductors*, R. H. Urli and J. W. Corbett, eds., Institute of Physics, London, 1977, p. 186.

36. F. J. Demond, S. Kalbitzer, H. Mannsperger, and H. Damjantschitsch, *Phys. Lett.* **A93**, 503 (1983).
37. H. Bracht, E. E. Haller, and R. Clarck-Phelps, *Phys. Rev. Lett.* **81**, 393 (1998).
38. A. Ural, P. B. Griffin, and J. D. Plummer, *J. Appl. Phys.* **85**, 6440 (1999).
39. J. C. Mikkelsen, *Appl. Phys. Lett.* **41**, 871 (1982).
40. R. H. Doremus, *J. Appl. Phys.* **66**, 4441 (1989).
41. T. Y. Tan and V. Gösele, *Appl. Phys. Lett.* **40**, 616 (1982).
42. G. K. Celler and L. E. Trimble, *Appl. Phys. Lett.* **53**, 2492 (1988).
43. C. S. Nichols, C. S. Van de Walle, and S. T. Pantelides, *Phys. Rev. Lett.* **62**, 1049 (1989); *Phys. Rev.* **B40**, 5484 (1989).
44. P. M. Fahey, P. B. Griffin, and J. D. Plummer, *Rev. Mod. Phys.* **61**, 289 (1989).
45. J. A. Van Vechten, *Phys. Rev.* **10**, 1482 (1974).

17

DIFFUSION OF GROUP III, IV, AND V ELEMENTS IN SILICON

17.1 INTRODUCTION

The elements of groups III and V are introduced into the silicon structure substitutionally to provide holes and electrons to the silicon; these different carriers of charge lead to *p* (positive) and *n* (negative)-type silicon, which are necessary for making *p*-*n* junctions in devices. These elements are introduced into the silicon by diffusion from various sources, so their diffusion behavior is very important in processing of integrated circuits and silicon devices.

The experimental measurements of diffusion of group III and V elements by different investigators do not agree well; a spread of over an order of magnitude is common. Apparently, no one has examined these differences and tried to resolve them. Some factors that appear to lead to these differences are (1) concentration dependence of effective diffusion coefficients D_e, (2) influence of other dopants on a diffusing dopant, (3) influence of other elements such as oxygen and carbon on a diffusing dopant, and (4) unknown factors.

In this chapter I first list diffusion coefficients and activation energies for various elements in silicon. I have tried to choose recent experiments that are carefully done, but I have not made an exhaustive survey of the literature and have probably omitted deserving work. I have relied particularly on two reviews of data [1,2]. Reference 1 covers the period from 1967 though 1987, and Ref. 2 from 1988 through 1998.

Next, the mechanisms of diffusion of these elements in silicon are discussed. There was some discussion of these mechanisms and their equations in Chapter 14, which will be referenced appropriately. The emphasis here is on interstitial diffusion of dopant atoms, which seems to be the dominant mechanism of their diffusion in silicon. Then effects of concentration of dopants on their diffusion coefficients are considered, and some experiments on internal friction of boron-containing silicon are reinterpreted briefly. Finally, some additional factors such as oxidation of the silicon and concentration of other dopants on dopant diffusion are mentioned.

17.2 DIFFUSION COEFFICIENTS AND ACTIVATION ENERGIES FOR GROUP III, IV, AND V ELEMENTS

Diffusion coefficients at 1156°C ($10^4/T = 0.7$) and activation energies for diffusion of group III, IV, and V elements at temperature above about 800°C are given in Tables 17.1, 17.2, and 17.3. These values were taken from the diffusion equations calculated by the authors of the references, most of which were reported in Refs. 1 and 2. The data in the table were limited to those measured at temperatures including 1156°C; values only at considerably lower temperatures were not included. No attempt was made to assess the validity of the measurements, although values of the diffusion coefficients that were very large or small were excluded from the calculations of the mean values of D.

The measured diffusion coefficients from different investigators showed a wide spread, more than an order of magnitude in most cases. The standard deviations of D values for most elements are large. The activation energies are somewhat spread, although not nearly as much as the D values. This variability of measured diffusion coefficients has not been explained; authors have compared their measured values with those that agreed with their own and have ignored other equally valid measurements. Some possible reasons for these differences will be explored in subsequent sections. The main conclusion is that there are unknown factors influencing the measurements. Influences of element concentration, oxidizing atmosphere, and other elements have all been explored, but even when these factors are the same, there are differences in measured diffusion coefficients.

Mean values of diffusion coefficients and activation energies are summarized in Table 17.4. These values show differences from values often assumed as correct: for example, in Fig. 11 of Ref. 47 and Fig. 1 of Ref. 48, and as used for computer programs for processing. Choice of a single particular reference for the diffusion coefficient of an element in silicon can lead to a significant difference with the mean values in Table 17.4.

The mean diffusion coefficients for aluminum and boron are about the same, and somewhat lower than for gallium. The activation energies for diffusion of these three group III elements are about the same. Indium and thallium diffuse significantly slower with higher activation energies. These differences possibly result from the larger atomic sizes of indium and thallium as compared to the other group III elements, as shown in Table 17.5. The mean diffusion coefficient of phosphorus is about 20 times greater than those of the larger group V atoms of arsenic, antimony, and bismuth. The activation energies of these elements are larger than for phosphorus diffusion. Nitrogen is quite anomalous, having a much higher

Table 17.1 Diffusion Coefficients at 1156°C and Activation Energies for Diffusion of Group III Elements in Silicon[a]

Diffusion Coefficient ($cm^2/s \times 10^{13}$)	Activation Energy [kJ/mol (eV)]	Refs.
Boron		
10	356 (3.69)	1, 3
18	355 (3.68)	1, 4
4.4	338 (3.50)	1, 5
16	356 (3.69)	1, 6
7.8	340 (3.52)	1, 7
[0.0015]	410 (4.25)	1, 8
7.1	162 (1.68)	1, 9
5.5	338 (3.50)	1, 10
4.6	357 (3.70)	1, 11
1.9	275 (2.85)	1, 12
5.4	346 (3.59)	1, 13
7.2	291 (3.02)	1, 14
[540]	314 (3.34)	1, 15
3.0	346 (3.59)	1, 16
4.7	362 (3.75)	17
7.4 ± 4.8	330 ± 56 (3.42 ± 0.58)	Mean
Aluminum		
6.0	335 (3.48)	3
6.8	324 (3.36)	1, 18
13.0	289 (3.00)	1, 19
1.3	329 (3.41)	1, 12
9.3	308 (3.20)	1, 20
5.0	314 (3.26)	21
6.9 ± 4.0	316 ± 16 (3.28 ± 0.17)	Mean
Gallium		
14	339 (3.52)	1, 3
[0.065]	[397] [(4.12)]	1, 22
[2600]	[201] [(2.09)]	1, 23
11	327 (3.39)	1, 12
16	362 (3.76)	1, 24
15	260 (2.70)	1, 25
14 ± 0.2	314 ± 71 (3.26 ± 0.74)	Mean
Indium		
2.8	378 (3.92)	1, 3
1.2	350 (3.63)	1, 12
2.0	363 (3.77)	Mean

(*Continued*)

Table 17.1 (Continued)

Diffusion Coefficient ($cm^2/s \times 10^{13}$)	Activation Energy [kJ/mol (eV)]	Refs.
Thallium		
2.8	377 (3.91)	1, 3
1.2	356 (3.70)	1, 12
2.0	366 (3.80)	Mean

[a][ ], Not included in the mean.

Table 17.2 Diffusion Coefficients at 1156°C and Activation Energies for Diffusion of Group V Elements in Silicon

Diffusion Coefficient ($cm^2/s \times 10^{13}$)	Activation Energy [kJ/mol (eV)]	Refs.
Phosphorus		
10	356 (3.69)	1, 3
25	364 (3.77)	1, 27
9.4	357 (3.70)	1, 28
99	306 (3.17)	1, 23
12	183 (1.90)	1, 29
45	260 (2.70)	1, 30
1.7	318 (3.30)	1, 15
4.6	373 (3.87)	1, 13
5.1	356 (3.69)	1, 31
20 Five data sets	354 (3.67)	2, 32
Mean 21 ± 24	321 ± 48	3.33 ± 0.50
Arsenic		
0.89	343 (3.56)	1, 3
0.83	408 (4.23)	1, 33
0.53	374 (3.88)	1, 34
0.38	502 (5.20)	1, 35
0.93	405 (4.20)	1, 36
0.48	332 (3.44)	1, 12
2.0	442 (4.58)	1, 37
2.7	386 (4.00)	1, 38
1.1 ± 0.8	399 ± 55 (4.14 ± 0.57)	Mean
Nitrogen		
$1.6(10)^6$	280 (2.9)	26
Antimony		
0.66	381 (3.95)	1, 3

Table 17.2 (Continued)

Diffusion Coefficient ($cm^2/s \times 10^{13}$)	Activation Energy [kJ/mol (eV)]	Refs.
1.12	384 (3.98)	1, 39
1.0	425 (4.41)	1, 40
[56][a]	336 (3.48)	1, 30
0.29	352 (3.65)	1, 12
0.73	384 (3.98)	1, 41
0.65	470 (4.87)	1, 42
2.4	376 (3.90)	1, 43
2.5	375 (3.89)	2, 44
0.9	390 (4.04)	2, 45
Mean 1.1 ± 0.8	388 ± 38 (4.02 ± 0.39)	
Bismuth		
0.45	448 (4.64)	1, 3
0.29	371 (3.85)	1, 12
3.05	241 (2.50)	1, 46
1.3 ± 1.5	353 ± 104 (3.66 ± 1.08)	Mean

[a][ ], Not included in the mean.

Table 17.3 Diffusion Coefficients at 1156°C and Activation Energies for Diffusion of Group IV Elements in Silicon

Diffusion Coefficient ($cm^2/s \times 10^{14}$)	Activation Energy [kJ/mol (eV)]	Refs.
Silicon		
[13.2][a]	469 (4.86)	1, 49
2.7	460 (4.77)	1, 51
0.73	455 (4.72)	1, 52
2.0	396 (4.1)	1, 53
0.29	484 (5.02)	1, 54
0.57	450 (4.66)	1, 55
2.8	396 (4.1)	1, 56
0.61	425 (4.4)	1, 57
0.94	458 (4.75)	58
Mean 1.5 ± 1.0	449 ± 34(4.65 ± 0.35)	
Germanium		
[15][a]	510 (5.28)	1, 59
4.1	454 (4.70)	1, 60
3.6	499 (5.17)	1, 61

(Continued)

Table 17.3 (Continued)

Diffusion Coefficient ($cm^2/s \times 10^{14}$)	Activation Energy [kJ/mol (eV)]	Refs.
0.74	480 (4.97)	1, 62
1.6	514 (5.33)	1, 63
1.4	520 (5.39)	2, 64
Mean 2.3 ± 1.5	496 ± 25 (5.14 ± 0.26)	
Carbon		
$1.7(10)^3$	282 (2.92)	1, 65
$1.4(10)^5$	287 (2.94)	1, 66
Tin		
3.3	410 (4.25)	1, 67
2.5	338 (3.50)	1, 68
1.4	474 (4.91)	2, 69
Mean 2.4 ± 1.0	407 ± 69 (4.22 ± 0.71)	

[a][ ], Not included in the mean.

effective diffusion coefficient than the other group V elements. Nitrogen apparently does not exchange into the silicon structure to the same high concentrations of these elements, so the rapid interstitial diffusion of nitrogen is less retarded by this exchange.

The diffusion coefficients of the group IV elements silicon, germanium, and tin are all about the same at 1156°C, and the activation energies for diffusion are similar to one another. The effective diffusion coefficient of carbon is much higher than that of the other group IV elements, probably because of limited substitution for silicon, just as for nitrogen. The effective diffusion coefficient for nitrogen is apparently greater than for carbon, although the measurements on these two elements are quite limited.

17.3 MECHANISMS OF DIFFUSION AND REACTION

To understand the experimental results in Section 17.2 and the influence of factors such as oxidation, element concentration, and other elements on diffusion, one must consider the mechanisms of diffusion. Diffusion of element atoms is considered to be interstitial; that is, these atoms dissolve in the large space of interstice in the middle of the diamond cubic structure of silicon or in other interstitial sites. See Chapters 14 and 16 for diagrams showing this central interstice. These interstitial atoms then diffuse by jumping through "doorways" between the interstitial sites. Atoms and molecules with radii smaller than about 1.4 Å can diffuse very rapidly through the open silicon structure (see Chapter 14 and Table 17.5). It is possible

Table 17.4 Summary of Means of Diffusion Coefficients and Activation Energies for Diffusion in Silicon

Element	D ($cm^2/s \times 10^{13}$)	Q [kJ/mol (eV)]	Element	D ($cm^2/s \times 10^{14}$)	Q [kJ/mol (eV)]	Element	D ($cm^2/s \times 10^{13}$)	Q [kJ/mol (eV)]
	Group III			*Group IV*			*Group V*	
B	7.4	330 (3.42)	C	10^4	283 (2.93)	N	$1.6(10)^6$	280 (2.9)
Al	6.9	316 (3.28)	Si	1.5	449 (4.65)	P	21	321 (3.33)
Ga	14	322 (3.34)	Ge	2.3	496 (5.14)	As	1.1	400 (4.14)
In	2.0	364 (3.77)	Sn	2.4	407 (4.22)	Sb	1.1	388 (4.02)
Tl	2.0	367 (3.80)				Bi	1.3	353 (3.66)

Table 17.5 Approximate Atomic Radii of Elements (Å)[a]

Element	Atomic Radius	Element	Atomic Radius	Element	Atomic Radius
Group III		*Group IV*		*Group V*	
B	0.97	C	0.91	N	0.71
Al	1.43	Si	1.32	P	1.28
Ga	1.35	Ge	1.37	As	1.39
In	1.66	Sn	1.62	Sb	1.59
Tl	1.71	Pb	1.75	Bi	1.70

[a]From crystal structure; see Section 5.2.

that larger atoms such as indium, thallium, antimony, and bismuth diffuse somewhat more slowly than smaller atoms, although there are anomalies in this view (arsenic and tin). Other factors, such as ease of replacement of silicon (see below), are probably more important than atom size in determining the effective diffusion coefficients and activation energies in Table 17.4.

Silicon lattice vacancies are sometimes invoked to explain diffusion in silicon. My view is that vacancies are highly immobile in silicon because of the large energy needed to break silicon–silicon bonds, as described also in Chapter 14. Additionally, because of this high bond energy, the concentration of these immobile vacancies will be small. Thus I will disregard vacancies in diffusion mechanisms in silicon. The silicon surface is often suggested as a source of silicon interstitial atoms. However, the experiments on diffusion of gold and platinum in silicon, described in Chapter 14, provide good evidence that the silicon surface is a sink, not a source, for silicon interstitial atoms. Lattice defects such as dislocations and grain boundaries can possibly also serve as sinks for silicon interstitials, but there is no direct evidence that they serve as sources of interstitials.

The much smaller effective diffusion coefficients and higher activation energies for diffusion for group III, IV, and V elements (except for carbon and nitrogen) as compared to other elements (see Chapter 14) can be understood as resulting from the exchange of these group III, IV, and V elements with silicon atoms in the silicon structure. Atoms of these groups are known to substitute for silicon atoms and provide conduction electrons and holes in the silicon. This exchange can be described by the reaction

$$\mathrm{Si}_n + M_i = \mathrm{Si}_i + M_n \tag{17.1}$$

In this equation Si_n and Si_i are network and interstitial silicon atoms, respectively, and M_n and M_i are similarly network and interstitial atoms of other elements (dopants). An equilibrium constant K for Eq. 17.1 is

$$K = \frac{IS}{NC} \tag{17.2}$$

in which I is the concentration of interstitial silicon (Si_i), S the concentration of exchanged dopant (M_n), N the concentration of network silicon (Si_n), and C the concentration of dissolved interstitial atoms of M (M_i). Concentrations are used instead of thermodynamic activities because the concentrations I, S, and C are small, so that interactions between atoms of these species are small. The concentration of network silicon is large and constant.

In the present view the network atoms (Si_n and M_n) are strongly bonded in the silicon structure and immobile. Thus there are two mobile or diffusing species to consider, interstitial silicon and interstitial dopant M. Experiments on diffusion of gold in silicon, described in Chapter 14, provide good evidence that gold interstitials diffuse rapidly in silicon, and the measured effective diffusion is controlled by interstitial silicon atoms formed by the exchange of Eq. 17.1. The concentration of interstitial gold is low and constant with space and time during diffusion heat treatments; the gold diffuses rapidly across the silicon wafer and then stays uniform. This interstitial gold exchanges with silicon atoms in the silicon structure by Eq. 17.1, forming interstitial silicon. The profile of exchanged immobile gold in the silicon structure is U-shaped; this profile results from the effective diffusion coefficient, which is concentration dependent (see Chapter 14 for more information). The reason that gold behaves in this way is that it exchanges with only a limited concentration of silicon atoms, about $10^{16}\,cm^{-3}$, as compared to 10^{18} to $10^{20}\,cm^{-3}$ for the group III and V dopants. This limited exchange distinguishes gold from other rapidly diffusing interstitial atoms, such as transition metals (copper, iron, etc; see Chapter 14), which do not exchange appreciably with network silicon.

The other limiting situation is control of the apparent diffusion process by the interstitial dopant atoms only. In this situation the interstitial silicon formed by Eq. 17.1 diffuses rapidly to the silicon surface, giving a low and uniform concentration of these silicon interstitial atoms. The flux of dopant interstitial atoms must provide the much larger concentration of exchanged dopant atoms in the silicon structure than for gold, so the resulting effective diffusion coefficient of dopant is low. From the diffusion–reaction model (see Chapters 3 and 14) the effective diffusion coefficient D_e is

$$D_e = \frac{C_T D}{S_T} \tag{17.3}$$

in which C_T is the total concentration of interstitially dissolved dopant at equilibrium and S_T is the total concentration of dopant exchanged into the silicon structure; D is the diffusion coefficient of the interstitial dopant atoms. Since $S_T \gg C_T$, this effective diffusion coefficient is much smaller than the interstitial diffusion coefficient. Since D is not concentration dependent, D_e does not depend on the concentration of exchanged dopant, and the profiles measured are those for a constant diffusion coefficient.

Another limiting case is that the concentration of silicon interstitials is proportional to the concentration of exchanged dopant, or $I = kS$, where k is a constant: Then from Eq. 17.2,

$$K^1 = \frac{S^2}{C} \tag{17.4}$$

in which S is the measured concentration of exchanged dopant in the silicon structure and K^1 is a constant, $K^1 = KN/k$. For Eq. 17.4 an effective diffusion coefficient directly proportional to concentration is found (see Chapters 3 and 14):

$$D_e = \frac{2SD}{K^1} \tag{17.5}$$

The profiles expected for this diffusion coefficient are given in Chapter 2.

17.4 CONCENTRATION DEPENDENCE OF EFFECTIVE DIFFUSION COEFFICIENTS OF GROUP III AND V ELEMENTS IN SILICON

An early study of the diffusion of boron and phosphorus in silicon showed constant effective diffusion coefficients of these elements from a concentration of 10^{16} to $5(10)^{21}\,\text{cm}^{-3}$ at temperatures from 1000 to 1250°C [3]. Concentration-independent effective diffusion coefficients for these elements at concentrations below about $10^{19}\,\text{cm}^{-3}$ were confirmed in an extensive study [12,13]. Concentration-independent effective diffusion coefficients were also found for arsenic below about $2(10)^{19}\,\text{cm}^{-3}$ [12,70], antimony below $3(10)^{18}\,\text{cm}^{-3}$, and bismuth below $2(10)^{16}\,\text{cm}^{-3}$. However, at higher concentrations of dopant the effective diffusion coefficients are found to increase as the concentration increases (see Refs. 24, 46, 71, and references within). This increase is linear with concentration in some studies [24,46]. Some results of Solmi and Noble [71] are shown in Fig. 17.1. They show increases for boron, phosphorus, arsenic, and antimony diffusion with increasing concentration. The decreases at high concentrations for phosphorus and arsenic at concentrations above the carrier concentrations are uncertain, because of possible precipitation of complexes or chemical compounds of the arsenic and phosphorus.

The contrast between profiles of phosphorus measured by Fuller and Ditzenberger [3] that fit the equations for a constant diffusion coefficient at surface concentration up to $6(10)^{20}\,\text{cm}^{-3}$ and later measurements that show concentration dependence above about $10^{19}\,\text{cm}^{-3}$ is surprising. The silicon used for Ref. 3 experiments was single crystal with a resistivity of 10 Ω-cm and *p*-type doped with boron. The silicon in more recent studies was quite similar; for example, Solmi and Nobli used *p*-type 10 Ω-cm single-crystal silicon. There must be some different unknown factor between the experiments of Fuller and Ditzenberger and later experiments. A possible explanation for this difference is given below.

To examine the experiments on concentration dependence of effective diffusion coefficients of dopants in silicon, I will adopt the razor of William of Occam:

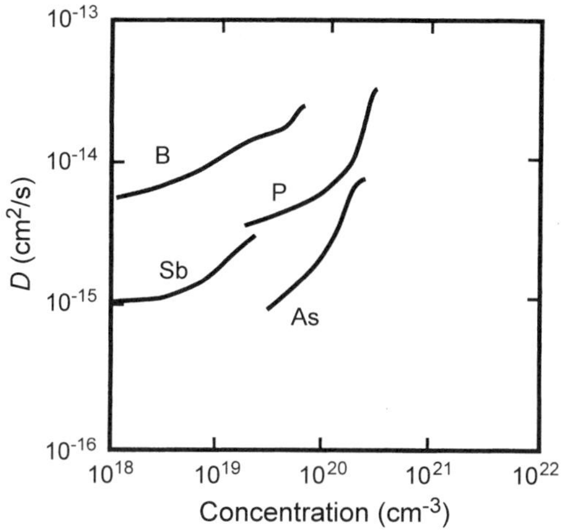

Figure 17.1 Effective diffusion coefficients of dopants in silicon as a function of dopant concentration. (From. Fig. 5 of Ref. 71.)

namely, that the simplest explanation is the best. I have suggested above and in Chapter 14 that lattice vacancies in silicon are present at low concentration and are effectively immobile, because of the strong covalent silicon–silicon bonds. Thus the only diffusing entities are interstitial atoms of silicon and dopants. As described in Section 17.3, the interstitial silicon atoms formed in reaction 1 diffuse rapidly to sinks, especially the silicon surface. Then the rate of diffusion is controlled by diffusion of the dopant atoms, and the effective diffusion coefficient D_e of Eq. 17.3 is not a function of concentration. This situation occurs at dopant concentrations below about $10^{19}\,\mathrm{cm}^{-3}$. The actual diffusion coefficient of dopant atoms, D in Eq. 17.3, is large, but the effective diffusion coefficient D_e, measured from atoms that replace silicon atoms in the silicon structure, is small, because the concentration S_T of replacing dopant atoms is much larger than the dissolved interstitial concentration C of dopant atoms.

A possible reason that Fuller and Ditzenherger found an effective diffusion coefficient independent of concentration for boron and phosphorus, even at concentrations above about $10^{19}\,\mathrm{cm}^{-3}$, could be that sinks for interstitial silicon, such as immobile silicon vacancies, stacking faults, or dislocations, were introduced into their samples during processing. These defects possibly lowered the concentration of interstitial silicon produced by Eq. 17.1, leading to diffusion control by interstitial dopant atoms. Why these defects were not introduced in later studies is not clear.

At concentrations above about $10^{19}\,\mathrm{cm}^{-3}$ of exchanged dopant atoms, the effective diffusion coefficient is often proportional to concentration. An example of this change of functionality of D_e is shown in the profiles of phosphorus measured by Yoshida *et al.* [72] after heating 2 h at 900°C, and shown in Figs. 17.2 and 17.3. Their profiles can be divided into two separate parts, one diffusing from a surface concentration of about $7.1(10)^{19}\,\mathrm{cm}^{-3}$ with a constant D_e value and a second with a

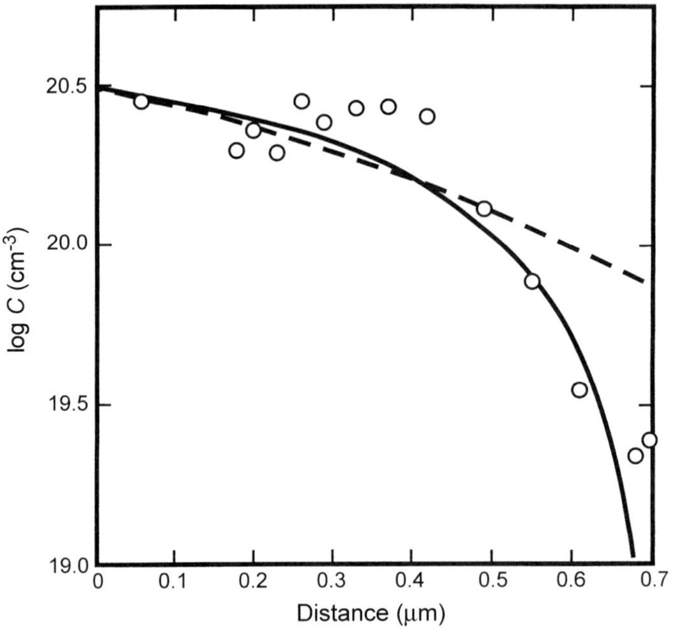

Figure 17.2 Profile of phosphorus in silicon after diffusion for 2 h at 900°C from a source of phosphorus in silicon dioxide. Points, data in a region near the silicon surface; line, diffusion coefficient directly proportional to concentration (Eq. 17.5); dashed line, error function solution. (From Fig. 1 of Ref. 72.)

surface concentration of about $3.2(10)^{20}\,\text{cm}^{-3}$ with an effective diffusion coefficient that is directly proportional to concentration. The constant D_e values for a surface concentration of $7.1(10)^{19}\,\text{cm}^{-3}$ was $1.6(10)^{-13}\,\text{cm}^2/\text{s}$, and for the variable D_e at the same concentration was $0.29(10)^{-13}\,\text{cm}^2/\text{s}$.

In a study of diffusion of arsenic in silicon, Masters and Fairfield [36] compared diffusion of arsenic tracer into silicon containing a uniform arsenic concentration, which they called *isocentration diffusion*, and diffusion of arsenic into silicon initially containing no arsenic. I have called these two types of experiments *exchange* and *reaction diffusion*. Their profile of arsenic diffusing into silicon containing no arsenic is shown in Figure 17.4. It does not fit the curve for constant diffusion coefficient and does fit the profile for a diffusion coefficient directly proportional to concentration, as shown in the figure, in the concentration range of about $2.6(10)^{20}$ down to $0.23(10)^{20}\,\text{cm}^{-3}$. At lower concentrations there is some indication, for only two data points, that the profile is changing to that for a constant diffusion coefficient, consistent with the findings described previously that below a concentration about $10^{19}\,\text{cm}^{-3}$ of dopant atoms the effective diffusion coefficient does not depend on concentration. In the experiments of Masters and Fairfield the diffusion coefficient for isocentration diffusion with a uniform arsenic concentration of about $2.6(10)^{20}\,\text{cm}^{-3}$ was $2.2(10)^{-14}\,\text{cm}^2/\text{s}$ at 948°C. For diffusion into intrinsic silicon at the same temperature and surface concentration

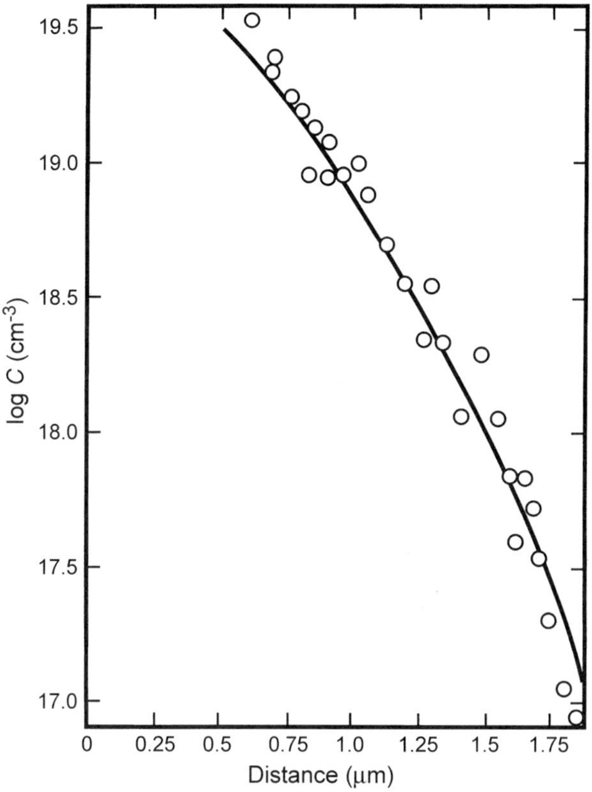

Figure 17.3 Profile of phosphorus in silicon; same conditions as in Fig. 17.2. Points, data at lower concentrations than in Fig. 17.2; line from a constant diffusion coefficient (Eq. 17.3).

was $1.8(10)^{-14}\,\mathrm{cm}^2/\mathrm{s}$, as calculated from the profile of Fig. 17.4 and the analysis of Crank [73, p. 120ff.]. The isocentration diffusion coefficients of Masters and Fairfield are plotted as a function of the uniform arsenic concentration in the silicon in Fig. 17.5. The slope of the log-log line in the figure is 1.20 by linear regression ($r = 0.982$), probably within experimental error of unity, down to a concentration of arsenic of about $10^{19}\,\mathrm{cm}^{-3}$. The isocontration diffusion coefficients of Makris and Masters for phosphorus are also plotted in Fig. 17.5; the linear regression slope of the log-log plot is 0.71 with $r = 977$; without the point at the lowest concentration the slope is 0.86. These data are therefore consistent with an effective diffusion coefficient of arsenic and phosphorus directly proportional to concentration above a dopant concentration of about $10^{19}\,\mathrm{cm}^{-3}$.

To understand these concentration dependencies, the first suggestion is that the constant diffusion coefficients result from control of diffusion by the interstitial dopant only; interstitial silicon formed in reaction 1 diffuses rapidly to surface or internal sinks, and the concentration of these silicon interstitial is low and uniform. It is possible that the diffusion coefficients of boron and phosphorus that are independent of concentration even at concentrations above $10^{19}\,\mathrm{cm}^{-3}$ in the

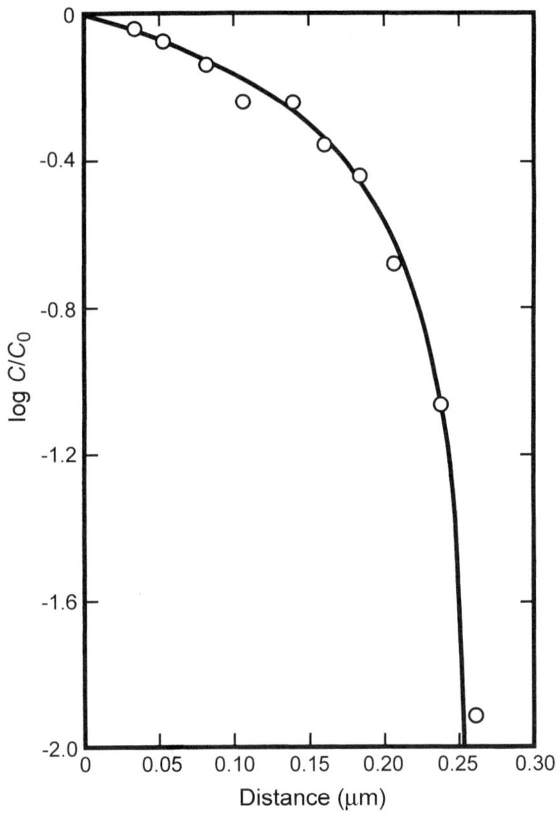

Figure 17.4 Profile of arsenic in silicon after diffusion for 4 h at 948°C from radioactive silicon–arsenic source on the silicon surface. Points, experimental data from Fig. 1 of Ref. 36; line from diffusion coefficient directly proportional to concentration (Eq. 17.5).

experiments of Fuller and Ditzenberger [3] result because there are sinks in the silicon produced during processing, as described above. These sinks could be immobile lattice vacancies or stacking faults. There is no independent evidence for these defects.

At higher concentrations than about $10^{19}\,cm^{-3}$, most experiments show that the effective dopant diffusion coefficients are directly proportional to concentration. I suggest that this functional dependence results because both dopant and silicon interstitials are influential in the diffusion. As described in Section 17.3, if the concentrations of these two interstitials are proportional to one another, the effective diffusion coefficient should be directly proportional to the total dopant concentration that replaces silicon atoms (Eq. 17.5). It will be difficult to test this possibility directly by experiment, because the interstitial concentrations are low and hard to measure separately from lattice silicon and dopant. Nevertheless, this possibility provides a relatively simple explanation for the different concentration dependencies of dopant atoms in silicon.

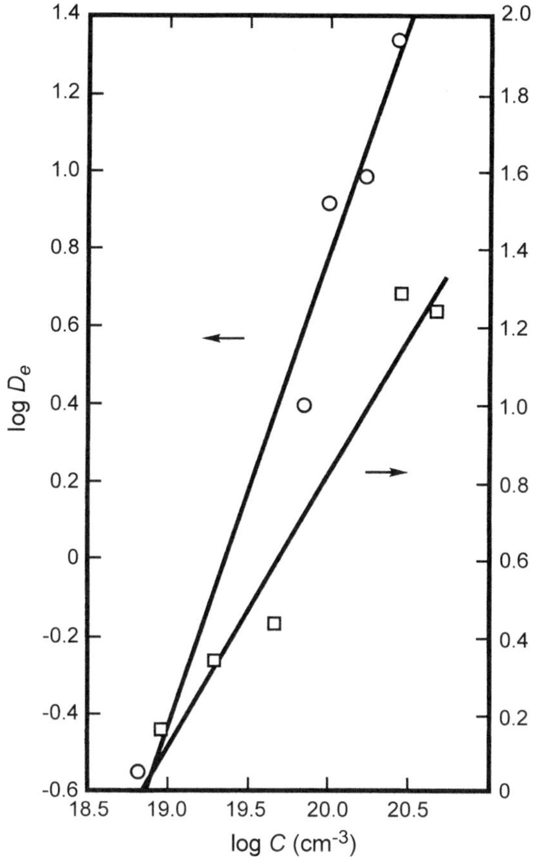

Figure 17.5 Logarithm of effective diffusion coefficients of phosphorus (○) and arsenic (□) as a function of total uniform concentrations of these dopants (isoconcentration experiments). (From Ref. 36.)

In the literature on dopant diffusion there is a confusing use of the terms *intrinsic* and *extrinsic silicon*. *Intrinsic silicon* traditionally has meant pure silicon with negligible dopants, so the concentrations of electrons and holes are equal, but in the dopant diffusion literature it is taken to mean an initial carrier concentration that is low compared to the concentration of diffusing dopant. *Extrinsic silicon* then means a condition of high dopant concentration. These terms are therefore contrary to traditional usage; I suggest they not be used in connection with dopant diffusion, and concentrations of dopants should be clearly described.

17.5 INTERNAL FRICTION OF SILICON

Tan and Berry measured the internal friction of silicon containing implanted boron, as reported by Frank [74]. They found peaks at temperatures of about −100, −64, and 73°C in the as-implanted silicon by flexural vibration at 438 Hz. After annealing for 1 h at 280°C, the 73°C peak disappeared, and the −100 and −64°C peaks were much reduced; a new peak at about 184°C developed with this heat treatment. The −100, −64, and 184°C peaks had reduced intensity when the stress was applied in a ⟨111⟩ direction; the 73°C peak was present for all orientations. Frank interpreted these peaks as resulting from the motion of covalently bonded pairs of boron–silicon and silicon–silicon pairs substituted for silicon atoms in the silicon structure. I feel that these pairs are highly unlikely to be present at appreciable concentrations, because of their high energies of formation. Furthermore, the activation energies corresponding to these peaks vary from 28 kJ/mol (0.29 eV) to 89 kJ/mol (0.72 eV), which are much too low for the motion of atoms covalently bonded in the silicon network.

I suggest that these internal friction peaks result from the motion of boron interstitial atoms in the silicon structure. Atoms with cubic symmetry will not show an internal friction peak. A possible position for a boron interstitial atom in silicon is at atomic positions [00 1/4] in the silicon unit cell. In this position the boron atom is between two silicon atoms and closer to 1 than the other as the result of a weak boron–silicon interaction. In this position the boron atom has approximate trigonal symmetry and so should show an internal friction peak with stress in a ⟨100⟩ direction but not in a ⟨111⟩ direction. Annealing at 280°C causes these implanted boron atoms to jump into more favorable sites with cubic symmetry, causing a reduction of the initial peaks. Some boron jump to other sites with noncubic symmetry to give the new peak at 184°C. The low activation energies for these motions show that interstitial boron atoms can diffuse with low interstitial activation energy in silicon, consistent with the assumption throughout this chapter of rapid interstitial diffusion of dopants. The much higher activation energies of the effective diffusion coefficients of dopants results because of the influence of the exchange reaction 17.1, which has a strong temperature dependence.

17.6 INFLUENCE OF OTHER FACTORS ON DOPANT DIFFUSION IN SILICON

There have been a very large number of studies of the influence of surface oxidation of silicon on dopant diffusion. Some of these recent studies are reviewed in Ref. 2; see also Refs. 75 to 77. Above about 1000°C, this influence of oxidation is limited to increases of factors of 2 or less in the effective diffusion coefficient of the dopant [75]; decreases in D_e are also sometimes found. However, at lower temperatures, increases in D_e over values of D_e extrapolated from Tables 17.1 and 17.2 can be quite large, up to a factor of 100 or more [76].

The favorite explanation of these effects is the generation of silicon interstitials at the silicon surface by the oxidation reaction. However, the experiments on gold diffusion described in Chapter 14 show that the silicon surface is a sink, not a source, for silicon interstitials. Furthermore, these interstitials at the silicon surface should be much more reactive than silicon atoms in the silicon structure, so the oxidation reaction should reduce, not increase, the concentration of silicon interstitials at the silicon surface. I feel it is much more likely that the small influence of oxidation rate at higher temperatures can probably be explained by correct treatment of the moving silicon–oxide interface into the silicon. The large increases in D_e at lower temperatures probably require the introduction of some additional species, such as water or dopant monoxides, into the silicon. See Chapter 5 for a discussion of the stability of dopant monoxides, and Chapter 16 for the role of SiO in oxygen and silicon diffusion in silicon. These monoxides or water could change the reaction equilibrium (Eq. 17.2) of Eq. 17.1, reducing the total concentration S_T of reacted dopant in the silicon, and thus increasing the effective diffusion coefficient of Eq. 17.3. More sensitive means for analyzing these monoxides and water at low concentrations are needed to test these possibilities.

Other dopants are also found to influence the diffusion of dopants. Addition of appreciable amounts of a third interstitial species leads to substantial complications in the analysis and calculation of diffusion coefficients. It is probably necessary to determine the relative roles of dopant and silicon interstitials in dopant diffusion before the influence of a third component can be explained.

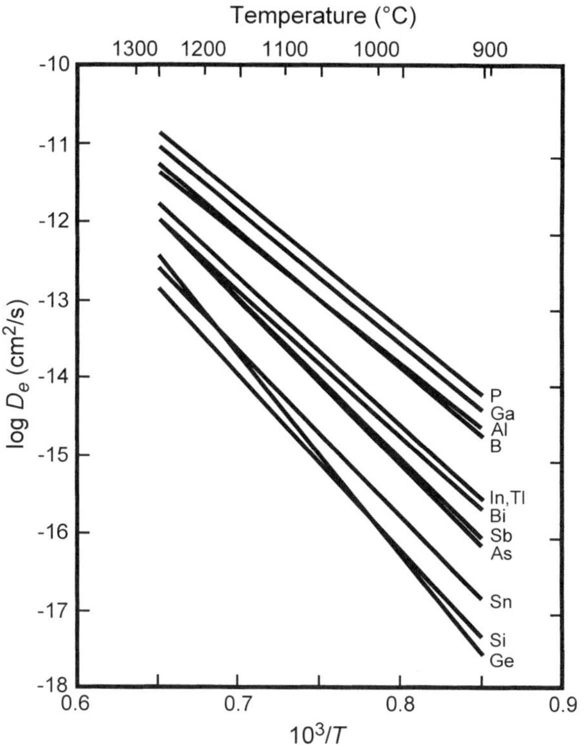

Figure 17.6 Logarithm of effective diffusion coefficients of group III, IV, and V elements in silicon as a function of $10^3/T$; from Table 17.4.

The diffusion of boron in silicon carbide [78] shows features expected for a diffusion–reaction, or kick-out, mechanism. The profiles of implanted boron can be explained [78] as resulting from diffusion plus the reaction

$$SiC + B = BC + Si \tag{17.6}$$

Features expected from a reaction not at equilibrium were found; for example, steps in boron concentration at the interface between the implanted boron and boron-free silicon. These steps have been described in Chapters 4 and 15, and arise when the reaction, such as in Eq. 17.6, is not at equilibrium. These results show that the diffusion–reaction model is valid for diffusion in other semiconductors or insulators, as well as silicon.

To summarize, the mean values of effective diffusion coefficients and activation energies given in Table 17.4, values of log D_e are plotted as a function of $10^3/T$ in Fig. 17.6. Some of these data are also compared with diffusion coefficients of other elements in Fig. 14.1.

REFERENCES

1. *Diffusion Defect Data*, **47** (1986).
2. D. J. Fisher, ed., *Diffusion in Silicon*, Scitec Publications, Zurich, Switzerland, 1998.
3. C. S. Fuller and J. A. Ditzenberger, *J. Appl. Phys.* **27**, 544 (1956).
4. J. Yamaguchi *et al.*, *J. Phys. Soc. Jpn.* **15**, 154 (1960).
5. A. D. Kurtz and R. Yee, *J. Appl. Phys.* **31**, 303 (1960).
6. E. L. Williams, *J. Electrochem. Soc.* **108**, 795 (1961).
7. J. Kato and Y. Nishi, *Jpn. J. Appl. Phys.* **3**, 377 (1964).
8. K. Nagamo, S. Jevauchi, and T. Tanaka, *Jpn. J. Appl. Phys.* **7**, 1361 (1968).
9. G. L. Vick and K. M. Whittle, *J. Electrochem. Soc.* **116**, 1142 (1969).
10. J. L. Alvarez, *An. Fis.* **65**, 299 (1969).
11. M. Okamura, *Jpn. J. Appl. Phys.* **8**, 1440 (1969).
12. R. N. Ghoshtagore, *Phys. Rev.* **B3**, 389, 397 (1971).
13. R. N. Ghoshtagore, *Solid State Electron.* **15**, 1113 (1972).
14. J. L. Price and F. N. Schwettmann, *J. Electrochem. Soc.* **12**, 705 (1974).
15. G. Masetti, S. Solmo, and G. Soncini, *Solid State Electron.* **19**, 545 (1976).
16. K. P. Frohmader and L. Baumbauer, *Solid State Electron.* **23**, 1263 (1980).
17. Y. M. Haddara *et al.*, *Appl. Phys. Lett.* **77**, 1976 (2000).
18. R. C. Miller and A. Savage, *J. Appl. Phys.* **27**, 1420 (1956).
19. Y. C. Kao, *Electrochem. Technol.* **5**, 90 (1967).
20. W. Rosnowski, *J. Electrochem. Soc.* **125**, 957 (1978).
21. Ch. Ortiz, D. Mathiot, Ch. Dubois, and R. Jerisian, *J. Appl. Phys.* **87**, 2661 (2000).
22. A. D. Kurtz and C. L. Gravel, *J. Appl. Phys.* **29**, 1456 (1958).

23. V. P. Prutkin, A. S. Lyutovich, and M. J. Kardzhaubaev, *Krist. Tonkickh Plenok* **139** (1970).
24. J. S. Makris and B. J. Masters, *J. Appl. Phys.* **42**, 3750 (1971).
25. S. Haridors, F. Beniere, M. Gauneau, and A. Rupert, *J. Appl. Phys.* **51**, 5833 (1980).
26. T. Itoh and T. Abe, *Appl. Phys. Lett.* **53**, 39 (1988).
27. J. C. C. Tsai, *Proc. IEEE* **57**, 1499 (1969).
28. T. Ouchiyama and M. Katsua, *Shin Nippon Denki Giho* **5**, 116 (1970).
29. Y. W. Hsueh, *J. Electrochem. Soc.* **117**, 807 (1970).
30. A. S. Lyutovich *et al.*, *Krist. Tonkickh Plenok* **91** (1970).
31. J. S. Makris and B. J. Masters, *J. Appl. Phys.* **42**, 3750 (1971).
32. V. Hartmann, *Wiss. Z. Tech. Hochsch. Ilmenau* **20**, 75 (1974).
33. W. J. Armstrong, *J. Electrochem. Soc.* **109**, 1065 (1962).
34. P. S. Raju, N. R. K. Rao, and E. V. K. Rao, *Indian J. Pure Appl. Phys.* **2**, 353 (1964).
35. Y. W. Hsueh, *Electrochem. Technol.* **6**, 361 (1968).
36. B. J. Masters and J. M. Fairfield, *J. Appl. Phys.* **40**, 2390 (1969).
37. D. P. Kennedy and P. C. Murley, *Proc. IEEE* **59**, 335 (1971).
38. C. Gontrand, P. Ancey, H. Haddab, and G. Chaussemy, *Semicond. Sci. Technol.* **7**, 181 (1992).
39. J. J. Rohan, N. E. Pickering, and J. Kennedy, *J. Electrochem. Soc.* **106**, 705 (1959).
40. S. Nakanuma and S. Yamaguchi, *J. Electrochem. Soc. Jpn.* **36**, 3 (1968).
41. V. A. Uskov and S. P. Svetlav, *Izv. Vyssh. Uchebn. Zaved. Fiz.* **15**, 145 (1972).
42. S. H. Song, S. Matsumoto, and T. Niimi, *Jpn. J. Appl. Phys.* **18**, 218 (1979).
43. F. H. M. Spit *et al.*, *Phys. Status. Solidi.* **A89**, 105 (1985).
44. F. H. M. Spit and H. Bakker, *Phys. Status. Solidi.* **A97**, 135 (1986).
45. A. N. Larsen and P. Krinaghøj, *Appl. Phys. Lett.* **68**, 2684 (1996).
46. Y. Ishikawa, K. Yazaki, and I. Nakamichi, *Jpn. J. Appl. Phys.* **28**, 1272 (1989).
47. W. Frank, U. Gösele, H. Mehrer, and A. Seeger, in *Diffusion in Crystalline Silicon*, G. E. Murch and A. S. Nowick, eds., Academic Press, San Diego, CA, 1984, p. 63.
48. W. Frank, *Defect Diffusion Forum* **75**, 121 (1991).
49. W. R. Wilcox and J. J. LaChapelle, *J. Appl. Phys.* **35**, 240 (1974).
50. R. F. Peart, *Phys. Status. Solidi.* **15**, K119 (1966).
51. R. N. Ghostagore, *Phys. Rev. Lett.* **16**, 890 (1966).
52. J. M. Fairfield and B. J. Masters, *J. Appl. Phys.* **38**, 3148 (1967).
53. I. R. Sanders and P. S. Dobson, *J. Matls. Sci.* **9**, 1987 (1974).
54. H. J. Mayer, H. Mehrer, and K. Maier, *Inst. Phys. Conf. Ser.* **31**, 186 (1976).
55. L. Kalinowski and R. Sequin, *Appl. Phys. Lett.* **35**, 211 (1979).
56. J. Hirvonen and A. Anttila, *Appl. Phys. Lett.* **35**, 703 (1979).
57. F. J. Demond *et al.*, *Phys. Lett.* **93A**, 503 (1983).
58. H. Bracht, E. E. Haller, and R. Clark-Phelps, *Phys. Rev. Lett.* **81**, 393 (1998).

59. D. A. Petrov, J. M. Shashkov, and J. P. Akimchenko, *Vopspsy Met. I. Fiz. Poluprovodnikov*, Akademii Nauk SSSR, Sbornik, Moscow, 1957, p. 130.
60. G. L. McVay and A. R. DuCharme, *J. Appl. Phys.* **44**, 1409 (1973).
61. P. V. Pavlov, V. I. Pashkov, and E. V. Dobrokhlov, *Fiz. Tverd. Tela* **15**, 3396 (1973).
62. H. Hettich, H. Mehrer, and K. Meier, *Inst. Phys. Conf. Ser.* **46**, 500 (1979).
63. P. Dorner *et al.*, *Philos. Mag.* **A49**, 557 (1984).
64. A. L. Bouchetout, N. Tabet, and C. Monte, *Mater. Sci. Forum* **10–12**, 127 (1986).
65. R. C. Newman and J. Wakefield, *J. Phys. Chem. Solids* **19**, 230 (1961).
66. P. L. Gruzin *et al.*, *Fiz. Tekh. Poluprovodnikov* **7**, 1853 (1973).
67. T. H. Yeh, S. M. Hu, and R. H. Kastl, *J. Appl. Phys.* **39**, 4266 (1968).
68. Y. Akasaka, *Jpn. J. Appl. Phys.* **13**, 1533 (1975).
69. P. Kringhøj and R. H. Elliman, *Appl. Phys. Lett.* **65**, 324 (1994).
70. T. L. Chiu and H. N. Ghosh, *IBM J. Res. Dev.* **15**, 472 (1971).
71. S. Solmi and D. Nobli, *J. Appl. Phys.* **83**, 2484 (1998).
72. M. Yohida, E. Arai, H. Nakamura, and Y. Terunuma, *J. Appl. Phys.* **45**, 1498 (1974).
73. J. Crank, *Mathematics of Diffusion*, 2nd ed., Oxford University Press, Oxford, 1975.
74. W. Frank, *Radiat. Effects* **21**, 119 (1974).
75. Y. Shibata, S. Hashimoto, K. Taniguchi, and C. Hamaguchi, *J. Electrochem. Soc.* **139**, 231 (1992).
76. R. E. Lever, P. B. Griffin, and W. A. Rausch, *J. Appl. Phys.* **78**, 3115 (1995).
77. F. Iacona *et al.*, *J. Electrochem. Soc.* **147**, 2762 (2000).
78. H. Bracht, N. A. Stolevijk, M. Laube, and G. Pensl, *Appl. Phys. Lett.* **77**, 3188 (2000).

18

DIFFUSION IN GERMANIUM

18.1 INTRODUCTION

In the 1950s germanium was considered as a possible device semiconductor; there were a number of studies done on diffusion in germanium. Since then silicon technology has become dominant for semiconductor devices, and the interest in diffusion in germanium has waned. Nevertheless, diffusion in germanium is of interest by way of comparison with diffusion in silicon, as discussed in the previous four chapters.

The lattice parameter of germanium in the cubic diamond structure is 5.66 Å, compared to 5.43 Å for silicon. It is not possible to decide which of these two structures is the most open, because of problems with the hard-sphere model of atoms in covalently bonded structures, as described in Chapters 5 and 16. The sizes of the germanium and silicon atoms in their solid structures are uncertain. One might expect the germanium atom to be larger than the silicon atom because germanium has more electrons. Whether this size difference results in more or less "open space" in the germanium crystal structure, with a larger lattice parameter, is uncertain. Diffusion coefficients of most elements that have been measured in both germanium and silicon are larger in germanium than in silicon, suggesting that the germanium structure is more open than that of silicon.

In this chapter measurements of diffusion coefficients in germanium are listed and compared with the diffusion–reaction and diffusion–exchange (kick-out) models. Finally, diffusion mechanisms in germanium and silicon are compared and discussed.

18.2 MEASUREMENTS OF DIFFUSION COEFFICIENTS IN GERMANIUM AND SILICON

Diffusion coefficients measured in germanium are given in Table 18.1, taken mostly from Ref. 1. Elements diffusing in germanium fit into the same categories as in silicon (see Chapter 14). Elements with a valence of plus one or two diffuse rapidly in germanium, just as they do in silicon. Gold, with plus one and three valences,

Table 18.1 Diffusion Coefficients in Germanium and Silicon at 800°C

	D, cm^2/s	
Element	Ge	Si
Fast Diffusers		
Li	$2(10)^{-5}$	$2(10)^{-6}$
Na	$2(10)^{-5}$	$4(10)^{-5}$
Cu	$8(10)^{-5}$	$1(10)^{-5}$
Ni	$5(10)^{-5}$	$2(10)^{-7}$
Fe	$6(10)^{-7}$	$2(10)^{-7}$
Co	$9(10)^{-7}$	$8(10)^{-6}$
Ag	$9(10)^{-7}$	$5(10)^{-10}$
Intermediate		
Au	$4(10)^{-10}$	$1(10)^{-15}$
Acceptors		
B	$8(10)^{-13}$	$7(10)^{-17}$
Al	$6(10)^{-14}$	$1(10)^{-16}$
Ga	$3(10)^{-14}$	$2(10)^{-16}$
In	$1(10)^{-13}$	$8(10)^{-18}$
Tl	$2(10)^{-13}$	$7(10)^{-18}$
Donors		
P	$6(10)^{-12}$	$3(10)^{-16}$
As	$1(10)^{-11}$	$2(10)^{-18}$
Sb	$1(10)^{-11}$	$2(10)^{-18}$
Bi	$3(10)^{-12}$	$7(10)^{-18}$
Group IV		
Ge	$4(10)^{-14}$	$3(10)^{-19}$
Si	$6(10)^{-15}$	$5(10)^{-20}$

Data for germanium mostly from Ref. 1. Data and references for silicon are in Chapters 14 and 17.

Table 18.2 Activation Energies for Diffusion of Groups III, IV, and V Elements in Silicon and Germanium

	Activation energy [kJ/mole (eV)]	
Element	Ge	Si
B	439 (4.55)	330 (3.42)
Al	286 (2.96)	316 (3.28)
Ga	302 (3.13)	322 (3.34)
In	302 (3.13)	364 (3.77)
Tl	328 (3.4)	367 (3.80)
P	241 (2.5)	321 (3.33)
As	230 (2.38)	400 (4.14)
Sb	227 (2.35)	388 (4.02)
Bi	248 (2.57)	353 (3.66)
Si	280 (2.9)	449 (4.65)
Ge	302 (3.13)	496 (5.14)

Data for germanium from Ref. 1; data and references for silicon are in Chapter 17.

diffuses more slowly in germanium ("intermediate diffuser"). Elements of groups III, IV, and V diffuse much more slowly in germanium than fast and intermediate diffusers, just as in silicon. Thus qualitatively the diffusion behavior of all these elements seems to be similar in silicon and germanium.

The fast diffusers diffuse more rapidly in germanium than in silicon, except for cobalt. The diffusion coefficients in silicon are about an order of magnitude slower than in germanium. Gold and elements from groups III, IV, and V diffuse much more slowly in silicon than in germanium; the difference is from two to seven orders of magnitude.

Activation energies for diffusion of group III, IV, and V elements for diffusion in silicon and germanium are compared in Table 18.2. For all elements, except boron, the activation energy for diffusion in silicon is greater than for germanium. It is possible that the one experimental result for boron in germanium is in error. These lower activation energies for diffusion in germanium correspond with the lower diffusion coefficients in germanium.

18.3 DIFFUSION PROFILES IN GERMANIUM

Experimental measurements of diffusion of various elements in germanium were made by depositing the element or its radioactive tracer on the germanium surface and analyzing for the element or tracer in successive slices of the germanium. The samples were thicker than the penetration of the element, so the geometry was semi-infinite with various surface boundary conditions. If the amount of element or tracer was large compared to the total amount diffused, then the surface boundary condition is constant concentration S_0 at distance $x=0$. If the effective diffusion coefficient D_e is not a function of concentration, then from Eq. 2.17 the profile of concentration S is

$$S = S_0 \operatorname{erfc}(x/2\sqrt{D_e t}) \tag{18.1}$$

where erfc is the conjugate error function, defined following Eq. 2.15, x is the distance, t the time, and D_e the effective diffusion coefficient. If the amount of material on the surface is so small that the surface concentration decreases with time, the profile is

$$S = \frac{M}{\sqrt{\pi D_e t}} \exp\left(-\frac{X^2}{4Dt}\right) \tag{18.2}$$

as given in Eq. 2.20, in which M is the total amount of material deposited on the surface at t=0.

The above equations are valid for a diffusion coefficient not dependent on concentration. If there is an exchange reaction as well as diffusion (sometimes called a *kick-out mechanism*), these profiles can be modified. If the element or tracer atom diffuses interstitially in the germanium and exchanges with network germanium atoms, one can write the exchange reaction as (see Eq. 14.1)

$$Ge_n + Cu_i = Cu_n + Ge_i \tag{18.3}$$

where subscript n means network and i interstitial. Then one of the three different conditions listed in Section 14.4, following Eq. 14.1, can hold. In the first condition it is assumed that the concentration of the impurity interstitial (in this case copper) is constant and uniform, because its diffusion coefficient is high. The profile for this condition is given by Eq. 4.12

$$\pm \operatorname{erf}\left(\ln\frac{S}{S_m}\right)^{1/2} = \frac{2x}{d} - 1 \tag{18.4}$$

where erf is the error function (defined in Section 2.2). For semi-infinite geometry d is a distance at which the reacted concentration S becomes uniform at the value S_m. (See Ref. 2 for more discussion of Eq. 18.4.) This equation is valid for profiles of gold and platinum in silicon, as described in Section 14.4. The second possible condition for diffusion and exchange with Eq. 18.3 is that the concentration of germanium interstitials is constant and uniform. In this condition the effective diffusion coefficient is independent of concentrations (Eq. 4.16) if local equilibrium exists. However, if there is no local equilibrium, different profiles are expected (see Section 4.5). The profile of reacted or network concentration S of impurities in the germanium (e.g., Cu_n from Eq. 18.3) can be found in Eq. 4.60

$$S(x,t) = S_T[\operatorname{erfc} r - e^{-r}u(\sqrt{kt}, r)] \tag{18.5}$$

where

$$r = \frac{x}{2}\left[\frac{1-\exp(-kt)}{D_e t}\right]^{1/2}. \tag{18.6}$$

In Eq. 18.5 S_T is the total concentration of copper in the germanium at equilibrium and k is the reaction rate coefficient for both the forward and backward reactions of Eq. 18.3. The function u arises in error functions of complex argument, as described in Chapter 4. (See Chapter 4 for discussion of these equations.)

Selected measurements of diffusion profiles in germanium are now compared to these equations. Fuller *et al.* [3] measured the diffusion of radioactive copper into germanium at temperatures from 680° to 900°C. Their profile after 10 minutes at 780°C is given in Fig. 18.1. The authors assumed that their profiles fit Eq. 18.1; in their words; "the curves of Fig. 3 descend normally." However, the comparison in Fig. 18.1 shows that the experimental concentrations deviated from the erfc profile of Eq. 18.1 at distances greater than about 1.5 mm into the germanium. Instead, they fit better in Eq. 18.5 with $\sqrt{kt} = 3$. The implication is that Eq. 18.3 is taking place. In the region near the germanium surface, local equilibrium has been reached so that the effective diffusion coefficient is independent of concentration; deeper into the germanium, local equilibrium has not been reached. This comparison also

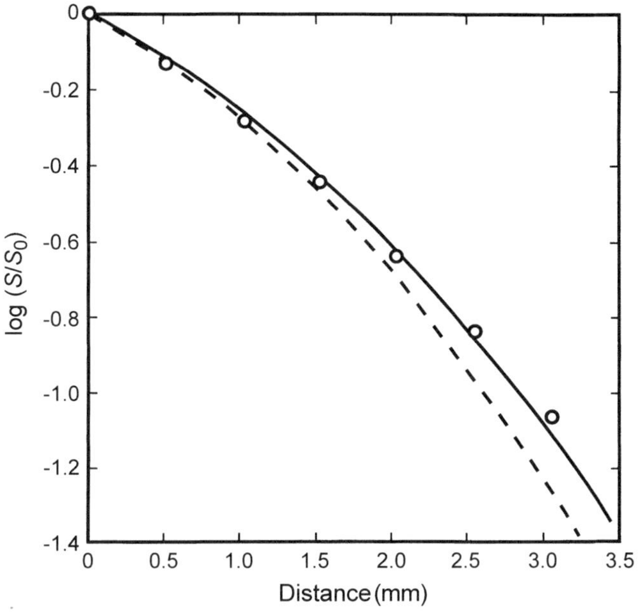

Figure 18.1 Log concentration of radioactive copper as a function of distance in single-crystal germanium after 10 minutes at 780°C. Data from Ref. 3; solid line from Eq. 14.5 with $\sqrt{kt} = 3$; dashed line from Eq. 18.1.

shows the usefulness of Eq. 18.5 in fitting diffusion–exchange profiles. Other profiles for different temperatures in Fig. 3 of Ref. 3 show similar fits.

Van der Maesen and Brenkman [4] measured profiles of elemental copper and nickel diffusing into germanium, apparently with chemical analysis of the metal. They concluded that the diffusion coefficient was dependent upon copper concentration. Their profile of copper after 20 min. at 800°C is compared with Eq. 18.5 with $\sqrt{kt} = 1$ in Fig. 18.2, and the fit is good except at low levels of copper, where the chemical analysis is possibly in error.

Wei [5] measured the diffusion profiles of radioactive silver, cobalt, and iron in germanium. The iron results were rather scattered and will not be discussed further. In germanium with a relatively high dislocation density (at about $10^5/\text{cm}^2$), both silver and cobalt showed exponential profiles; the profiles did not fit the profile of Eq. 18.1 for a diffusion coefficient independent of concentration. Furthermore, the surface concentrations changed with time, although rather erratically. Exponential profiles are expected in the diffusion–reaction model at surface concentrations less than about half he equilibrium value, as discussed in Sections 4.2 and 4.5; thus these profiles of silver and cobalt in germanium are consistent with Eq. 18.5 with $\sqrt{kt} \leq 1$.

Two profiles of silver measured by Wei in dislocation-free germanium had an unusual shape. The silver concentration dropped sharply at near the germanium surface, and then leveled off at longer distances from the surface. This shape is reminiscent of the U-shaped profiles of gold and platinum in silicon described in Section 14.4, which fit Eq. 18.4. A plot of a profile of silver in dislocation-free germanium as compared to Eq. 18.4 is given in Fig. 18.3, and fits Eq. 18.4 well.

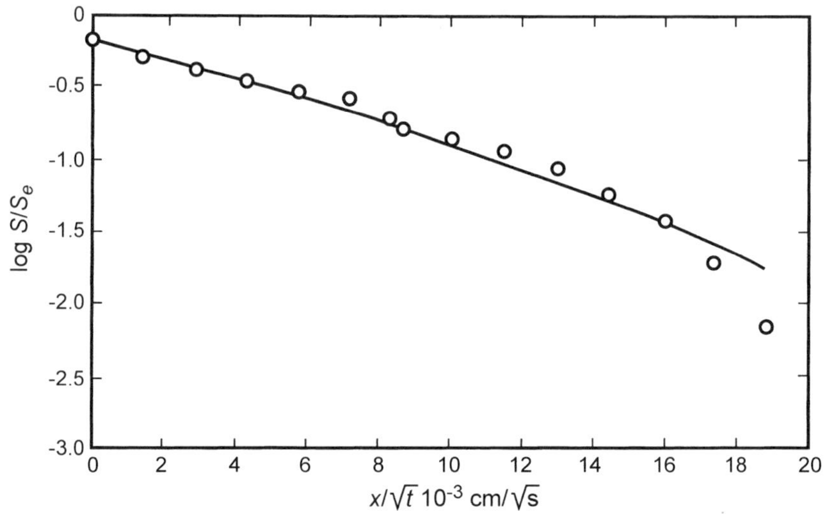

Figure 18.2 Log concentration of copper as a function of distance in single-crystal germanium after twenty minutes at 800°C. Data from Ref. 4; line from Eq. 14.5 with $\sqrt{kt} = 1$.

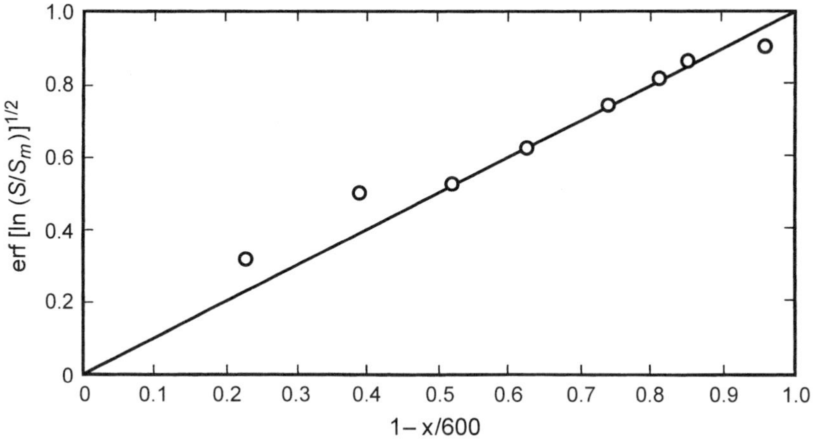

Figure 18.3 Diffusion profile of silver in dislocation-free germanium after 1000 *s* at 710°C. Data points are from Ref. 5; line from Eq. 18.4. See Section 14.4 for additional discussion and plots.

This equation results from condition one above, in which the interstitial silver diffuses rapidly into the germanium until it is constant and uniform in the region of measurements; the process is controlled by the slower diffusion of interstitial germanium, formed in Eq. 18.3, out of the germanium sample. This mechanism is directly related to the interpretation of gold and platinum diffusion in silicon (Chapter 14 and Ref. 2).

Profiles of arsenic in germanium after heating at 650°C and 580°C for twenty hours, fit the equation for a constant diffusion coefficient [6]. Profiles of radioactive antimony in germanium after $2.59(10)^5\,s$ at 900°C and 837°C fit Eq. 18.2 for a diffusion coefficient independent of concentration [7].

These comparisons of profiles of diffusing atoms in germanium with various equations were possible because the authors of the papers from the decades of 1950 and early 1960 reported their profiles in tables or clear graphs. It is striking that many measurements of diffusion profiles in silicon are not available; for example, there have been many measurements of copper diffusion in silicon, but I have found no reports that contain usable profiles. Profiles have valuable information about mechanisms of diffusion and reaction, and it is sad that many measurements are lost to further analysis.

18.4 MECHANISMS OF DIFFUSION IN GERMANIUM

The diffusion of copper in germanium was one of the first semiconductor systems for which a diffusion–exchange mechanism was suggested. Frank and Turnbull [8] noted that copper had a diffusion coefficient about 50 times greater in deformed germanium than in dislocation-free germanium. They recognized that the copper could be either interstitial or substituted for germanium in the germanium structure, as also proposed by van der Maesen and Brenkman [4]. Frank and Turnbull claimed that the dislocations in deformed germanium could not increase the interstitial diffusion coefficient of copper by a factor of 50, so they postulated that vacancies resulting from the deformation caused an increase in the effective diffusion coefficient. In their model the effective diffusion coefficient is that of the vacancies. This dissociative model has been discussed in detail by W. Frank and co-workers; they show that it cannot explain the result on diffusion of gold in silicon [2, 9]. In my view vacancies in silicon and germanium are effectively immobile, because of the strong covalent bonding in the diamond structure. Thus it seems unlikely that the dissociative model of Frank and Turnbull applies to interstitial diffusion in silicon or germanium.

The profiles discussed in Section 18.3 and shown in Figs. 18.1–18.3 provide good evidence that copper and silver diffuse by the diffusion–exchange or kick-out mechanism of Eq. 18.4. The diffusion can be controlled by either interstitial impurity (copper or silver) or by interstitial germanium (self-interstitial). In Sections 14.4 and 14.5 there are detailed discussions of diffusion of intermediate diffusers in silicon. Silver in germanium also fits this pattern. In dislocation-free germanium the fast-diffusing silver atoms reach a steady-state of a uniform and constant profile, so that the diffusion and reaction are controlled by the slower diffusion of interstitial germanium atoms. Dislocations in the germanium increase the effective diffusion rate of the interstitial germanium, so that the overall diffusion is controlled by silver diffusion. Mixed impurity-germanium diffusion is also possible, as discussed in Chapter 14 for silicon.

In Section 14.5 this faster effective diffusion of interstitial germanium (or silicon) with higher dislocation density is attributed to two factors. The first is enhanced diffusion along or near the dislocation (pipe diffusion). The second is removal of germanium interstitials by formation of stacking faults or by dislocation climb; in this mechanism the dislocations or deformation defects (possibly immobile vacancies) effectively act as a sink for germanium interstitials, thus reducing their effective diffusion distance.

It is not clear whether the same considerations apply to copper diffusion in germanium as for silver. An increase in the dislocation density in germanium increases the effective diffusion coefficient of copper in germanium (see the first paragraph of this section) and, again, this result could be caused by both pipe diffusion and sinks for interstitial germanium. However, the profiles of copper diffusion in dislocation-free germanium do not fit Eq. 18.4. It is possible that the concentration of reacted copper (Cu_n) is so high that the diffusion and reaction of interstitial copper is controlling even at low dislocation densities. (See Sections 14.4 and 14.5 for pertinent discussion.)

There are few reports of profiles of group III, IV, and V diffusers in germanium. Presumably the interstitial dopants control the rate of diffusion because their high network concentrations. (See Section 17.4.) In Chapter 16 it was proposed that the diffusion of silicon in silicon could take place by transport of silicon monoxide. It is also possible that germanium monoxide transports oxygen and germanium in germanium. (See Section 5.6 for discussion of monoxides.)

In summary it appears that there are many similarities between diffusion in silicon and germanium, with a few differences. Silver and copper in germanium exchange with network germanium, leading to diffusion by the diffusion–exchange (kick-out) mechanism. In silicon there is no evidence for appreciable exchange of these atoms with the silicon network.

Many of the mechanistic ideas expressed here are speculative and require further experiments for confirmation. Perhaps in the future there will be renewed interest in germanium for specialized device applications, and renewed interest in diffusion-exchange mechanism in germanium.

REFERENCES

1. B. L. Sharma, *Defect and Diffusion Forum* **70** & **71**, 1 (1990).
2. U. Gösele, W. Frank, and A. Seeger, *J. Appl. Phys.* **23**, 361 (1980).
3. C. S. Fuller, J. D. Struthers, J. A. Ditzenberger, and K. H. Wolfstein, *Phys. Rev.* **93**, 1182 (1954).
4. F. van der Maesen and J. A. Brenkman, *J. Electrochem. Soc.* **102**, 229 (1955).
5. L. Y. Wei, *J. Phys. Chem. Solids* **18**, 162 (1961).
6. R. Wölfe and H. Dorendorf, *Solid State Electron.* **5**, 98 (1962).
7. W. C. Dunlap, *Phys. Rev.* **94**, 1531 (1954).
8. F. C. Frank and D. Turnbull, *Phys. Rev.* **104**, 617 (1956).
9. W. Frank, *Defect and Diffusion Forum* **75**, 121 (1991).

Appendix

Conversion Factors between Various Length Units[a]

	Meters (m)	Centi- meters (cm)	Micro- meters (m)	Nano- meters (nm)	Angstron units (Å)	Pico- meters (pm)
Meters	1	100	10^{6}	10^{9}	10^{10}	10^{12}
Centimeters	10^{-2}	1	10^{4}	10^{7}	10^{8}	10^{10}
Micrometers	10^{-6}	10^{-4}	1	10^{3}	10^{4}	10^{6}
Nanometers	10^{-9}	10^{-7}	10^{-3}	1	10	10^{3}
Angstrom units	10^{-10}	10^{-8}	10^{-4}	0.1	1	100
Picometers	10^{-12}	10^{-10}	10^{-6}	10^{-3}	10^{-2}	1

[a]To convert a parameter in units of the row to that in the column, multiply by the factor in the table.

Conversion Factors between Various Volume Units[a]

	m^3	cm^3	μm^3	nm^3
m^3	1	10^{6}	10^{18}	10^{27}
cm^3	10^{-6}	1	10^{12}	10^{21}
μm^3	10^{-18}	10^{-12}	1	10^{9}
nm^3	10^{-27}	10^{-21}	10^{-9}	1

[a]To convert a parameter in units of the row to that in the column, multiply by the factor in the table.

AUTHOR INDEX

The book of Crank (*Mathematics of Diffusion*) and my own papers are not listed in this index because they are referred to in almost every chapter. If a reference has more than three authors, usually only the first author is listed.

Abe, M., 65, 72
Abe, T., 256, 271
Abraham, F., 174, 183
Agius, B., 46, 57, 164–165, 168, 182, 231, 240
Akasaka, Y., 258, 272
Akermark, T., 164–166, 182
Alpert, D., 18, 22
Alvarez, J. L., 255, 270
Anderson, O. L., 63–64, 66, 73, 238, 240
Ando, K. 176, 183
Anttila, A., 248, 251, 257, 271
Araujo, F. M., 155, 161
Artacho, E., 245, 251

Bansal, N. P., 138, 146
Barber, W. D., 20–22
Barrer, R. M., 18, 22, 65, 73, 79, 95, 131–132, 146
Bartholomew, R. F., 129–130, 134–136
Barton, J. L., 158–159, 161
Bates. J. B., 160, 162
Beales, K. J., 155, 161
Begeal, D. R., 65, 72
Behrens, H., 113–114, 118, 130, 133–135, 137, 139–140, 146, 188, 198
Belle, J., 176, 183
Berghoz, J. L., 241, 246, 251
Berry, B. S., 242, 251
Beshers, D. N., 2, 8
Beyer, W., 234, 240
Birnie, D. P., 172–173, 182
Bluman, G., 209, 217
Booth, F., 30, 32–35, 154, 161
Bosomworth, D. R., 244, 246, 251
Bouchetout, A. L., 258, 272
Bracht, H., 248–249, 252, 257, 270–272
Brady, J. B., 175, 183
Brandt, W. W., 65, 72
Branz, H. M., 233, 235, 240
Brenkman, J. A., 278, 280–281
Briscoe, J., 78, 94
Burlitz, R. S., 18, 22
Burn, J., 81–82, 84–85, 95, 125, 127
Burnham, C. W., 134, 141, 143, 147
Burrus, C. A., 27, 35
Burt, R., 65, 73
Büttner, H., 65, 67, 73

Canil, D., 180, 184
Cardona, M., 159, 161
Carlson, D. E., 234, 240
Carroll, M. R., 63, 67, 73
Carslaw, H. C., 9, 13, 22, 44, 55, 57
Carter, R. E., 172, 174, 176, 182
Cathcast, J. V., 160, 162
Cawley, J. D., 164–167, 170, 182
Celler, G. K., 249, 252
Chakraborty, S., 180, 184
Chen, W. K., 176, 183
Cherniak, D. J., 125, 128, 176, 183
Chevallier, J., 219–220, 239
Chiang, Y. M., 172–173, 182
Chiu, T. L., 262, 272
Choudhury, A., 101–103, 107
Christian, J. W., 28, 215,
Coffa, S., 207–211, 213, 215–218
Connolly, C., 176, 183
Contamin, P., 176, 183
Cook, G. B., 197–198
Cooper, D. M., 155, 161
Cooper, R. F., 197–198
Cooper, A. R., 165, 176, 183
Corbett, J. W., 222, 239, 242, 251
Costello, J. A., 164–167, 169, 182
Cottrell, A. H., 2, 8
Cronemeyer, D. C., 192, 198
Cullity, B. D., 61, 72

Dalle, C., 155, 161
Darken, L. S., 71, 73
Davis, E. A., 219, 220, 239
Davis, K. M., 91, 95
Davis, N. F., 134, 141, 143, 147
Deak, P., 222, 239
Deaton, R. L., 176, 183
Delaney, J. R., 108, 111–114, 118, 125, 127, 188, 198
Demond, F. J., 248, 22, 257, 271
Denk, J., 172, 184
Dennis, P. F., 98, 101–103, 107, 125, 127, 176, 183
Derdau, D., 176, 183
Derry, D. J., 176, 184
Dersch, O., 37, 57, 104–105, 107
Dingwell, D. B., 130, 133–134, 138–139, 146
Ditzenberger, J. A., 255–257, 262–263, 270, 277–278, 281
Dorendorf, H., 279, 281
Dorner, P., 258, 272
Douglas, R. W., 195, 198
Draper, N. R., 118
Doyle, M. V., 2, 8
Drowart, J., 72–73
Drury, J., 80, 82, 85, 95
DuCharme, A. E., 257, 272
Dümbgen, G., 103–104, 107, 164–165, 169, 176, 178, 182–183
Dunken, H. H., 244, 251
Dunlap, W. C., 279, 281
Dunn, T., 180, 184

Eagan, R. J., 67, 73
Einstein, A., 187, 198
Elphick, S. C., 176, 183
Ericson, J. E., 37, 57
Ernsberger, F. M., 129, 145, 222, 239
Essene, E. J., 5, 8, 123, 125, 128
Etsell, T. H., 172, 174, 176, 182

Fahey, P. M., 2, 8, 199, 217, 250, 252
Fairfield, J. M., 248, 251
Farver, J. R., 2, 8, 98, 101–103, 107, 123, 125–127, 178, 183
Fisher, D. J., 199, 201–202, 204, 217, 219, 220, 222, 239, 253–255, 257, 268, 270
Flenglas, S. N., 172, 174, 176, 182
Fortier, S. M., 124, 126–127
Fowler, R. H., 79, 95, 131, 146
Fox, M., 155, 161
Frank, F. C., 279–281
Frank, R. C., 65, 72, 149–150, 152–153, 160
Frank, W., 199–201, 106–209, 211–213, 215–218, 254, 267–268, 271–272, 276, 280–281
Franz, A. J., 238, 240
Freer, R., 175–176, 183
Frenkel, J., 64, 73, 180, 184, 246, 251
Friebele, E. J., 154, 161
Frischat, G. H., 67, 73
Frohmadev, K. D., 255, 270
Fry, D. L., 149, 153, 160, 236, 240
Fukata, N., 220, 239
Fuller, C. S., 255–257, 262–262, 270, 277–278, 281
Furuichi, A., 222, 236–237, 239

Gaines, R. V., 123, 127
Gallagher, C. J., 204, 217
Ganem, J. J., 164, 167, 182
Gautason, B., 175–176, 183
Gaworzewski, P., 241, 246, 251
Ghoshtagore, R. N., 248, 251, 255–256, 262, 270
Giletti, B. J., 67, 73, 98, 101–102, 106–107, 124–127, 176, 178, 183
Gland, J. L., 238, 240

Glicksman, M. E., 94–95
Gontrand, C., 256, 271
Goodenough, J. B., 174, 184
Gösele, V., 199–200, 206–209, 217, 247–249, 251–252, 254, 271, 276, 280–281
Greene, B. I., 155, 161
Greenway, D. L., 159, 161
Griffin, P. B., 2, 8, 199, 217, 250, 252, 268, 272
Grimley, D. J., 78, 94
Grinrod, P., 7–8
Griscom, D. L., 154, 161
Gruzin, P. L., 258, 272
Guggenheim, E. A., 79, 95, 131, 146
Gurry, R. W., 71, 73
Gusev, E. P., 170, 182

Haddara, Y. M., 255, 270
Hagel, W. C., 176, 183
Han, C. J., 164–168, 182
Haridors, S., 255, 271
Harrop, P. J., 175, 182
Hart, E. W., 215, 218
Hartwig, C. M., 79, 95
Harwig, H. A., 174, 183
Hass, C., 242, 246, 251
Hauber, J., 207, 217
Haul, R., 103–104, 107, 164–165, 169, 176, 178, 182–183
Hayashi, T., 176, 183
Heiser, T., 204, 217
Helmich, M., 88–89, 91–92, 95, 125, 127
Helms, C. R., 164–168, 182
Hercule, S., 160, 162
Herring, C., 220, 222, 235–236, 239
Herzberg, G., 245, 251
Hess, K. C., 124–125, 127
Heuer, A. H., 176, 184
Hirronen, J., 248, 251, 257, 271
Hettich, H., 257, 272
Hocine, S., 213, 218
Hoffman, J. W., 176, 184
Holloway, J. R., 108, 111–114, 118, 125, 127, 188, 198
Holtz, F., 130, 133–134, 136–140, 146
Hrostowski, H. J., 241–242, 245, 251
Hseuh, Y. W., 256, 271
Huntington, H. B., 2, 8
Hurley, C. N., 60, 72

Iacona, F., 268, 272
Ichimiya, T., 222, 236–237, 239
Ihinger, P. D., 109, 110, 118, 124–125, 130, 134–136, 146, 174, 183
Ingrin, J., 160, 162

Ishihara, T., 174, 183
Ishikawa, Y., 257, 262, 271
Itoh, H., 155, 161
Itoh, T., 256, 271

Jackson, W. B., 54, 57, 219–220, 222, 239
Jaeger, J. C., 9, 13, 22, 44, 55, 57
Jain, S. C., 15, 22, 231–232, 240
Jaoul, O., 176, 183
Jeans, J., 63, 72, 181, 184
Jeffer, J. H. E., 71, 73
Johnson, J., 65, 73
Johnson, N. M., 219–220, 222, 235–236, 239
Johnson, O. W., 160, 162

Kaiser, R. H., 241–242, 245, 251
Kalen, J. D., 164–168, 182
Kalinowski, L., 248, 251, 257, 271
Kao, Y., 255, 270
Karsten, J. L., 108, 111–114, 118, 125, 127, 188, 198
Kato, J., 255, 270
Kats, A., 97–103, 107, 125, 127
Kennedy, D. P., 256, 271
Keppler, H., 28, 35, 114, 118, 130, 132–135, 137–138, 145–146
Kiesling, Y., 172, 176, 184
Kim, B. K., 176, 183
Kingery, W. D., 172–173, 182
Kittel, C., 61, 72
Kohlic, J. T., 158, 161
Kohn, S. C., 129, 146
Kotomin, E., 7–8
Kramer, S. A., 174, 183
Kringho, P., 258, 272
Kuge, S., 21, 218
Kumei, S., 209, 217
Kurtz, A. D., 255, 270
Kuzovkov, V., 7–8

Labotka, T. C., 172, 176, 178, 183, 190, 198
Lacorre, P., 174, 183
Lanford, W. A., 91, 93, 95
Lange, R. A., 134, 136, 142–144, 147
Langmuir, I., 130, 136, 146
Lapham, K. E., 108, 111–114, 118, 125, 127, 188, 198
Larsen, A. N., 257, 271
Lee, R. W., 65, 72, 149, 150, 152–153, 160, 236, 240
Lee, S., 241, 250
Lerch, W., 212–213, 215, 218
Lesher, C. E., 180–181, 184
Lever, R. E., 268, 272

Lidiard, A. B., 2, 8, 172, 182
Ligenza, J. R., 2, 8, 86, 95
Lowe, J. P., 60–61, 72
Lyutovich, A. S., 256–257, 271

MacDonald, R. S., 121, 127, 242, 251
Magee, C. W., 234, 240
Maier, K., 248, 251, 257, 271
Major, L. D., 164–165, 182
Makris, J. S., 255–256, 262, 271
Marin, J. F., 176, 183
Maruse, D., 27, 35
Masaryk, J. S., 79, 95, 131, 146
Masetti, G., 255–256, 270
Masters, B. J., 248, 251, 255, 256, 262, 264–267, 271
Masterton, W. L., 60, 72
Matano, C., 22
Mathiol, D., 213, 218
Matsudo, J. J., 66, 73
Matzke, H., 65, 73
Mayer, H. J., 248, 251, 257, 271
McQuaid, S. A., 241, 247, 251
McVay, G. L., 257, 272
Mehrer, H., 199, 207–209, 212–213, 215, 217–218, 248, 251, 254, 257, 271
Mesli, C. J., 204, 217
Mikhailov, M. D., 9, 22
Mikkelson, J. C., 164, 169, 182, 241, 249, 250, 252
Miller, R. C., 255, 270
Monson, T. K., 207, 210, 217
Moore, D. K., 125, 128, 176, 183
Moore, W. J., 176, 184
Morain, M., 158–159, 161
Mott, N. F., 145, 147, 180, 184
Moulson, A. J., 80–85, 95, 125, 127, 139, 146, 174, 182
Mozzi, R. L., 76–77, 94
Muehlenbachs, K., 175–176, 178, 180, 182–183
Mulfinger, H. O., 125, 127

Nagamo, K., 255, 270
Nakamura, K. G., 238, 240
Nakanuma, S., 257, 271
Nakashima, H., 213, 218
Nakayama, G. S., 63, 73, 78, 95
Newman, R. C., 246, 251, 258, 272
Nichols, C. S., 250, 252
Nickel, N. H., 219–220, 222, 239
Nishi, Y., 255, 270
Nobli, D., 262–263, 272
Norberg, R. E., 221, 239
Norton, F. J., 65, 73, 78, 95, 150, 161, 164, 169, 182
Novak, M., 113–114, 118, 130, 134–135, 137, 146
Nowick, A. S., 242, 251

Ochs, F. A., 134, 136, 142–144, 147
Oel, H. J., 67, 73
Ohring, M., 169, 182
Oishi, Y., 176, 183
Okamura, M., 255, 270
Ortiz, C., 255, 270
Ouchiyama, T., 256, 271
Özizik, M. N., 9, 22

Pajot, B., 245, 251
Palik, E. D., 159, 161
Pankove, J. I., 219–220, 235, 239
Pantelides, S. T., 250, 252
Pavlov, P. V., 257, 272
Paul, A., 188, 191–192, 198
Pauling, L., 60–61, 72
Peart, R. F., 248, 251, 271
Perkins, R. A., 160, 162
Perkins, W. G., 65, 72
Petrov, D. A., 257, 272
Pfeffer, R., 169, 182
Phillips, B. L., 179, 183
Pichaud, B., 214, 218
Pitzer, K. S., 137, 146
Plessner, K. W., 155, 161
Plummer, J. D., 2, 8, 199, 217, 250, 252
Polian, A., 129, 146
Powers, R. W., 2, 8
Price, J. L., 255, 270
Prutkin, V. P., 255–256, 271

Rammensee, W., 65, 67, 73
Rauch, B., 65, 72
Rauch, F., 37, 57, 88–89, 91–92, 95, 104–105, 107, 125, 127
Reddy, K. P. R., 176, 183
Rice, S. A., 7, 8
Richardson, F. D., 71, 73
Richet, P., 129, 146
Rigo, S., 46, 57, 85–90, 95, 164–168, 170, 182, 231, 240
Ritter, G., 241, 246, 251
Roberts, G. J., 65, 73, 80–82, 85, 91, 95, 125, 127
Roberts, J. P., 80–85, 95, 125, 127, 139, 146, 174, 182
Robin, R., 176, 184
Rochet, F., 46, 57, 85–90, 95, 164–165, 168, 170, 182, 231, 240

Rogers, W. A., 18, 22
Rohan, J. J., 257, 271
Roselieh, K., 65, 67, 73
Rosenhauer, M., 65, 67, 73
Rosnowski, W., 255, 270
Rubie, D. C., 180, 184
Rush, J. D., 155, 161
Ryerson, F. J., 176, 178, 183

Sadoh, T., 213, 218
Santos, P. V., 54, 57
Schachtner, R., 103–104, 107
Schick, H. L., 181, 184
Scholze, H., 125, 127
Schurman, M. K., 155, 161
Seeger, A., 199, 206–209, 217, 254, 271, 276, 280–281
Seitz, F., 2, 8
Semet, M. P., 67, 73, 125, 127
Sequin, R., 248, 251, 257, 271
Shackelford, J. F., 63, 72, 78–79, 95, 131–132, 146
Shannon, R. D., 61, 64, 69, 72, 244, 251
Sharma, B. L., 274–275, 281
Sharp, Z. D., 106–107, 176, 183
Shelby, J. E., 64–65, 67, 72–73, 79, 83, 95, 131–132, 137, 146, 149–150, 153–154, 158–159, 161, 175, 183, 236, 240
Shen, A., 114, 118
Shewmon, P., 94–95
Shibata, Y., 268, 272
Shimizu, N., 180, 184
Shinar, J., 234–235, 240
Silver, L. A., 124–125, 127, 130, 134–136, 146
Simpson, L. A., 172, 174, 182
Smith, A. W., 176, 183
Smith, H., 118
Snyder, L. C., 222, 239
Sickel, H. G., 103–104, 107
Solmi, S., 262–263, 272
Song, Z., 224–225, 230, 238–239, 257, 271
Soules, T. F., 76–77, 94
Southgate, P. D., 242, 251
Sowerby, J. R., 130, 132–135, 137–138, 145–146
Spit, F. H. M., 257, 271
Spitzer, W. G., 2, 8, 86, 95
Srivastava, K. R., 65, 73
Stannard-Powell, S. J., 155, 161
Stavola, M., 246, 251
Sterner, S. M., 137, 146
Stolper, E. M., 42, 44, 57, 63, 67, 73, 108–110, 114–118, 124–125, 129–130, 134–136, 146, 174, 182, 188, 198
Stolwijk, N. A., 207–208, 212–213, 215, 217–218
Stone, J., 27, 35, 155–157, 161
Straboni, A., 46, 57, 164–165, 168, 182, 231, 240
Street, R. A., 219–220, 235, 239
Strong, R. L., 176, 183
Stuart, D. A., 63–64, 66, 73, 238, 240
Stutzmann, M., 219–220, 239
Sucov, E. W., 164–165, 169, 182
Swets, D. E., 65, 72, 149–150, 152–153, 160
Sykes, D., 129, 146

Takahashi, T., 174, 183
Takeno, H., 241, 246, 251
Tan, T. Y., 247–249, 251–252
Thompson, B. A., 176, 183
Tilquin, J. Y., 188, 198
Tokuda, Y., 247, 251
Tomlins, G. W., 175–176, 183–184
Tomozawa, M., 90–91, 95, 125, 127, 155, 161
Tress, H. J., 194, 198
Tressler, R. E., 164–167, 169, 182
Trimble, L. E., 249, 252
Tsai, C. C., 54, 57, 256, 271
Tuller, H. L., 174, 183
Turnbull, D., 279–281
Tuzzolo, M. R., 158, 161

Uchida, N., 155, 161
Ural, A., 249, 252
Uskov, V. A., 257, 271

Van de Hulst, H. C., 159, 161
Van der Maesen, F., 278, 280–281
Van de Walle, C. S., 250, 252
Van Vechten, J. A., 207, 217, 250, 252
Van Wieringen, A., 222, 236–237, 239
Vaughn, D. E. W., 18, 22, 79, 95, 131–132, 146
Vick, G. L., 255, 270
Vitko, J., 158–159, 161
Von Hippel, 222, 239

Wagman, D. D., 72, 673, 194, 198, 221, 239
Wagner, C., 19, 22
Wagner, H., 234, 240
Wakabayashi, H., 90, 95, 125, 127
Wakefield, J., 258, 272
Walker, J., 219–220, 222, 239
Wall, F. T., 131, 141, 146, 186, 194, 198
Wang, L., 5, 8, 123, 125, 128
Warmoltz, N., 222, 236–237, 239
Warren, B. E., 76–78, 94

Wasserburg, G. J., 42, 44, 57, 67, 73, 108, 110, 114–118, 125, 127, 130, 146, 188, 198
Watkins, G. D., 242, 244, 251
Watson, E. B., 125, 128, 176, 183
Webb, S. L., 130, 146
Wei, L. Y., 278, 279, 281
Widmer, A. E., 221, 239
Wiesenfeld, J. M., 7, 35, 155–157, 161
Wilcox, W. R., 257, 271
Wilke, M., 188, 198
Wilkins, J. E., 19, 22
Williams, E. L., 164–165, 169, 182, 255, 270
Wilson, E. B., 60–61, 72
Wölfe, R., 279, 281
Wood, S. C., 160, 162
Woods, K. N., 18, 22
Wright, A. C., 78, 94

Xu, Z., 63–65, 73, 244, 251

Yamaguchi, J., 255, 270
Yang, S., 27, 35
Yeh, T. H., 258, 272
Yesugi, N., 155, 161
Yoder, H. S., 98, 106–107, 176, 183
Yoshida, M., 263–265, 272
Yund, R. A., 2, 8, 67, 73, 98, 101–103, 107, 125, 127

Zamouche, A., 204, 217
Zellama, K., 230–234, 238, 240
Zener, C., 94–95
Zeng, Q., 155, 161
Zhang, Y., 5, 8, 42, 44, 57, 63–65, 67, 73, 108–110, 114–118, 123, 127, 130, 135–137, 146, 174, 182, 188, 198, 244, 251
Zouine, A., 37, 57
Zotov, N., 28, 35, 145, 147

SUBJECT INDEX

Activation energy of diffusion
  in oxides, 64, 125, 175–177
  in silicon, 200, 255ff, 269
Activity, thermodynamic, 10, 131
Aluminum–oxygen hole center, 154
Atom sizes
  covalent, 61, 62, 205, 274
  gaseous, 62–67
  ionic, 61, 68–69
  metallic, 61, 205, 260

Bond enthalpies, 173, 204

Calcite, diffusion of water in, 122
Carbon monoxide, carbon dioxide reaction, 190
Chemical potential, 10, 131, 186
Coefficient of solubility, 78
Concentration profiles
  arsenic in silicon, 266
  calculated, 26–27, 40, 45, 47, 56, 210
  copper in germanium, 277–278
  gold in silicon, 208–209
  hydrogen in silicon, 224–225, 233
  phosphorous in silicon, 264–265
  platinum in silicon, 213
  schematic, 86, 153, 226
  silver in germanium, 279
  water in obsidian, 42, 112, 115–116
  water in quartz, 37
  water in silica glass, 81, 89
Copper diffusion in germanium, 277–278
Covalent bonding, 61

Desorption, 3
Diffusion and exchange
  hydrogen in silicon, 227
  in silicon, 6, 205, 260
  local equilibrium 24, 27–29, 52ff, 165, 205
Diffusion and reaction
  derivation of equations, 24–26, 37ff, 43–44
  local equilibrium, 23ff
  nonequilibrium, 36ff
  surface reaction, 32, 55–56, 179
Diffusion coefficient
  definition, 10
  function of concentration, 19–22, 26, 207ff, 262
  Matano calculation, 22
  pressure effects, 29
  tracer, 52
Diffusion equations
  derivation, 9
  Fick's first law, 10
  Fick's second law, 12
  time range of validity, 54

Diffusion geometries
  infinite, 12, 48
  semi-infinite, 13, 45
  plane sheet, 14–18, 232
  surface reaction, 32, 55–56, 179. *See also* Surface concentrations
Diffusion-limited reaction, 7
Diffusion mechanisms
  in germanium, 280
  in silicon, 200, 258
  ionic, 172
  molecular, 64ff, 170, 177
  silicon monoxide, 181, 242ff
Dislocations, in silicon, 214

Electron distribution around atoms, 60
Ellingham diagram, 71
Error function of complex argument, 44, 47, 53, 57

Fiber optic cables
  germanium in, 155
  phosphorous in, 155–158
  reaction of hydrogen in, 154
Free energy of oxidation, reduction reactions, 191–194
Frenkel defect, 172

Gas solubility
  in silica glass, 78
  of water, 136ff
Germanium
  diffusion in, 273ff
  diffusion in silicon, 257
Gold diffusion in silicon, 208ff

Hard sphere model of atoms, 59–62
Hydroxyl ions, 174
Hydrogen
  atoms, 221
  diffusion in oxides, 148ff
  diffusion in silica glass, 65, 149
  diffusion in silicon, 219ff
  molecular size, 63
  reaction with silicates, 148, 152
  to water ratios, 188

Inert gas atoms
  diffusion in rhyolitic and albitic glasses, 67–68
  diffusion in silica glass, 65–66
  sizes, 63
Internal friction
  of boron in silicon, 267
  of oxygen in silicon, 246
Interstitial atoms, 2, 202
Ionic porosity, 62, 68, 70, 125–126, 238, 273

Kick-out mechanism, *see* Diffusion and exchange

Langmuir model of limited sites, 79, 130ff
Local equilibrium, 4, 23ff

Matano method, 22
Monoxides, 70ff
Moving boundary equations, 30–34. *See also* Tarnishing reaction

Obsidian (rhyolite glass)
  composition, 109
  diffusion of water in, 108ff
Openness of structure, *see* Ionic porosity
Ostwald solubility, 78
Oxidation–reduction reactions, 158, 185ff, 190–191
  and ionic transport, 197
  and the phase rule, 193
  cerium, 196
  copper, 195
  iron, 148, 160, 193, 196
Oxides
  diffusion of hydrogen in, 148ff
  diffusion of oxygen in, 163ff
  diffusion of water in, 120ff
  free energy of formation, 71, 72
  ionic porosity, 70, 125
Oxygen
  diffusion in dry oxides, 175ff
  diffusion in silica glass, 65, 151, 164
  diffusion in silicon, 200, 241ff
  fast diffusion in silicon, 247
  network (ionic) diffusion, 169, 172–174
  partial pressure control, 71, 188

Partial molar volume, 140ff
Permeation
  equations, 16
  of hydrogen in silicon, 236
Phase rule, and oxidation–reduction reactions, 193ff
Platinum diffusion in silicon, 213
Porosity, *see* Ionic porosity
Pyrope (garnet), diffusion of water in, 122

Quartz
  diffusion of water in, 96ff
  structure, 97–98

Reaction rates
  experimental, 44, 88–89, 231–232
  second order, 49, 84ff
Rutile (TiO_2), diffusion of hydrogen in, 160

Schottky defect, 172
Silica glass
  diffusion of gases in, 65–66
  diffusion of water in, 74ff, 151, 164
  gas solubility in, 78
  names, 74
  structure, 76–77
  types and compositions, 75, 153
Silicon
  acceptor diffusion in, 201, 253ff
  amorphous, 215
  arsenic diffusion in, 266
  dislocations in, 214
  donor diffusion in, 201, 253ff
  fast diffusion in, 199ff
  gold diffusion in, 213
  hydrogen diffusion in, 219ff
  intermediate diffusion in, 199ff
  openness 61–62, 203
  oxygen diffusion in, 201, 253ff
  phosphorous diffusion in, 264
  (110) plane, 62, 203
  platinum diffusion in, 213
  self-diffusion in, 200, 248, 257
Silicon monoxide, 71–72, 181, 242ff
Sorption, 3, 13
Speciation of water, 129ff
Surface concentrations
  nonequilibrium, 39, 42, 47, 49–52, 56, 87
  with surface exchange, 55–56, 179

Tarnishing reaction, 30–34, 154

Vacancies, 2, 172, 177, 202, 260
Viscosity, Mott theory, 180

Water
  diffusion in
    obsidian, 42
    oxides, 120ff, 163ff
    quartz 7, 96ff
    silica glass, 74ff, 151, 164
    silicon, 247
  molecular size, 63
  residual, 163ff, 186ff
  solubility, 136ff
  speciation of, 129ff